Behavior, Health, and Environmental Stress

Behavior, Health, and Environmental Stress

Sheldon Cohen
Carnegie-Mellon University
Pittsburgh, Pennsylvania

Gary W. Evans
and
Daniel Stokols
University of California at Irvine
Irvine, California

and
David S. Krantz
Uniformed Services University of the Health Sciences
Bethesda, Maryland

Plenum Press • *New York and London*

Library of Congress Cataloging in Publication Data

Main entry under title:

Behavior, health, and environmental stress.

Bibliography: p.
Includes indexes.
1. Medicine and psychology. 2. Environmentally induced diseases. I. Cohen,
Sheldon, 1947-   . [DNLM: 1. Behavior. 2. Environment. 3. Stress. Psychological
—etiology. WM 172 B4186]
R726.5.B38   1986   616.07′1   86-480
ISBN 0-306-42138-0

First Printing — April 1986
Second Printing — June 1991

© 1986 Plenum Press, New York
A Division of Plenum Publishing Corporation
233 Spring Street, New York, N.Y. 10013

Printed in the United States of America

For our wives—Mary Miller, Merrie Wilent, Jeanne Stokols, and Marsha Douma—and our children— Tristan Evans-Wilent, Eli Stokols, Michael Douma, and Della Elizabeth Krantz.

This book is also dedicated to David C. Glass, whose teaching and research significantly influenced our intellectual perspective.

Finally, this volume is a tribute to the collegiality and friendship among the four authors, which, despite the demands of the project, sustained our efforts to produce something of lasting value.

Preface

Eight years ago, four psychologists with varying backgrounds but a common interest in the impact of environmental stress on behavior and health met to plan a study of the effects of aircraft noise on children. The impetus for the study was an article in the *Los Angeles Times* about architectural interventions that were planned for several noise-impacted schools under the air corridor of Los Angeles International Airport. These interventions created an opportunity to study the same children during noise exposure and then later after the exposure had been attenuated. The study was designed to test the generality of several noise effects that had been well established in laboratory experimental studies. It focused on three areas: the relationship between noise and personal control, noise and attention, and noise and cardiovascular response. Two years later, a second study, designed to replicate and extend findings from the first, was conducted.

The experience of conducting the Los Angeles Noise Project and the varied experiences as active scientists in the fields of environment, health, and social psychology we have had since its completion have provided us with a broad and in-depth perspective on stress and human well-being. This volume is our attempt to summarize and pass on some of what we have learned during this period. It includes critical reviews of research on environmental stress and control, cognitive performance, and health; discussions of methodological issues in the study of environmental stress, including many practical issues often not raised in print; the presentation of two theoretical approaches (*cost-of-coping* and *contextual analyses*) that we feel are essential in understanding the effects of environmental and psychosocial stresses in the everyday lives of people; and data from the Los Angeles Noise Project, presented in the context of each of the preceding discussions.

The volume has multiple goals and is directed at multiple audiences. The targeted audiences include those actively investigating the effects of stress on behavior and health, interested researchers in related fields, professionals and policymakers interested in the effects of environmental stress, and graduate students in psychology, sociology, and epidemiology. The research reviews (Chapters 3, 4, & 5) provide state-of-the-art information for policymaker and researcher alike. Potentially, our analysis of the costs of coping (Chapter 1) could help organize a broad literature on stress and coping and help direct future research in the stress field. The contextual analyses (Chapter 6) are designed to sensitize researchers to the importance of context in stress analyses and provide an organization to aid in future contextually based research. We have also made a special attempt to describe the important issues in designing and evaluating correlational field research both as a teaching tool and so that the uninitiated could appreciate (and criticize) the work we present (Chapter 2). Finally, the data from the noise project help clarify many of our arguments and answer and raise a number of questions about the stress and coping process.

We owe a great deal to the institutions that helped fund this research. An initial grant-in-aid was received by the Society for the Psychological Study of Social Issues that allowed us to get the project started in time to take advantage of the natural intervention. Subsequent grants were received from the National Science Foundation (BNS 77-08576 and BNS 79-23453), the National Institute of Environmental Health Sciences (ESO176401 OBR), the University of Oregon Bio-Medical Fund, and the Focused Research Program on Stress at the University of California at Irvine.

We would like to thank Sheryl Kelly, Laura Martin, and Laurie Poore for assistance with various aspects of this project. We also appreciate the expert technical accomplishments of Al Murphy, Nick Garshnik, and Fran Renner. Sybil Carrere, David Glass, Richard Lazarus, Ed Lichtenstein, M. N. Palsane, and Ronald Kessler provided comments on specific chapters; we sincerely appreciate their help, and note that it is we and not they who are responsible for any errors or misinterpretations. Finally, we are indebted to the children, parents, teachers, and administrative staff of the schools participating in this research. Without their interest and cooperation, this research program would not have been possible.

SHELDON COHEN
GARY W. EVANS
DANIEL STOKOLS
DAVID S. KRANTZ

Contents

Stress Processes and the Costs of Coping

Increased public attention in the late 1960s and early 1970s to the importance of the physical environment in shaping our health and behavior stimulated a renewed interest in these issues on the part of social and health scientists. To a great extent, work during this period was stressor-specific. Independent groups of researchers studied noise, population density, temperature variations, and the like with little cross-stressor integration of theory or data. This work was also conducted in relative isolation from theoretical and empirical work on similar problems. For example, even though Lazarus (1966), Janis (1958), and others had described and empirically demonstrated the role of cognition in mediating the associations between stressor exposure and behavior and/or health, those studying environmental stressors tended to focus on stressor intensity alone with only a handful of researchers considering cognitive factors. Since the mid-1970s research on the impact of environmental stressors on behavior and health has become increasingly integrated into the larger field of psychological stress research. This book documents that integration and discusses its implications for the study of stress and coping processes. We develop a broad theoretical framework that emphasizes the commonalities between psychosocial stress and environmental stress as well as the commonalities between various environmental stressors. We focus on issues that we view as central to understanding the stress and coping process and present original data that exemplify and occasionally provide answers to the questions we raise. In order to place this discussion in context, we include critical reviews of the literatures on the impact of environmental stress on cognition, control, and health.

The data that we report are from the Los Angeles Noise Project—a set of naturalistic studies of the effects of aircraft noise on children. Noise project data are presented in the context of discussions of the issues we view as central to the understanding the generic stress-coping process. Our theoretical perspective places particular emphasis on the deleterious effects of the process of coping, the importance of the context in which a stressor occurs, and the role of one's perceptions of the stressful situation as a predictor of health and behavioral outcomes. In many instances, the scope of our theoretical formulations is considerably broader than that of the data. Our hope is that our perspective will stimulate others to fill in missing evidence.

Plan of the Book

This chapter presents an overview of our major theoretical perspective. We begin by outlining the physiological and psychological stress traditions and then present our views on the "costs of coping" with stressful events. Chapter 2 presents a discussion of methodological approaches employed in the study of environmental and psychosocial stressors. We outline a set of criteria for evaluating correlational research and discuss a number of practical problems that arise in the design and implementation of such work. This discussion is followed by a description and rationale for the designs employed in the noise project studies. Statistical techniques used in this work are also summarized.

Each of the remaining chapters of this book reports data from the studies of the Los Angeles Noise Project in the context of broadly based discussions of the relationship of environmental and psychosocial stressors to the outcome measures being considered. Chapters 3, 4, and 5 focus on the effects of the stress process on personal control, health, and attentional processes respectively. These chapters reflect our concern with the cost of the process of coping and provide evidence relevant to a number of the issues discussed in Chapter 1. The chapters include selective/critical reviews of existing research on the effects of stress on control, health, and attention. The noise project data presented in these chapters examine the question of whether mere exposure to the stressor, for example, attending a noisy as opposed to a quiet school, is associated with the outcome variable under consideration. Evidence on the possibility that children habituate or otherwise adapt to noise with prolonged exposure is also examined in this section of the book.

Chapter 6 argues for the importance of viewing stressors within their broader social and physical environmental contexts. We discuss issues one should consider in doing contextual research and suggest criteria for designing contextual based studies. Analyses of noise project data that reflect the role of contextual factors in determining response to noise are presented.

The final chapter summarizes the theoretical points made in the book and the major findings of the noise project. Implications of this work are drawn for a broad range of environmental psychosocial stressors and for policy decisions regarding environmental stress.

Models of Stress

Some Preliminary Definitions

Although the term *stress* is used widely by psychologists, sociologists, epidemiologists, medical practitioners, and lay persons, there appears to be little agreement either within or across disciplines in regard to a scientifically precise definition. Numerous definitions have been provided that vary in the extent to which they emphasize stimulus, response, or intervening mechanism (e.g., Appley

& Trumbull, 1967; Mason, 1975; McGrath, 1970). They also differ in the processes under consideration with some concerned with cognition, others with motivation, and others with physiology.

Like a number of contemporary writers (e.g., Monat & Lazarus, 1977; Singer, 1980), we avoid a semantic controversy by using the term *stress* to refer to a broad area of study. In particular, it is used to refer to the study of situations in which the demands on individuals tax or exceed their adaptive capabilities. Our own emphasis will be on specifying several mechanisms that may be responsible for deleterious effects on behavior and health that sometimes occur in these situations.

Actually, there are not one but several fields of study that are identified as stress research (cf. Singer, 1980). That is, there are a number of research traditions, each with a different emphasis or focus on the problem. Our own conceptions of the stress process have been influenced by work in what can be termed the physiological and psychological stress traditions. These traditions, although related, provide distinct clues as to the effects of psychosocial and environmental stressors on people. The following pages will provide a short description of the orientations of these two traditions and attempt to distill the central assumptions of each. We will then focus on a number of stress processes we feel are important in understanding response to chronic stress.

Physiological Model of Stress

Many of the pathophysiological concomitants of stress are believed to result from activation of the sympathetic-adrenal medullary system (SAM) and the pituitary-adrenocortical axis (PAC). Although detailed descriptions of these two systems and their relationships to one another are beyond the scope of this chapter (see Baum, Singer, & Baum, 1981; Levi, 1972), each will be discussed in brief in order to provide a basic understanding of their roles in the stress process.

Sympathetic-Adrenal Medullary System

Interest in the impact of SAM activation on bodily reactions to emergency situations may be traced to Cannon's early work on the fight or flight response (Cannon, 1932). Cannon proposed that the SAM system reacts to various emergency states with increased adrenalin (epinephrine) secretion. Althouth 25 years passed before this increased secretion was actually demonstrated, there is now a large body of evidence indicating increased output of epinephrine and norepinephrine in response to a wide variety of psychosocial stressors (Levi, 1972).

It has been claimed that if SAM activation is excessive, is persistent over a period of time, or is repeated too often, it may result in a sequence of responses that culminate in illness. The responses include functional disturbance in various organs and organ systems (cf. Dunbar, 1954) and ultimately permanent structural changes of pathogenic significance at least in predisposed individuals (e.g., Raab, 1976). Particularly culpable in this regard is the secretion of the catecholamines, epinephrine and norepinephrine by the adrenal medulla and/or sympathetic nerve endings. Catecholamine discharge is believed to induce many of the pathogenic states associated with psychological stress, including (a) hemodynamic effects, such as increased blood pressure and heart rate (McCubbin *et al.*, 1980); (b) induction of myocardial lesions (Raab, 1971; Raab, Chaplin, & Bajusz, 1964); (c) increased cardiac demand for oxygen (Raab, 1971); and (d) provocation of ventricular arrhythmias believed to lead to sudden death (Herd, 1978).

Pituitary-Adrenocortical Axis: Selye's Model

The hormonal responses of the PAC axis were emphasized in Selye's (e.g., 1956, 1974) influential description of a nonspecific (general) physiological reaction that occurs in response to aversive stimulation. Selye argued that pathogens, physical stressors (e.g., shock or noise), and psychosocial stressors all elicit the same pattern of physiological response. This response is said to proceed in a characteristic three-stage pattern referred to as the general adaptation syndrome (GAS). During the first stage of the GAS, the alarm stage, the organism's physiological changes reflect the initial reactions necessary to meet the demands made by the stressor agent. The anterior pituitary gland secretes ACTH, which then activates the adrenal cortex to secrete additional hormones (cortical steroids). The hormone output from the adrenal cortex increases rapidly during this stage. The second stage, resistance, involves a full adaptation to the stressor with consequent improvement or disappearance of symptoms. The output of cortical steroids remains high but stable during the resistance stage. Finally, the third stage, exhaustion, occurs if the stressor is sufficiently severe and prolonged to deplete somatic defenses. The anterior pituitary and the adrenal cortex lose their capacity to secrete hormones, and the organism can no longer adapt to the stressor. Symptoms reappear, and, if stress continues unabated, vulnerable organs (determined by genetic and environmental factors) will break down. This breakdown results in illness and ultimately death.

Recall that Selye argued the *generality* position that any noxious agent, physical or psychological in nature, could mobilize the GAS. In contrast, recent critiques of Selye's model (Lazarus, 1977; Mason, 1975) argue a more specificitylike position, that is, that particular stressors have their own distinct physiological reactions. These authors agree that there is a nonspecific physiological response to stressors. They argue, however, that the response is a concomitant of the emotional reaction to stressful situations that occurs only when situations are appraised as stressful. In short, their position is that the nonspecific response is psychologically mediated (e.g., occurs when there is a cognitive appraisal of threat). When conditions are designed to reduce the psychological threat that might be engendered by laboratory procedures, there is no nonspecific reaction to the physical stressor (Mason, 1975). For example, by minimizing competitive concerns and avoiding severe exertion, the danger that young men would be threatened by treadmill exercise was reduced, and the GAS pattern was not found. It is noteworthy that, in Selye's more recent work (1974, 1980), he acknowledges that there are both specific as well as general (nonspecific) factors in one's physiological response to a stressor but maintains that the nonspecific response is not always psychologically mediated. He also suggests that the GAS does not occur (or is at least not destructive) under all kinds of stress. For example, he suggests that there may be a pleasant stress of fulfillment and victory and a self-destructive distress of failure, frustration, and hatred.

Two aspects of Selye's model have exerted a profound influence on the development of behavioral and psychological conceptions of the stress process. The first is his assumption that there is a finite amount of adaptive energy that can be invested in one's transaction with a stressor or stressors. A depletion of this energy results in a deleterious impact on the organism. Although Selye's energy was physiological in nature, a number of other theorists (e.g., Cohen, 1978; DuBos, 1965; Glass & Singer, 1972) have proposed similar limitations of psychic energies. A second influential aspect of Selye's theory is his argument that pathogenic effects occur as a result of the body's attempt to cope with the stressor. This assertion is often termed the "adaptive-cost hypothesis." Specifically, the hypothesis suggests that the process of adaptation itself causes deleterious effects that occur either during or after exposure to a stressor. This notion has been borrowed by those interested in behavioral adaptations and their consequences for both behavior and health (e.g., Cohen, 1980; Glass & Singer, 1972; Holmes & Rahe, 1967). Some specific implications of these assumptions for behavioral and psychological formulations of the stress process will be discussed later.

Psychological Model of Stress

The psychological stress tradition places emphasis on the organism's perception and evaluation of the potential harm posed by a stimulus. The perception of threat arises when the demands imposed upon an individual are perceived to exceed his or her felt ability to cope with those demands. This imbalance gives rise to the experience of stress and to a stress response that may be physiological and/or behavioral in nature. Psychological stress then is not defined solely in terms of the stimulus condition or solely in terms of the response variables but rather in terms of the transaction between the person and the environment. It involves interpretation of the meaning of the event and the interpretation of the adequacy of coping resources. In short, the psychological stress tradition assumes that stress arises totally out of persons' perceptions (whether accurate or inaccurate) of their relationship to their environment.

The most influential model of psychological stress has been the one proposed by Lazarus (1966, 1977). The model is depicted in Table 1. In the original formulation of his model, Lazarus (1966) argued that an appraisal of a stimulus as threatening or benign (primary appraisal) occurs between stimulus presentation and stress reaction. For a situation to be deemed threatening, the stimulus must be evaluated as harmful. In his later writings, Lazarus (1977, 1980) argued that a situation will also result in a stress reaction if it is evaluated as a harm/loss, threat, or challenge.

Table 1. Lazarus's Model of Stressor Appraisal

Appraisal stage	Process	Determinant factors
Primary	Determining whether a situation poses a threat, challenge, or potential harm or loss	Perceived features of the stimulus situation Psychological structure of the individual
Secondary	Evaluating resources in order to determine whether one can cope with the situation	Perceived availability of coping resources from either within the individual or from the environment
Reappraisal	Repetition of primary and secondary appraisal as perception of the stressor or available resources changes over time	Change in the situation or person

Primary appraisal is presumed to depend on two classes of antecedent conditions: the perceived features of the stimulus situation and the psychological structure of the individual. Some stimulus factors affecting primary appraisal include the imminence of harmful confrontation, the magnitude or intensity of the stimulus, the duration of the stimulus, and the potential controllability of the stimulus. Factors within the individuals that affect primary appraisal include their beliefs about themselves and the environment, the pattern and strength of their values and commitments, and related personality dispositions.

When a stimulus is appraised as requiring a coping response, the individuals evaluate their resources in order to determine whether they can cope with the situation, that is, eliminate or at least lessen the effects of a stressful stimulus. This process is termed "secondary appraisal." Coping responses may involve actions designed to directly alter the threatening conditions (e.g., fight or flight) or thoughts or actions whose goals are to relieve the emotional impact of stress (i.e., bodily or psychological disturbances). The latter group of responses, referred to as "emotionally focused" coping, may be somatically oriented, for example, the use of tranquilizers, or intrapsychic in nature, for example, denial of danger (Lazarus, 1975).

If one perceives that effective coping responses are available, then the threat is short-circuited, and no stress response occurs. If, on the other hand, one is uncertain that she or he is capable of coping with a situation that has been appraised as threatening or otherwise demanding, stress is experienced. It is important to note that this process of evaluating the demands of a situation and evaluating one's ability to cope does not occur only at the onset of a stressful event but will often recur during the course of the event (cf. Folkins, 1970; Lazarus, 1980). Thus, an event that is initially appraised as threatening may be later reappraised as benign, and coping strategies that are initially found to be lacking may later be found to be adequate. Conversely, events that one initially evaluates as nonthreatening may be later reevaluated as stressful.

Although it is recognized that certain events are almost universally appraised as stressful, for example, the death of a loved one, the impact of even these events can be expected to depend on the individual's appraisal of the threat entailed and ability to cope with it. For example, the death of a spouse for someone with neither family nor friends may be experienced as more severe than the same event for someone with close ties to family and friends.

Unlike the physiological stress tradition in which stressor stimuli presumably lead to a restricted range of physiological responses, the psychological tradition has posited a broad range of outcomes as indicators of stress. Typically, response variables in this work include self-reported annoyance or stress, negative-toned affect, changes (usually deficits) in performance of complex tasks, and alterations in interpersonal behavior as well as the physiological changes discussed earlier. Unfortunately, cognitive stress models tend to be vague in their predictions of the particular measures that will be affected in any instance, and the nature of the relationships among these outcome measures.

Effects of the Coping Process

We noted earlier that one of the influential aspects of Selye's model is his argument that there is a cost of the adaptive process. Specifically, the cost refers to the deleterious effects of an encounter with a stressor that occurs as a consequence of the coping processes that are employed (cf. Cohen, 1980; DuBos, 1968; Glass & Singer, 1972). We will call these *secondary,* or *indirect,* effects of one's exposure to a stressor in that they occur because of the coping process rather than because of the stressor itself.

In the following section, we suggest a number of conditions under which both behavioral and psychological costs of the adaptive process occur. Although, to some degree, these theoretical speculations have been shaped by the physiological and psychological stress traditions discussed previously, they diverge from these traditions in a number of significant ways. First, let us consider our divergence from Selye's model. The adaptive process discussed by Selye is both automatic and nonspecific. By automatic, we mean that the PAC responses are not voluntary strategies chosen by the subject but are rather part and parcel of a biologically programmed reaction to threatening events. In contrast, many of the coping strategies that will be discussed in the following pages are voluntary.

The nonspecific nature of the physiological reactions studied by Selye was discussed earlier. As noted before, the nonspecific aspects of one's physiological response to a stressor may be elicited by an emotional reaction to the stressor rather than by the physical stressor itself. Unlike Selye's model, several of the strategies discussed in this section are specific in nature. That is, they are strategies that are developed to cope with a particular stressful event. It is possible, however, that these various coping strategies share some common underlying physiological outcome(s).

Finally, we note an important divergence we will make from the psychological stress tradition. Recall that the psychological tradition suggests that people experience stress when they appraise a situation as threatening or otherwise demanding and when they perceive that they lack the ability to adequately cope. We will argue, however, that secondary effects of stressor exposure can occur when a situation is appraised as potentially threatening and one perceives adequate coping responses are available. Specifically, we will propose that successful (as well as unsuccessful) coping can have a deleterious impact on both one's behavior and health.

Successful Coping

First, consider situations in which a person is successfully coping, that is, the direct (primary) effects of the stressor are neutralized by an apparently adjustive coping strategy. There are (at least) three types of secondary effects that can occur for those involved in an apparently successful adaptation to their environment. These effects can be labeled (a) cumulative fatigue effects; (b) overgeneralization of a coping strategy; and (c) coping side effects.

Cumulative Fatigue Effects. Many previous theoretical discussions of the adaptive cost hypothesis have proposed what we term a *cumulative fatigue effect*. This concept assumes that a person has a limited amount of energy (whether biological or psychological in nature) and that prolonged coping demands deplete this energy supply. Recall, for example, that Selye (1975) asserted that after prolonged exposure to a stressor, one's adaptive biological reserves are drained, resistance breaks down, and exhaustion sets in. Symptoms appear, and, if stress continues unabated, vulnerable organs will break down, and ultimately death will ensue. Glass and Singer (1972) similarly suggest that the process of adaptation requires cognitive effort. This work includes searching for appropriate coping responses and/or attempting to define the stimulus. Moreover, they assumed that the work required to adapt to unpredictable and uncontrollable stressors was substantially greater than that required to adapt to predictable and controllable stimulation. Glass and Singer further argued that increased adaptive effort depletes one's available energies and thus results in deficits on subsequent task performance.

Cohen (1978) similarly argued that there were psychic costs of prolonged exposure to unpredictable and uncontrollable stressors, but he described these loses in terms of one's attentional capacities or information-processing abilities. Specifically, he suggested that exposure to unpredictable, uncontrollable stressors substantially increases demands on one's attentional capacity. This increased demand might occur because individuals are required to continually monitor threatening stimuli to evaluate their adaptive significance and decide on appropriate coping responses (cf. Lazarus, 1966). Increased demand may also occur because of effort required in inhibiting response to a distracting stimulus. Cohen further asserts that an individual's attentional capacity is not fixed but shrinks when there are prolonged demands. This shrinkage, or "cognitive fatigue," presumably increases with both the attentional load (information rate) of an activity and the duration of an activity. Thus, prolonged exposure to an environmental stressor and/or to a high information rate task should result in an insufficient reserve of attention to perform demanding tasks.

Others (Basowitz, Persky, Korchin, & Grinker, 1955; Dubos, 1965; Wohlwill, 1966) make similar claims concerning the depletion of psychological reserves over the course of one's exposure to a stressor. These various models have been used to explain poststress deficits in task performance (e.g., Cohen, 1980; Glass & Singer, 1972), poststress insensitivity toward others including increased aggression and decreased helping (e.g., Cohen, 1978; Cohen & Spacapan, 1978), and the increased deficits that occur over time on demanding tasks performed under stress (Cohen, 1978; Wilkinson, 1969).

The reduced capacity to cope with environmental demands for those *already coping* with life stressors may also be interpreted as a result of cumulative fatigue. An example is provided by a study of effects of smog exposure on a large sample of Los Angeles residence (Evans *et al.,* 1982). Evans and his colleagues found that higher smog levels resulted in increased risk of mental health problems *only* for those who had recently experienced a stressful life event.

Overgeneralization. Another secondary effect of successful coping is the overgeneralization of a coping strategy. This effect occurs when a strategy employed to deal with a stressor persists even in situations where a person is not confronted with the offending stressor. Overgeneralization results in deleterious effects if the coping responses are inappropriate in other situations in which they are employed. The persistent use of a particular strategy may be due to an overlearning of the coping response.

Overgeneralization has been shown to occur both in laboratory and field settings. For example, Epstein and Karlin (1975) reported that the competitiveness and withdrawal displayed by men during crowding and the cooperativeness and cohesiveness displayed by women during crowding persisted into the poststress session. Baum and Valins (1977) similarly reported that students from dormitories whose architectural designs fostered high levels of forced interaction made more active attempts to avoid the possibility of contact with a stranger outside of the dormitory than students from dormitories with lower levels of interaction. Thus, an avoidance response that presumably developed as an attempt to cope with dormitory life persisted outside of the dormitory setting.

Evidence on the use of denial in coping with heart attacks (myocardial infarctions) similarly suggests that the persistence of a coping strategy can be detrimental after the stressor is terminated (cf. Krantz & Schulz, 1980). Several studies have found that deniers show less anxiety when first admitted to the coronary care unit (CCU) than those not employing a denial strategy (e.g., Froese, Hackett, & Cassem, 1974; Gentry, Foster, & Haney, 1972). These patients deny that they have had a heart attack, often protest detention in the CCU, and insist on returning to normal activities. Although useful during the initial phases of the attack, use of denial has been related to rehabilitation problems due to long-term resistance to following medical instructions (e.g., Croog, Shapiro, & Levine, 1971; Garrity, 1975). In sum, use of denial may make for better coping with the early stress of illness in the CCU. However, in the long term, patients may endanger their chances of recovery by ignoring medical recommendations that are important for satisfactory rehabilitation.

An example of overgeneralization of a strategy employed to cope with noise was proposed by Deutsch (1964). She suggested that children reared in a noisy environment eventually become inattentive to acoustic cues; that is, they learn to "tune out" sound. In tuning out his or her noisy environment, a child is not likely to distinguish between speech-relevant and speech-irrelevant sounds. Thus, he or she will lack experience with appropriate speech cues and will generally show an inability to recognize relevant sounds and their referents. The inability to discriminate sound could account, in part, for subsequent problems in learning to read. A child who cannot readily discriminate basic speech sounds faces a difficult task in learning to associate these sounds with their appropriate signs. A number of studies have found auditory discrimination and reading deficits on the part of children living and/or attending school in noisy environments (e.g., Bronzaft & McCarthy, 1975; Cohen, Glass, & Singer, 1973; Cohen, Evans, Krantz, Stokols, & Kelly, 1980). However, it is still unclear whether these deficits are attributable to a tuning-out strategy or to the noise masking parent speech (cf. Cohen & Weinstein, 1982).

Although the data just described are limited to situations in which the coping strategy is one that develops as a response to a particular stressor (e.g., withdrawal as a response to crowding), it is possible that nonspecific strategies that are used to cope with a wide range of stressors persist outside of stressor exposure (cf. Milgram, 1970). For example, the strategy of focusing one's attention on the essential aspects of a task during stressor exposure (e.g., Broadbent, 1971; Hockey, 1970; Wachtel, 1968) may persist even after exposure is terminated. Deficits on complex tasks administered after exposure to unpredictable and uncontrollable stressors including noise (e.g., Glass & Singer, 1972; Sherrod, Hage, Halpern, & Moore, 1977), electric shock (e.g., Glass *et al.*, 1973), and crowding (e.g., Evans, 1979; Sherrod, 1974) are explicable within this context (cf. Cohen, 1980).

Coping Side Effects. The third form of the secondary effect occurs when coping behaviors, which were successful in ameliorating the possible effects of the stressor, are detrimental to the individual in other ways. We have labeled this phenomenon *coping side effects.* We will discuss two ways in which the act of coping itself may be detrimental: (a) the coping process directly produces pathogenic effects and (b) coping behaviors interfere with health maintenance.

The first approach suggests that direct coping may be effective in initially removing a threat, but at the same time, it may itself produce pathogenic physiological reaction (Obrist, 1981). For example, Obrist reported that situations in which active coping or behavioral adjustment are required result in cardiovascular responses, for example, elevated blood pressure and increased water retention by the kidneys, above the level that is efficient for the body's metabolic needs. Active coping situations are those in which the organism can exert some control over a noxious stimulus by some aspect of his or her behavior, for example, attempting to escape or avoid shock by pressing a key. By contrast, these physiological responses do not seem to be elicited in similar intensity or kind in stressful situations where the individual copes passively or intrapsychically and does not take direct action to attempt to control the situation. In related studies, active coping with a stressor has also been found to lead to increased specific discharge of norepinephrine (e.g., Elmadjian, 1963; Elmadjian, Hope, & Larson, 1958). Moreover, norepinephrine levels in blood and urine remain elevated in subjects engaged in active efforts to escape or avoid stressors (Contrada *et al.,* 1982; Frankenhaeuser, 1971; Frankenhaeuser & Rissler, 1970; Weiss, Stone, & Harrell, 1970).

Incidentally, this notion of active coping seems applicable to descriptions of those psychosocial conditions (e.g., rapid cultural change, migration, socioeconomic mobility) shown by epidemiologic research to be associated with blood pressure elevations in some human populations (Gutmann & Benson, 1971; Krantz *et al.,* 1981). As Gutmann and Benson have pointed out, this research literature similarly suggests that situations that require active coping, that is, continuous behavioral adjustment, likely play a causal role in hypertension. Although there is no convincing evidence at this point, it has also been proposed that the effects of stressful life events on the etiology of disease are attributable to the effort of coping with life change (cf. Holmes & Rahe, 1967).

Recent work (e.g., Manuck, Harvey, Lechleiter, & Neal, 1978; Solomon, Holmes, & McCaul, 1980) suggests that the detrimental effects of active coping occur under conditions of high- but not low-effort coping. In particular, when control is easy to exercise, subjects report a level of anxiety and display physiological arousal patterns that are similar to subjects who are not threatened by an aversive event. However, when control is difficult to exercise, subjects report a level of anxiety and display physiological response patterns that are like those of subjects threatened with an aversive event who are unable to exercise control over it.

There is recent evidence that there are predictable *individual differences* in physiological (sympathetic) reactivity for those involved in high-effort coping. Specifically, these data come from experiments comparing Type A's—those who respond to challenge by showing competitive drive, impatience, hostility, and vigorous voice characteristics—with those not showing this behavioral style (Type B's). First, a number of studies have found that A's respond with higher systolic blood pressure than B's when performing a challenging task (e.g., Dembroski *et al.,* 1978; Glass *et al.,* 1980). If we assume that challenge leads to high-effort coping, the greater elevation of blood pressure on the part of A's may be explicable in effort terms. Second, two recent studies compared A's and B's in an experimental situation where subjects exerted effort to control an aversive stimulus. Pittner, Houston, and Spiridigliozze (1983) found that under high-effort control conditions, Type A subjects had greater elevations in blood pressure and pulse rate than did Type B subjects. In low-effort control conditions these differences were absent or weaker. Contrada *et al.* (1984) similarly reported evidence of greater plasma nor-epinephrine secretion among A's than among B's when effortful control was re-quired. Hence, Type A's may show greater physiological reactivity when effortfully coping, and hence coping A's may be under greater risk for illness than coping B's.

A further refinement of the conditions under which active coping results in pathogenic physiological responses is suggested by Weiss's (e.g., 1977) discussion of the conditions under which stress produces ulcers. Like the theorist discussed previously, Weiss suggested that effortful coping, in his terms a high rate of coping responses, is related to increased pathology. He also suggests, however, that patho-genic effects will occur only when one's coping responses do not continually produce feedback that one is manipulating his or her environment. The theory's first proposition is that the more coping responses one emits, the greater is the ul-cerogenic stress. Note that Weiss does not say that increased responses cause increased ulceration, only that they rise together. Although coping may cause ulceration, a strong alternative would view both increased response rate and in-creased ulcerations as a consequence of a high level of anxiety or fear. The theory's second proposition is that if the response immediately produces appropriate feed-back—brings about stimuli that are not associated with the stressful situation—ulcerogenic stress will not occur. Weiss's emphasis is on immediate changes in one's environment following each coping attempt. Hence, those coping attempts that are successful in avoiding an aversive situation are considered stressful unless there is continuous feedback signaling the effectiveness of each coping response.

Evidence supporting Weiss's theory is provided by a set of experiments in which he manipulated whether or not coping (lever-pressing) rats received relevant feedback. Ulceration was low for those rats who received feedback, higher for those who coped and received no feedback, and highest for those who coped but received inappropriate feedback (short, low-intensity shocks). Studies in which number of coping responses and rate of feedback are independently manipulated by the experi-menter would determine whether the effort involved in coping is the cause of the pathogenic response (as mentioned before there is evidence that this is true), and whether feedback does cancel out this effect.

Cassel's (1975) analysis of social stress and its implications for the etiology of disease resulted in a similar emphasis on coping and feedback. In Cassel's words: "Stressful social situations might be those in which the actor is not receiving adequate evidence [feedback] that his actions are leading to anticipated conse-quences" (p. 543). He suggests that these stressful situations are particularly likely to occur when individuals are unfamiliar with cues and expectations of the society in which they live. Thus, like Guttman and Benson (1971), Cassel would expect that migrants, those undergoing rapid social change or social disorganization, would be at risk for disease. He argues, however, that it is not only the effortful adaption that results in increased pathology but rather the adaptive efforts in the absence of information about their effectiveness.

A second side effect of successful coping occurs when the coping behaviors one engages in are themselves detrimental to the maintenance of one's health and well-being. For example, increased coping efforts may result in increased allocation of time to a particular problem. Thus, one might spend long hours at work attempting to complete a demanding project. Although furthering short-term goals, this kind of coping is detrimental to one's health to the extent that it interferes with proper nutrition, exercise, or the practice of hygienic behaviors. Cigarette smoking, coffee drinking, drug consumption, and overeating, which often increase in times of stress, are salient behaviors in this category (e.g., Conway, Vickers, Ward, & Rahe, 1981; Rahe & Arthur, 1978). Smoking, while often resulting in self-reported decreases in anxiety or fear (actually sympathetic increases, Schachter, Kozlowski, & Silverman, 1977), has been implicated as a risk factor for three leading causes of death in the United States (heart disease, cancer, stroke). Similarly, alcohol consumption, although providing a temporary release from stress, is also implicated as a risk factor for physical diseases and social pathology.

Summary of the Effects of Successful Coping. In sum, we have outlined three kinds of effects of successful coping. First, the coping process may drain one's cognitive or physiological energies. Second, a strategy employed to deal with a stressor may persist even in situations where a person is not confronted with the offending stressor. Finally, coping behaviors that are successful in ameliorating the possible effects of a stressor can be detrimental to individuals in other ways.

When Coping Fails

There are also secondary effects that occur when one's attempts to cope with an aversive event fail. These effects occur either as a cost of actively coping with the stressor or because of one's perception of uncontrollability and failure *per se*. They are separate from any primary effects that may occur as a result of being unable to cope with the demands of a specific stressor.

Costs of Active Coping. The cumulative fatigue effects and coping side effects discussed before in the context of the costs of successful coping are also potential effects of failure to cope. These effects would occur to the extent that there is a prolonged active and effortful attempt to cope. Clearly, cognitive fatigue, pathogenic effects of active coping, and the interference of coping processes with health maintenance should occur when one engages in prolonged effortful coping irrespective of whether the coping behavior is successful or not. In cases where the ineffectiveness of coping attempts becomes immediately apparent and active coping is terminated, these mechanisms would not come into play.

A form of overgeneralization may also be operative in situations where coping fails. Instead of overgeneralizing successful coping strategies, it is also possible to overgeneralize the expectation that coping is ineffective. This perspective has been developed in learned helplessness theory and is discussed in that context next.

Learned Helplessness Theory. Specific effects of a continual inability to control important events are suggested by Seligman's (1975) learned helplessness theory. Seligman has proposed that when persons (or animals) learn that their reinforcements are independent of their responses (i.e., that they lack control over their outcomes), this learning undermines their motivation to initiate further instrumental responses, interferes with learning that other outcomes are controllable, and causes a depression of mood. Extreme effects of helplessness presumably include fear, clinical depression, disease, and even death. A later modification of learned helplessness theory suggests that the impact of helplessness on these outcomes increases when the individual lacks a requisite controlling response that is available to others (personal helplessness), and when persons attribute their helplessness to long-lived and recurrent (stable) causes and to causes that are important for a wide range of outcomes (global) (Abramson, Seligman, & Teasdale, 1978).

In general, laboratory studies of learned helplessness in humans provide strong evidence that depriving someone of control over important (in the context of the experiment) outcomes ultimately (often after an initial period of attempting to reassert control, e.g., Roth & Kubal, 1975; Wortman & Brehm, 1975) results in deficits on subsequently administered tasks (e.g., Cohen, Rothbart, & Phillips, 1976; Hiroto & Seligman, 1975; Krantz, Glass, & Snyder, 1974). These deficits presumably manifest negative transfer that inhibits one from learning that their outcomes are under his or her control as well as a decrease in motivation. Similar studies have indicated that those lacking control over noxious stimulation exhibit increases in reported levels of depression, anxiety, and hostility (Gatchel, Paulus, & Maples, 1975; Miller & Seligman, 1975). This depressed mood, however, apparently occurs only when subjects perceive that they have failed the task (Griffith, 1977). Although we will not pursue alternative explanations here, there is evidence suggesting the possibility that the poor performance in the helplessness group may occur because of the distracting effects of anxiety (Coyne, Metalsky, & Lavelle, 1980; Lavelle, Metalsky, & Coyne, 1979), or that poor performance is the result of a low-effort strategy used to avoid making a self-attribution of low ability (Frankel & Snyder, 1978; Snyder, Smoller, Strenta, & Frankel, 1981).

Helplessness has been manipulated in the laboratory by both the administration of insoluble tasks and by exposing subjects to inescapable noxious stimuli. These two paradigms are presumed to have the same effect because in both cases the probability of various response outcomes (correct or incorrect, noxious stimulation or no noxious stimulation) is independent of one's responses. Both paradigms use a triadic design including groups in which a person's performance on a test phase task is assessed after a pretreatment task in which either (a) reinforcement is contingent on a subject's behavior; (b) reinforcement is not contingent on a subject's behavior; and (c) there is no pretreatment.

An experiment by Hiroto (1974) serves as an example of the helplessness paradigm. Subjects in an escape group received loud noise that they learned to turn off by button pushing. The subjects in the inescapable group received the same noise, but the noise was independent of their responding. A third group received no noise. After the pretreatment session, subjects were again exposed to noise but could escape the loud bursts by mastering a hand shuttle-box task. In order to escape noise in this situation, an individual had only to move his or her hand from one side to the other. Both the no-noise group and the escape group learned readily to shuttle with their hands. However, those in the inescapable group generally failed to escape and avoid, with many sitting passively and taking the aversive noise.

Deficits on tasks administered after an experience of uncontrollability are not typically limited to the original situation in which responding was ineffective but rather may generalize to a wide variety of new situations. For example, failure to control noxious stimuli can lead to later difficulties in solving mental tasks, and failures on cognitive tasks can interfere with instrumental responding to noxious stimuli (Hiroto & Seligman, 1975).

Most research on the effects of helplessness on physiology and behavior has emphasized deficits in task performance and indications of "giving up" that occur subsequent to a helplessness treatment. In real life, this apparent reluctance to cope with new demands could be detrimental to both health maintenance and to the achievement of challenging goals. In terms of health maintenance, one could react to the onset of illness and even one's own symptoms as if they were uncontrollable. This may well result in failures to perform routine health care behaviors, for example, brushing one's teeth, seeking medical care, or complying with treatment and rehabilitation regimes (e.g., Krantz & Schulz, 1980; Taylor, 1979). Challenging situations on the job in school or at home may similarly be met by a lack of active coping. Performance on difficult tasks would thus be expected to suffer, possibly reinforcing one's previous inappropriate expectancy that his or her outcomes are not contingent on his or her behavior.

Physiological Concomitants of Helplessness. The psychological state of helplessness may play a more direct role in the etiology of disease states. For example, it has been suggested that helplessness inhibits the operation of the immune system. This hypothesis is supported by experiments in which tumors have been found to grow more quickly and to be rejected less often in mice given inescapable shock than in mice given escapable shock or no shock (Sklar & Anisman, 1979; Visintainer, Volpicelli, & Seligman, 1982). Helplessness has also been proposed as a possible cause of sudden coronary death (cf. Engel, 1968; Green, Goldstein, & Moss, 1972) and of increased risk of heart disease for those displaying Type A behavior patterns, that is, competitive drive, hostility, and impatience (Glass, 1977). However, the physiological bases for these assertions have not been clearly spelled out, and further research is needed in order to determine mechanisms translating psychological states of helplessness into physiological effects that culminate in disease.

There is, however, research on animal response to hopeless situations that may provide a clue to a physiological mechanism linking helplessness to sudden death. Richter (1957) found that wild rats whose whiskers were cut off apparently gave up any attempt to survive an emersion in a water bath. The dewhiskered rats swam for less than 5 minutes before dying, whereas rats whose whiskers were not removed swam for 6 to 8 hours before drowning. Richter reasoned that the wild rats' whiskers were an important source of contact with their environment and that their removal totally deprived the rats of the perception that they could control their outcomes. This assertion was lent credence by Richter's demonstration that rats who were given a "hopelessness pretraining," in which they received several experiences of being immersed in water, subsequently did not give up and die but swam for 6 to 8 hours. Richter found that death in the hopeless rats was preceded by a decrease in arousal, including a slowing of heart rate and lowering of body temperature. He interpreted this as evidence that death was due to *underarousal,* possibly brought on by an overstimulation of the parasympathetic system, rather than to overarousal due to the *overactivation* of the pituitary-adrenal cortical axis (as outlined by Selye) and/or the sympathetic-adrenal-medullary system. Others have similarly found that swimming deaths in rats are associated with changes in cardiovascular function indicative of overstimulation of the parasympathetic system (e.g., Binik, Theriault, & Shustack, 1977). However, the interpretation that these changes are induced by helplessness has been challenged and a number of viable alternative explanations suggested (cf. Binik, Deikel, Theriault, Shustack, & Balthazard, 1979; Hughes & Lynch, 1978). Studies directly manipulating helplessness in rats are mixed in their support for a helplessness-linked reduction of arousal. Although Weiss and his colleagues (Weiss, Glazer, & Pohorecky, 1977) found that giving up in the face of an uncontrollable stressor was associated with substantial depletions in norepinephrine, Binik *et al.* (1979) failed to find any increase in swimming deaths or decrease in heart rate during swimming trials for rats pretreated with uncontrollable shock.

Human studies in which helplessness was manipulated in the laboratory have found a similar pattern of decreased sympathetic arousal (after an initial period of increased response) among people deprived of control over their outcomes. These studies indicate that subjects who initially expect that they can control an aversive agent but who learn over the course of the experiment that they are helpless show lower levels of arousal after the initial trials than subjects provided with an effective coping response. For example, as discussed earlier, Obrist *et al.* (1978) found that subjects with a hard coping task showed higher arousal than those with an easy task. They also report that after the first few trials, subjects working on an impossible coping task, designed to foster "giving up," had levels of systolic blood pressure, heart rate, and carotid dP/dt (how quickly a heart beat is transmitted as a pulse rate through the body) equivalent to the easy-task group. Apparently, after learning that their coping attempts were unsuccessful, both their coping efforts and associated sympathetic activation decreased. Further support for lowered physiological arousal in helpless subjects is reported by Gatchel and Proctor (1976). After the first few trials, subjects treated with a standard helplessness pretreatment—a series of inescapable tones—demonstrated lower tonic skin conductance and smaller phasic skin-conductance responses than a group pretreated with escapable tones. The helplessness group did, however, show more spontaneous electrodermal activity, the measure most closely correlated with self-reported arousal. Similar reports of relatively low levels of phasic skin-conductance response on the part of subjects who initially expected that they could control an aversive stimulus but were unable to do so (helplessness condition) as opposed to those who were able to escape have been reported by Gatchel, McKinney, and Koebernick (1977) and Krantz, Glass, and

Snyder (1974). The levels of arousal reported in the studies discussed here were not below those expected in normally functioning people. Thus, this work cannot be viewed as evidence for a pathogenic response but only as supportive of the hypothesis that arousal decreases after prolonged exposure to a helplessness-inducing task or situation.

In sum, there is accumulating evidence that the experience of helplessness is associated with relatively low levels of arousal, which is possibly attributable to reaction of the parasympathetic system. Moreover, extreme instances of such a response may be associated with sudden death as well as susceptibility to disease. This work is, however, merely suggestive, and further research on physiological response to helplessness is required before any definitive conclusions are possible. Overall, it is clear that one's perceived failure to cope with a stressor may result in behavior and physiological changes inimical to the well-being of the organism. Hence we apparently suffer when coping fails as well as when it succeeds.

Helplessness Resulting in Alternative Coping Modes. Learned helplessness theory views helplessness as a state resulting from the inability to *directly* cope with a stressful event. This approach does not address the possibility that persons unable to directly cope may still be able to handle the affective consequences of that event (e.g., Lazarus & Folkman, 1984). It seems likely that the perception of helplessness, although inhibiting further problem-focused coping, may alternatively elicit emotional-focused coping. For instance, Evans, Jacobs, and Frager (1982) found that long-term residents of air polluted neighborhoods did not directly confront the problem but instead tended to adopt emotionally focused coping strategies such as rationalization or denial. In comparison to recently exposed persons, they were more likely to underestimate the extent of the smog problem as well as its impact on their own health. These same residents also felt that community smog abatement programs like car pooling would not do any good. Moreover, long-term residents were less likely to follow health advisories during smog episodes and did not restrict outdoor activities. Hence, many of their actions were consistent with the helplessness model and at the same time suggest that they did not become totally passive but rather adopted alternative emotionally focused modes of coping. An interesting question that is as yet uninvestigated is whether health outcomes differ for persons who just fail at direct coping and those who fail at both problem-focused (direct) and emotionally focused coping.

Passivity in the Face of Stressful Events

Our discussons of successful and unsuccessful coping were concerned with situations where a person actively attempts to cope with a stressful event. We can also ask what happens when a situation is structured such that one responds passively to a stressor, that is, does not engage in any direct coping behavior. The issue of passive exposure to a stressor is both conceptually and empirically controversial and, as is indicated later, the present state of the literature does not lend itself to clear conclusions.

Let us first consider research on the aftereffects of stress. This work focuses on the important difference between instances of passive exposure in which there is an expectancy that, if active coping was engaged in, it would be successful and instances in which there is no reason to expect that coping efforts would be fruitful. Specifically, Glass and Singer (1972) and others (see review by Cohen, 1980) have found that persons passively exposed to a stressor who are not led to expect that it is possible to escape or avoid it (no perceived control group) perform more poorly on tasks administered after stressor termination than a group not exposed to the stressor. However, when a passive exposure group is led to expect that they can terminate the stressor if they desire (perceived control group), even though they do not actually implement that control, the deleterious effects of stressor exposure are partly or wholly ameliorated. These data suggest that passive exposure is deleterious, at least in terms of performance on subsequent task, only if one's passivity is forced on him or her by the structure of the situation, that is, there is no choice of whether or not one copes.

Seligman (1975) has interpreted these data as supportive of learned helplessness. Before discussing Seligman's analysis, it is important to emphasize the differences between the "aftereffects" and "learned helplessness" paradigms. In the helplessness studies, subjects in the control and no-control groups both expect (because of the instructions) that it is possible to control the situation. Subjects with control actively and successfully cope. Those without control cope initially but eventually learn that they are not effectively manipulating their environment. In the aftereffect studies, no one actually copes. The perceived control group is told that there is an effective coping strategy available to them, but they do not actually use it. The no-perceived-control group is never given the expectation that coping is possible. According to Seligman (also Glass & Singer, 1972), passive exposure without perceived control presumably results in the deterioration of poststress task performance because of the motivational and cognitive deficits associated with helplessness (see Cohen, 1980, for alternative explanations of aftereffects).

It is paradoxical that the helplessness literature itself provides data that are apparently inconsistent with the helplessness interpretation of the aftereffects phenomenon. Several learned helplessness studies, in fact, demonstrate that helplessness is not produced in subjects who are passively exposed to noxious stimuli. Moreover, these studies employ a passive exposure condition that is strikingly similar to Glass and Singer's no-perceived-control group. In work reported by Gatchel *et al.* (1977), Gatchel and Proctor (1976), and Hiroto and Seligman (1975), a condition is included in which subjects are passively exposed to a noxious stimulus without being told that it is possible to escape or avoid it. In all cases, subjects in this condition do not show the poststimulation deficits in performance found in the inescapable stressor (helplessness) condition but in fact perform like subjects in the escapable condition.

Because of the difference between the aftereffects and helplessness paradigms, it is difficult to assess whether failure to find poststimulation deficits in the passive exposure condition are inconsistent with results from aftereffect studies. For example, subjects may merely perform better under perceived control than under the escape condition in helplessness studies. Unfortunately, there are no experiments at this point that include all of the relevant conditions for comparison of the after-effects and helplessness paradigms. (These would include two groups from the helplessness paradigm—those who actively cope and succeed, those who actively cope and fail, and two groups from the aftereffects paradigm—those who do not cope but expect coping would be successful and those who do not cope and do not expect that coping would be successful.) Hence, it is still uncertain whether a passive response to a stressor is adjustive or not.

When Not Coping Is Adjustive. There are some important implications of the research and theory discussed previously that deserve further attention. Earlier, we pointed out that effortful coping without feedback is associated with increased pathogenesis in rats and increased sympathetic response (possibly a precursor of pathogenesis) in humans. One corollary of this finding is that in the face of an objectively uncontrollable stressor, it is better to remain passive than it is to actively cope. Because there would be little if any feedback indicating that one is successfully manipulating his or her environment in an uncontrollable situation, the active coper would presumably suffer, whereas the passive coper would not. On the other hand, accepting a passive strategy may itself be a sign of helplessness or may eventually result in feelings of helplessness. As noted earlier, such feelings are themselves linked to disease outcomes and behavioral deficits.

The preceding analysis suggests that one cannot win in the face of an uncontrollable stressor. That is, if you actively cope, you increase your risk of disease because of an overarousal of the sympathetic system. If you fail to cope, you become (or already are) helpless and are at higher risk for the behavioral and health consequences of such a state. The possibility does exist, however, that in some cases passivity in the face of an uncontrollable stressor is adjustive (not increasing one's risk for disease or behavioral disorder). These are situations in which one chooses to remain passive as opposed to having his or her passivity forced on them because they have no effective active strategies. One may choose a passive strategy because not responding is perceived to be the most effective (or at least temporarily useful) way of dealing with the situation. In other words, when not responding is perceived by the subject as a coping strategy or chosen as a temporary state that the subject perceives can be replaced at any time with an effective coping strategy (e.g., perceived control conditions in be aftereffects paradigm), it is an effective means of coping with an uncontrollable stressor. When not responding is perceived by a person as a state of not coping, a state that if one had a choice would be avoided (e.g., no-escape condition in the helplessness paradigm), it has the detrimental effects associated with helplessness. Another way of viewing this distinction is whether or not passivity is a preferred state. When it is, it is an effective means of coping and when it is not, it results in helplessness-related effects.

Choosing not to directly cope with a stressor does not necessarily imply total passivity. One strategy that may be employed by those choosing not to directly cope is to accept the event and cope instead with its emotional impact (e.g., Lazarus &

Folkman, 1984). For example, Baum and his colleagues (Baum, Fleming, & Singer, 1983; Collins, Baum, & Singer, 1983) studied persons living near the Three Mile Island nuclear plant. These persons were viewed as confronting an uncontrollable stressor because a sharp decline in property values made moving (the direct response with the most likelihood of being effective) impossible for most. Baum found that over the long run, those using emotionally focused coping strategies reported less distress and fewer psychological symptoms than those using problem-focused strategies. Hence, it is possible that the effectiveness of passivity (a lack of direct coping attempts) in the face of an uncontrollable stressor may be partly or wholly due to refocusing efforts on an aspect of the experience that is controllable. It is to be hoped that future research on the relationship between passivity in the face of uncontrollable stressors will test the predictive validity of the distinction between preferred and nonpreferred states and examine the role of emotionally focused coping for those preferring not to directly cope with the stressful event.

Dispositional Factors Mediating Coping Costs

Some individuals are more likely than others to cope in ways that accrue costs. It is beyond the scope of this chapter to outline the relationship between personality factors, individual coping styles, and the coping costs discussed here, but the mention of a few salient possibilities should clarify our point. We mentioned earlier that Type A's may respond to challenge with greater effort and the concomitant physiological reactivity. Hence, A's may be more susceptible to arousal induced by effortful coping. Coping flexibility may be similarly important in preventing other coping costs. The importance of flexibility has been raised by Pearlin and Schooler (1978) who found that successful copers are those who are able to adjust their behavioral and emotional repertoires to fit the changing demands of their diverse life domains. Perhaps individuals who can bring a variety of coping strategies to bear on a particular stressor may be less prone to an overgeneralization of inflexible coping behaviors than those persons who have a more limited repertoire. Lazarus's distinction between problem-focused and emotional-focused coping strategies may also be important in determining coping costs. For example, individuals who characteristically combine problem-focused and emotional-focused strategies in dealing with various stressors may be less prone to fatigue effects than people who rely on one or the other coping style exclusively. Persons who tend to use emotional-focused forms of coping may also avoid the potentially pathogenic effects of active problem-focused-coping described before.

Cost of Coping: Summary

As outlined here, it is clear that the process of active coping, whether successful or unsuccessful, can have a severe impact on one's health and behavior. Table 2 provides an outline of the secondary effects of coping discussed previously.

In some cases this impact is initially mediated by physiological processes, whereas in other cases it is mediated by behavioral ones. Deleterious effects may

Table 2. Costs of Coping with a Stressful Event

	Mechanism	Effects
Successful coping		
Cumulative fatigue effect	Adaptive processes drain one's cognitive (psychic) energies	Poststress deficits in task performance, poststress insensitivity toward others, and increased deficits over time on demanding tasks performed under stress
Coping side effect	When coping behaviors, which were successful in ameliorating the possible effects of the stressor, are detrimental to individuals in other ways	Physiological effects that are above the level efficient for the body's metabolic needs. Possible interference with health maintenance behaviors.
Overgeneralization of a coping strategy	A strategy employed to cope with a stressor persists even in situations where a person is not confronted with the offending stressor	Deleterious effects on task performance, interpersonal behavior, and health maintenance behavior occur to the extent that the coping responses are inappropriate in other situations in which they are employed
Unsuccessful coping		
Cumulative fatigue effect	Adaptive processes drain one's cognitive (psychic) energies	Poststress deficits in task performance, poststress insensitivity toward others, and increased deficits over time on demanding tasks performed under stress
Coping side effect	When coping behaviors, which were employed in an attempt to ameliorate the possible effects of the stressor, are detrimental to individuals in other ways.	Physiological effects that are above the level efficient for the body's metabolic needs. Possible interference with health maintenance behaviors.
Helplessness	When an expectation that one cannot effectively cope with a stressor persists even in situations where coping is possible.	Undermines motivation to initiate further instrumental responses, interferes with learning that other outcomes are controllable, causes depression of mood, possibly results in a level of underarousal
Passivity in the face of stressful events	One does not engage in any coping behavior.	When not responding is a chosen (preferred) state, there are no secondary effects. When not responding is not the preferred state effects associated with helplessness are expected.

occur when (a) one engages in effortful coping; (b) one persists in using a coping response in situations where the strategy is not adaptive; (c) coping responses have deleterious side effects; and (d) one perceives that his or her efforts to cope are fruitless. Data and theory on passivity in the face of a stressor allow only tentative conclusions at this time.

On Studying Children under Stress

Our discussion up to this point has not distinguished between the stress-appraisal processes or coping strategies that are used by adults as opposed to those used by children. This issue is critical to our presentation because the data that we report have been gathered exclusively from third- and fourth-grade schoolchildren, whereas our theoretical discussion and research reviews are primarily based on work with adults. It is possible that children in this age group (8 to 9 years old) employ appraisal styles or coping strategies that are peculiar to their stage of development. Although an assessment of these processes, evaluation of their efficacy, and description of the roles they play in child development would be a worthy pursuit, it is one more suited to those concerned with developmental processes *per se* than to those interested in generic forms of coping with stress (see Garmezy & Rutter, 1983). Thus, our own emphasis is on determining whether various processes demonstrated in the adult literature are similarly operative in children.

A central question about the effects of stress on children is whether children are more or less vulnerable than adults. There are a number of reasons to expect that children are more susceptible to pathogenic and behavioral effects of the stress process. Young children (a) may have less cognitive capacity and thus lower thresholds for information overload (Cohen, 1978; Evans, 1978b); (b) may be unable to adequately anticipate stressors and consequently plan coping strategies; (c) may lack well-developed coping repertoires and thus have less flexibility in meeting various adaptive challenges and threats from their surroundings; and (d) have organs that are not fully developed that may be more vulnerable to the physiological effects of stress and/or the coping process than those of physically mature adults.

In addition, the lack of *control* that children have over their physical and social environments may be especially important in determining susceptibility. Unlike adults, children usually cannot regulate their initial choice to expose themselves to stressors nor can they regulate subsequent environmental conditions. For example, children are seldom able to make choices about moving away from a noisy neighborhood, adding sound insulation to their house, or otherwise attempting to attenuate noise exposure. They are also seldom able to directly complain about the noise source. As a result, they must frequently rely on adults to regulate their exposure to stressful conditions. Unfortunately, adult caregivers may also suffer from exposure to the same aversive conditions and be unable or unwilling to attend to the child's needs (Cohen, Glass, & Phillips, 1979; Evans, 1978b; Saegert, 1981).

In sum, there are many reasons to think that children may be more susceptible to environmental and psychosocial stressors than adults. Hence children may be viewed as a high risk population who can provide an idea of the potential effects of chronic stressor exposure.

Making the Most of this Volume

As we noted earlier, this volume was designed to serve multiple purposes. Because we felt that some persons would be interested in specific topics but not in others, each chapter is written to stand on its own. (The exception is that the details of the design of the Los Angeles Noise Project studies are included only in Chapter 2). However, we have also attempted to produce a volume that holds together as a whole, representing an integration of our theoretical views, previously existing literature, and the data provided by the Los Angeles Noise Project studies. We encourage even those who are primarily interested in one or two of the three categories of outcomes we discuss (control, health, and attention) to read the theoretical discussions in Chapters 1 and 6 and the methodological and practical insights of Chapter 2 because we think that these chapters provide new and interesting ways of viewing these literatures. Our hope is that this organization allows efficient and optimal use of the volume for both those who are broadly interested in the topics we discuss and those who are interested in one or more specific areas.

Correlational Field Methodology in the Study of Stress

The following chapter is an attempt to accomplish two interrelated goals: (a) to summarize methodological and practical issues that are central to designing and interpreting correlational field research on the relationships between stress, health, and behavior; and (b) to provide a description of the research designs, procedures, and statistical analyses used in the Los Angeles Noise Project. Our intent in raising methodological and practical issues is to present a context for viewing the noise project and to provide those not intimately familiar with correlation field research with the tools necessary to evaluate this project and other studies. We hope this section will also be used as an outline for the design of future studies. The purpose of the description of the noise project is to provide an overview of the methodological strategies used in these studies and to aid in the interpretion of noise project data presented throughout this volume.

Designing Naturalistic Studies of Environmental Stress

This section provides an introduction to correlational field research for the uninitiated and hopefully some insight for the initiated as well. It is targeted at behavioral scientists whose methodological expertise is grounded in the laboratory experiment and graduate and advanced undergraduate students who are interested in understanding how one designs and interprets correlational field studies.

In an *experimental* study, investigators use two techniques that allow them to eliminate alternative causal explanations for their results. First, subjects are randomly assigned to conditions. Assuming a large enough sample of subjects is used, random assignment assures the investigator that differences between experimental conditions are not due to differences between groups on any individual characteristic. In other words, random assignment is equivalent to matching experimental groups on *all possible individual difference factors*. Second, other than the experimental (independent) variable, all characteristics of the social and physical environments are held constant across conditions. For example, in an experimental study of the effects of noise on performance, the quiet and noise groups work in the same room, at the same temperature, with the same lighting, same instructions, same experimenter, and so forth. Hence, differences between conditions on the outcome (dependent) variable cannot be attributed to any environmental factor other than the (independent) variable under consideration. *Neither of these tools is available when studying naturally occurring phenomena.* There are, however, correlational research techniques that provide some control over individual and environmental factors. Later, we discuss these techniques in the context of the study of the relationship between exposure to environmental stressors and health and behavior.

There are a variety of correlational research designs that can be employed in the study of environmental stressors in naturalistic settings. These designs vary in the number of alternative causal explanations that remain when the study is completed as well as in their monetary costs, practical difficulties, and the amount of time and effort needed to conduct the study. They also vary in terms of the amount of information they provide in regard to the effects of stressor exposure. To no one's surprise, the costlier, more difficult, and more time-consuming research designs are the one's that typically provide the most information. In the next few pages, we describe a number of design alternatives and discuss the costs and benefits of each. In each case, we use the study of the effects of noise on health to exemplify the implications of the design under consideration. Following this discussion, we present a detailed description of the design and methodology employed in the Los Angeles Noise Project studies.

Individual versus Aggregate Data

The first decision made in designing a correlational field study is the level of analysis. Correlational studies of stressor exposure can be conducted at an individual or aggregate level. In studies on an individual level, measurements are obtained from a sample of individuals. For example, in the longitudinal study of the Los Angeles Noise Project, 262 children *each* completed a number of questionnaires, a variety of tasks, and had their blood pressures taken. In an aggregate-level study, groups are used as the unit of analysis instead of individuals; the means, medians, or rates of the group are used instead of individuals' scores. Hence analyses deal with samples of groups rather than samples of individuals. One might compare, for example, the death rate of census tracts around a noisy airport or noisy industrial area to a sample of census tracts that are similar demographically but are not located in noise-impacted communities. One problem with aggregate analysis is that the variable under study, for example, noise exposure, is often not equally distributed across the entire geographic area that defines the aggregate. In our example, an area of a census tract adjacent to the airport may have ambient noise levels that are considerably higher than areas of the same tract that are further away from the airport. Because measurements, like rates of illness and mortality, are aggregated over the entire geographic area, it cannot be determined to what degree such rates in experimental census tracts are influenced by those exposed to the factor under consideration and to what degree they are influenced by others living in the same geographic area but not so exposed. Aggregate-level analyses also preclude estimates of variability in individual exposures due to daily travel habits, occupational conditions, and the like. For example, one might live in a quiet census tract but spend 8 hours a day in a noisy factory and commute on a busy expressway. As we will show in Chapter 6, interactions of home and school noise levels are important for some measures of children's health and behavior.

Another limitation of aggregate data is that they do not allow examination of the influence of dispositional or attitudinal differences on the impact of the stressor or examination of the role of psychological mediators in stressor effects. For example, in the case of community noise exposure, we know from survey data that attitudes toward the source of the noise, for example, its purpose and whether it is viewed as necessary, are important determinants of annoyance. Moreover, there is reason to think that noise-induced effects on health and well-being are mediated by anger about the noise and its source (Cohen & Weinstein, 1982). Such hypotheses cannot, however, be addressed in aggregate data sets.

As noted, before, aggregate analysis is usually considered less desirable than individual-level analysis but is sometimes employed because it is a less costly method. There are cases, however, when aggregate-level analysis is preferred to individual level. These are situations where the effect of a risk factor is on a group process. Consider, for example, the effect of noise on the performance of something learned in the classroom. This effect may occur because of disruption of the classroom (group) process such as interference in teacher–student communication, student–student communication, or changes in teaching style. Because the impact of the risk factor is presumed to be on the group rather than at an individual level, an aggregate data point (classroom) is used as the unit of analysis rather than individual students. The analysis of school achievement data from the Los Angeles Noise Project employs an aggregate analysis of this type.

Cross-Sectional versus Longitudinal Designs

Cross-Sectional Designs. A cross-sectional study is one in which predictor and outcome variables are all assessed at the same point (a cross-section) in time.[1] Differences between experimental and control groups in cross-sectional studies can be attributed to any of three alternative explanations. Assume, for example, that workers exposed to noise have higher blood pressure levels than a group not so exposed. One possibility is that their increased blood pressure is caused by the noise exposure. A second possibility is that persons with high blood pressure levels tend to choose or tend to be assigned to noisy jobs. A final possibility is that some *third factor* on which noise and quiet groups differ may actually be responsible for both job assignment and blood pressure levels. Examples of individual-level third factors on which noise and quiet groups may differ include education, race, age, seniority with company, and job difficulty. Environmental third factors on which noise and

---

[1]We use the terms *predictor* and *outcome variables* to refer to a *hypothesized* causal chain. As we will discuss later, causal inferences are not possible in cross-sectional work.

quiet groups may differ include job-related exposure to air pollution, vibration, and heat. These factors tend to covary with noise exposure and could be the actual cause of elevated blood pressure. Select third factor explanations can be eliminated if the experimental and control groups are closely matched on the factor in question. For example, if noise and quiet groups have equal levels of education, blood pressure differences between groups cannot be attributed to educational level. However, even if the investigator matches noise and quiet control groups on several key factors, it is always possible that some unknown variable, not measured or controlled for, is responsible for both job assignment and blood pressure level.

Longitudinal Designs. In a longitudinal design, data are collected from the same individuals at more than one point in time. In many cases both the predictor and outcomes are assessed at each measurement point, but it is not uncommon for the predictor to be assessed only initially and for the outcomes to be measured at each subsequent measurement point. Longitudinal designs allow the investigator to view *changes* in one or more variables in relationship to one or more other variables. Data from longitudinal studies are generally subject to the same three categories of alternative explanations that apply to cross-sectional studies.

There is one form of longitudinal study, the *prospective* study, that allows the elimination of one category of explanation—the possibility that the predictor is caused by the outcome. A prospective study is oriented toward the future effects of the predictor(s) under consideration. The defining characteristic of a prospective study is that it assesses the relationship between a predictor or predictors at one point in time and an outcome or outcomes at a later point. For example, one could look at whether working in a noise-impacted or quiet section of a factory now is related to the incidence of hypertension 2 years from now. In order to eliminate the possibility that the outcome causes the predictor, one must also equate the experimental and control groups on the outcome variable (e.g., blood pressure) at the initial testing. If the experimental and control groups are identical in terms of the outcome variable at the start of the study, any relationship between the predictor as assessed initially and outcomes assessed at later testings cannot be due to the outcomes causing the predictor. For convenience, we will call a prospective study that equates the experimental and control groups on the outcome variable at the onset of the study a "true" prospective study. In our example, a true prospective study would eliminate the possibility that initial blood pressure level was responsible for assignment to noise or quiet job. Groups can be equated either by restricting the sample to those who are similar at the beginning of the study (e.g., including only those with low blood pressure levels) or by using statistical procedures, such as partial correlation and multiple regression, to equate mathematically the groups. Hence, the true prospective study functionally matches the experimental and control groups on the outcome under consideration and then looks at changes in the outcome over time as a function of the initial value of the predictor variable (see the discussion by Kasl, 1983, on additional complexities and limitations of prospective designs in the study of stress).

Although desirable, prospective analysis of longitudinal data is not always appropriate. An underlying assumption of such an analysis is that the predictor variable remains relatively stable over the period of prediction. For example, if noise exposure is the predictor variable, the prospective[2] analysis assumes that persons are exposed to a specified range of noise from the original point of prediction until the final point of measurement. If, for example, noise levels varied randomly for different people in the sample, the probability of finding changes in health or behavior associated with *initial* level of exposure would be small. The stability assumption is not usually a problem in studying exposure to environmental stress but can be a serious problem when predicting changes in behavior or health from unstable psychological variables like emotional states. It is advisable, however, when doing prospective analyses to check on the stability of the predictor across the period of prediction.

Although longitudinal studies provide information not available from cross-sectional work, they are more difficult to execute. One of the major difficulties is getting data on the *entire* sample at each measurement point. People move, become ill, change their minds about participating, have changes in their schedules, and the like, that make it difficult to get information on the same sample at several points in time. The longer the duration of the study, the greater this problem.

Subject attrition is an especially difficult problem if leaving the study is related to outcome variables. For example, what could one conclude if people with high blood pressure dropped out of a longitudinal study on the effects of noise more often than those with low blood pressure. If blood pressure levels in one group (e.g., persons living in noisy neighborhoods) are higher than in another (e.g., equivalent persons living in quiet neighborhoods), the attrition bias will lead to the groups looking more similar than they actually are.

Attrition biases can also differ across groups. For example, in the longitudinal noise project study, persons in the noise group with high blood pressure were more likely to drop out of the study than those with low pressure. However, in the quiet group, those dropping out and those remaining in the study did not differ in blood pressure levels. A bias of this sort can make groups look either more similar than they actually are (as in the example given before) or make them look different when they actually are not. This occurs because an attrition bias in one group but not the other results in an increase or decrease (depending on the nature of the bias) in one group mean relative to the other. In order to determine whether attrition biases were acting in the longitudinal noise project study, we used a regression procedure to determine whether attrition itself, or the interaction of attrition and the noise-quiet condition were related to any of the outcome variables as measured at Time 1. A more detailed description of this method is provided later.

A new procedure (not employed in our analyses) is also available for correcting

---

[2]The exception to this rule is when one is testing the hypothesis that a very short exposure to the risk factor is sufficient to put a person under long-term risk.

such biases in attrition when they occur (Berk, 1983; Heckman, 1979). In short, this procedure creates a model to predict who drops out of the study. It then uses this model to assign weights to the data of remaining subjects based on whether they are like other remainers or like those who dropped out. The utility of this analysis depends on the ability to predict attrition from variables that have been collected in the study.

The best way of dealing with attrition problems, however, is avoiding them. In any longitudinal study, one should use all available means to find and collect measurements from as much of the sample as possible at each measurement point. A procedure employed in many longitudinal studies is to ask each respondent for the name and address of a person who will always know where he or she lives. This information allows the investigator to continue to collect data from people who ·move. Strategies that are useful in keeping people involved in a longitudinal study include (a) gaining a written or verbal commitment from them at the beginning of the study; (b) assigning a specific interviewer or tester to each respondent so that each respondent establishes a relationship with someone connected to the study; (c) increasing incentives (pay) over the course of the study and providing a bonus for those who follow through to completion. The most important thing is that all research personnel treat respondents as intelligent human beings whose participation in the study is a personal favor (even if they are being paid).

Basic Requirements for Sound Field Research

The previous discussion describes two common field research designs and their potential strengths and limitations. Proper implementation of any design requires an understanding of the phenomenon under study and a resulting sensitivity to possible alternative explanations. In other words, all studies using the same generic design are not equal. For example, two cross-sectional studies of the same phenomenon can be quite different in regard to the kinds of inferences one can make from the results.

Later, we list a number of criteria for judging (and designing) field correlational studies. In creating this list, we have borrowed from a personal communication with Tom Wills outlining the criteria for sound epidemiological work. The impact of psychosocial and environmental factors on health is the balywick of epidemiologists. The methodological issues they face are the same as those that confront psychologists or other social scientists interested in the relationships between stress and behavior in human populations.

In our description of the criteria, we employ the epidemiological term *risk factor* to refer to any factor that might result in an increased probability of developing diseases, behavioral disorders, or other forms of behavioral or physiological disruption. For example, in the case of the noise project, exposure to intense noise levels is the risk factor under consideration.

Criteria for Judging Field Correlational Studies

Our discussion focuses on five criteria that are especially relevant to interpreting field studies. These include (a) inclusion of an experimental (or case) group exposed to the risk factor and a nonexposed control group; (b) respondents are unaware of the hypotheses of the research; (c) careful verification of exposure to the risk factor; (d) data are collected from the total population at risk or a representative sample from that population; and (e) the risk factor outcome relationship is documented in laboratory studies.

1. Data are collected from an experimental (or case) group, which is known to be exposed to the risk factor and from a control group not exposed to the risk factor. The experimental and control groups should be comparable on factors that would be expected to be related to the outcome variable under consideration. For example, consider studying the mortality rate of those living in communities impacted by aircraft noise. Because age, race, and socioeconomic status are associated with death rates, it is essential that the noise and quiet control groups are adequately matched on these factors. When matching is not possible, there are a number of statistical techniques that allow artificial control of extraneous variables. These techniques vary from stratification of the sample on the variable under consideration, for example, analyzing the data separately for persons less than 20 years old, 20 to 40, and so forth, to the use of partial correlation and multiple regression. These techniques can only be used, however, when the data on the variable in question are available. Hence careful consideration of potential alternative explanations for data must occur in the planning stage of the study. Factors that are often important to consider as possible candidates for "controlled" variables in studies of the relationship between stress, behavior, and health include age, sex, race, income, and education. However, the factors controlled for in any particular study depend on the outcome variable(s) under consideration. This is where an understanding of the phenomenon under study and sensitivity to possible alternative explanations are crucial.

2. Aside from the essential requirements of informed consent, respondents should be unaware of the hypotheses of the research. This is particularly important in research involving psychological variables, where respondents may be motivated to respond in such a way as to confirm what they perceive to be the investigator's hypotheses or to support a result that is consistent with their own political values. For situations in which it is probable that respondents are aware of the purposes of the research—which generally occurs when there is strong and generalized community sentiment about an issue—it is advisable to guard against response biases to the greatest possible extent. Such precautions include careful questionnaire construction, with multiple items and dimensions, imbedding critical questions in a more general context, use of nonobvious or unobtrusive measures, and use of physiological measures (cf. Webb, Campbell, Schwartz, & Sechrest, 1966). Optimally, one would use several types of measures and examine the degree to which the results converge.

Community noise surveys, where respondents are questioned about how annoyed they are with aircraft, traffic, or industrial noise, provide a prime example of this potential bias. In order to emphasize their concern with noise, respondents will often report that their level of annoyance is more extreme than it actually is. Commonly, this is prevented by presenting questions in a context that hides the purpose of the study, for example, imbedding noise questions in surveys about neighborhood satisfaction. One can also avoid leading the respondents by allowing them to raise the noise issue themselves if they think it is important. This can be done by starting interviews with open-ended questions in which respondents are asked to list the annoying aspect of their neighborhood. In the noise project, we collected multiple measures including ones that were not obviously concerned with noise, ones imbedded in larger contexts, and a physiological measure.

3. There should be careful verification of exposure to the risk factor. Probable dose should be determined if possible. *Dose* refers to a quantitative measurement that combines intensity of the risk factor (e.g., sound level, amount of CO in the air or level of density) with the duration of exposure. Dose-response relationships are a particularly valuable form of evidence because they can be used to determine the level of exposure that is necessary to put persons under risk. For example, the Occupational Safety Hazards Administration's industrial noise-level regulations are based on evidence for the dose-response relationship between sound level and hearing loss. Workers are allowed exposure to 90 dB(A) sound for 8 hours, 5 days a week. However, if they are exposed to noise at levels of 95 dB(A), they are allowed only 4 hours of exposure, and if they are exposed to noise levels of 100 dB(A), they are allowed only 2 hours exposure per day.

It is worth raising two methodological issues that are important for the establishment of accurate dose–response curves. First, as noted before, careful attention to exposure outside of the research cite, for example, measurement of daily habits, occupational exposure, and the like, can be important. Second, it is preferable that research designs include a variety of exposure levels. A wide range of values is important because some environmental agents may have nonlinear relationships to health and behavior. For example, many pollutants need to reach a certain threshold before aversive effects are noted.

4. The criteria discussed so far address the internal validity of a study— whether the research was done competently enough to allow any conclusions about the relationship between the variables under study in that sample. Assuming internal validity, it is also important to collect data in a manner that allows external validity—the ability to generalize one's results to the target population. Data should be collected from either the *total population at risk* or a *representative (usually random) sample* from that population. Self-selection by respondents into the study is not appropriate. Representative sampling allows one to generalize from the sample to the population that it represents. One way of estimating the probability that a particular sample is representative is the percentage of the target sample who decline participation. The greater the number who decline, the greater the probability of bias. Although few field studies have participation rates over 90%, it is a

reasonable guideline to express serious concern if the participation rate drops below 80% to 85%. If possible, it is desirable to compare nonparticipants with participants on demographic variables used to describe the sample. Optimally, the nonpartici-pants will not differ from the participants.

5. It is desirable if the presumed risk factor outcome relationship can be documented in laboratory studies. As discussed earlier, laboratory experiments can be used to establish a causal link between a risk factor and the outcome under consideration. Converging evidence from both laboratory experiments and field studies is compelling, even if the designs of the field studies are less than ideal. In cases where human laboratory experiments are not feasible (particularly if the disease or behavior under study has a long developmental course), or it would be unethical to expose humans to the risk factor under consideration, studies of in-frahuman species are invaluable.

There are other methodological and statistical criteria that should be applied in designing and evaulating field research. These include the importance of using psychometrically valid instruments that are truely measuring the concept under consideration (construct validity), using statistical analyses appropriate to the data (statistical validity) (cf. Cook & Campbell, 1979), and of replicating results across samples and measures. Because these criteria are as important in laboratory experi-mentation as in field correlational studies, we will not elaborate on them here.

To the extent that a particular study meets Requirements 1 to 4 (as well as having adequate construct and statistical validity), reasonably sound conclusions usually can be drawn from the data. Violation of any requirement demands serious consideration of its implications for interpreting the results. Violation of more than one requirement is often regarded as sufficient to invalidate the study in terms of contemporary scientific standards. To the degree that a *set* of studies reaches the same conclusions and meets Requirements 1 to 5, it is possible to make general inferences about the relationship between the risk factor and outcome.

Practical Considerations in Field Research Design

Methodological criteria are not the only determinants of how a field study is designed. Study design is also affected by practical considerations. Primary among these considerations is obtaining cooperation to carry out the study. There are two essential design issues involved in obtaining and maintaining the cooperation of a sample. The first concern is limiting the intrusiveness of the study procedures on the lives of the subjects. Important areas in this regard are the amount of time required of subjects and the degree to which their privacy is violated. The second concern is convincing those who make the decision regarding access to the sample that the study can provide them or individual respondents with some valuable information or other benefit.

Some critical judgment is required in determining how much time will be required of each respondent and what will be done during that time. An estimate of the maximum amount of time one can ask for should be made in light of the nature of the subject population and the nature of the situation. For example, in the noise project, we were concerned with the attentional span of the children and minimizing the amount of class time a child missed and the number of intrusions in any particular classroom. In the case of parent questionnaires, we wanted to ask only the essential questions so that the time commitment would be minimal and participation would be maximal. Time limits imply limits in the amount of information that can be collected from a respondent. A good guideline for choosing measures is to give the highest priority to those questions and tasks that provide you with information necessary for rigorous methodological control (e.g., education, age, gender) and the information essential for testing the theoretical propositions or practical questions that are being posed. There is always a tendency to add an extra measure or two because the study is set up, subjects are available, and so forth. When considering adding tasks, one should remember that additional task time increases the probability of boredom, inattention, response bias, and attrition. In short, increased subject time often constitutes a threat to study validity.

Questions perceived as intimate or violating one's privacy pose similar threats. Responses to these questions are often invalid, and asking such questions can detrimentally affect respondent cooperation. If not essential, such questions should not be asked. If necessary, the least intrusive form of the question should be chosen. For example, in the noise project we felt it was necessary to get some valid indication of social class of the children's parents. Although knowledge of parent income would have increased the validity of our estimate, we felt that such a question would be viewed as inappropriate by some respondents. Hence, we used two less obtrusive questions—years of education and number of children in the family.

Although investigators usually view the theoretical or practical questions that they are posing as more than justifying their intrusion on the lives of respondents, the respondents and those who control access to respondents may find such justification irrelevant and/or inadequate. Before entering negotiations with a group of potential subjects, it is essential to seriously consider the possible benefits of the project for the participants. If existing measures do not provide such benefits, it is worth considering adding measures that provide potentially beneficial information. In the case of the noise project, we were able to offer both the school and parents information on hearing loss and blood pressure levels in cases where it seemed advisable for a child to be examined by a physician.

In sum, field researchers must be sensitive to the feelings, attitudes, and needs of their respondents as well as to the methodological criteria discussed earlier. Often, such sensitivity requires compromises in study design. In most cases, if the investigator carefully weighs the costs and benefits of potential compromises, a well-controlled study can be carried out. There are cases, however, where the necessary compromises will seriously damage the purpose or methodological validity of the study. These are cases where the investigator needs to seriously consider whether the study is worthwhile.

Los Angeles Noise Project Studies

The Los Angeles Noise Project includes both cross-sectional and longitudinal designs. Moreover, the studies and analyses are conducted in a manner consistent with the criteria we have proposed as desirable for sound field correlational research. Unlike previous work in this area, the project places much emphasis on confirming results of laboratory experiments and on replication and extension of project results. As with theoretical issues discussed in this book, our understanding of some of the more subtle methodological issues grew with our involvement in the noise project. Hence, the noise project studies do not perfectly fulfill all of our proposed criteria.

From the Laboratory to the Field

Recent laboratory research on the impact of high-intensity noise has directed attention to the possible effects of community and industrial noise on a number of nonauditory systems. For example, noise is associated with alterations in task performance (see Chapter 5) and elevation of a number of nonspecific physiological responses (see Chapter 4). Exposure to noise that is unpredictable and uncontrollable can also result in *aftereffects*—deficits in performance and social sensitivity that occur after the noise is terminated (see Chapter 5). One difficulty with this research is that it emphasizes acute rather than long-term noise effects. Thus, its implications for those suffering prolonged exposure in their homes or at work are unknown.

Although investigators have also begun to take a closer look at the nonauditory effects of noise in naturalistic settings (see Chapter 4 and reviews by Cohen, *et al.*, 1979; Cohen & Weinstein, 1982; Kryter, 1970; Miller, 1974; Thompson, 1981), methodologically rigorous studies are rare. This research also tends to be atheoretical and thus difficult to compare with existing laboratory work. Moreover, there are few *longitudinal* studies of people living and/or working under noise. Thus, it is unknown whether prolonged noise exposure results in increasingly deleterious effects or whether those exposed for prolonged periods adapt to noise, with effects disappearing after awhile. Studies comparing measures of health and behavior of the same person before exposure, immediately after exposure begins, and at set intervals for one or several years would allow us to determine the long-term course of stress and adaptation. In addition, longitudinal studies in situations in which the environmental stressor is removed or attenuated would make it possible to determine whether there are long-term aftereffects of prolonged noise exposure.

This book reports data from the Los Angeles Noise Project—a set of studies of the impact of aircraft noise on elementary-school children. The purpose of the project was to study the course of adaptation and the impact of a noise-abatement intervention on a variety of physiological, cognitive, and motivational measures. It is particularly concerned with exploring the generality of laboratory work on noise-induced shifts in attentional strategies, feelings of personal control, and nonauditory physiological responses related to health.

Methodological Overview

Subjects in the noise project were children attending the four noisiest elementary schools in the air corridor of Los Angeles International Airport and students from three control (quiet) schools. Peak sound-level readings in the noise schools are as high as 95 dB(A), and the schools are located in an air corridor that has over 300 overflights a day—approximately one flight every 2 1/2 minutes during school hours (Lane & Meecham, 1974). Noise-abating architectual interventions were constructed (during summers) both before and during the course of the study. Hence, we were able to assess the effectiveness of the noise abatement and to determine whether effects of noise that occurred during the first year of the study would persist after a child was assigned to a quieter classroom.

The study focuses on effects of noise that occur during periods when the children are not exposed—effects that persist outside of exposure. Thus all tasks and questionnaires (except the achievement test records gathered from school files) were administered in a quiet setting—a noise-insulated trailer parked directly outside the school. In the longitudinal study, students were tested first in the spring of 1977 (T1) and again in the spring of 1978 (T2). Data will also be presented from a replication of this study run 1 year later using only third-graders. The third grade replication involved only one testing session.

Tasks administered during the test periods were designed to assess feelings of personal control and to determine whether the children employed some common

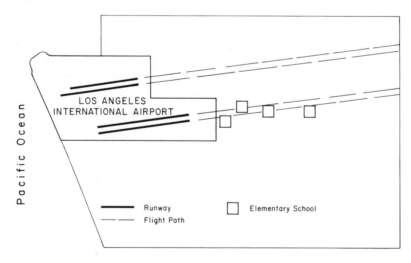

Figure 1. Location of noise-impacted elementary schools in relation to Los Angeles International Airport.

attentional coping strategies. Children were also asked a number of questions about their response to home and classroom noise and had their blood pressures measured. At the time of the first (but not the second) testing session, a parent questionnaire dealing with parent response to noise, mother and father's level of education, and the number of children in the family was sent home with each child. Scores on standardized reading and math tests and data on absenteeism were also collected from school files at the time of the first (but not second) session. Detailed description of the instruments and procedures are provided in subsequent chapters.

Cross-sectional data comparing children from noisy schools to children from quiet schools are drawn from both the initial and replication study in order to establish the relationship between noise exposure and the various outcomes under consideration. Data from subjects who were tested at both testing sessions (longitudinal data) were analyzed in order to determine if adaptation to noise occurs, that is, initial noise effects decrease or disappear, over the 1-year interval between sessions. Separate analyses (both cross-sectional and longitudinal) were conducted to evaluate the effectiveness of the noise-abatement interventions. These cross-sectional data were collected during the first testing session and compare children who were in noise-abated classrooms to those in noisy (nonabated) rooms and those from quiet schools. The longitudinal analyses look at the changes in response of children who moved from a noise to noise-abated classroom in contrast to those children who spent both years in noise-impacted rooms. The answers to questions about the effects of adaptation to noise and abatement effectiveness each require different blockings (or groupings) of the noise variable and analyses of different subsets of the sample. Table 3 provides an overview of these analyses. The table indicates for each analysis which sample is involved (Time 1, Time 2, attrition, and replication samples), when data for the sample(s) under consideration were collected, which of the groups are being compared (noise, quiet, abated), and whether there are any additional variables entered into the regression. It may be useful for readers to refer back to this table while reading the results sections of each chapter.

This chapter will provide descriptions of the procedure used to match noise and quiet samples and descriptions of subject characteristics, the noise measures, and the statistical analysis model. There is also a discussion of subject attrition and the analysis employed to determine the effects of attrition on the interpretation of longitudinal data. As noted before, other experimental tasks and procedures are described in subsequent chapters. The reader is referred to Cohen, Evans, Krantz and Stokols (1980) and Cohen, Evans, Krantz, Stokols, and Kelly (1981) for additional detail.

Matching

Three control schools (quiet schools) were matched with the experimental schools for grade level, ethnic and racial distribution of the children, the percentage of children whose families received assistance under the Aid to Families with Dependent Children program, and the occupations and education levels of parents.

Table 3. Overview of the Analyses Used in Original Study

Title of analysis	Sample	Classroom noise condition			Additional independent variable conditions
		1977 (T1)	1978 (T2)	1979	
I. Noise effects: Cross-sectional analyses	T1 & replication	(long-study) Noise vs. quiet		(replication) Noise vs. quiet	None
II. Attrition bias	T1	Noise vs. quiet			Retested at T2? Yes No
III. Adaptation to noise	Attrition (T1 & T2)	Noise^a ———— Noise / Quiet — vs. — Quiet			None
IV. Blood pressure: Habituation or attrition?	T1	Noise vs. quiet			Migration Not enrolled in school 1 year after T1 vs. enrolled in school in 1 year after T1 but not 2 years later vs. still enrolled 2 years after T1
V. Noise abatement: Cross-sectional analyses	T1 & replication	Noise vs. abated vs. quiet			None
VI. Noise abatement: Longitudinal analyses	Attrition (T1 & T2)	Noise ———— Noise / Noise — vs. — Abated			None

^aThe few classrooms that had had noise-abatement work completed prior to T1 are included as noisy classrooms in these analyses. This was justified by our findings suggesting little if any effect of abatement.

Initial school matching was based on aggregate archival data collected from the school districts and 1970 census records. In the case of census data, the census tract in which the school was located was used to provide estimated parent education levels. Mean values on each of these factors for noise and quiet schools are reported in Table 4. Thus, we were able to compare samples of children attending noise and quiet schools who were relatively similar in terms of age, social class, and race. Data on race, social class, and mobility were also obtained from the children in the sample and from their parents. These data allowed additional information on the adequacy of the matching procedure and were used in a regression analysis procedure (described later) that provided additional statistical control over these factors.

Subjects

The study included children from all noise-impacted third- and fourth-grade classrooms in each noise school as well as children from an equivalent number of classrooms in quiet schools. Eighty-seven percent of the students received parent permission and participated in the study. To ensure that performance differences between children from noise schools and those from quiet schools could not be attributed to noise-induced losses in hearing sensitivity, an audiometric pure-tone threshold screening was administered to each child. Children were screened at 25 dB(A) for select speech frequencies (500, 1000, 2000, and 4000 Hz). Children failing to detect 25 dB(A) tones at any one of these frequencies in either ear were not included in the study. Six percent of the noise-school children and 7% of the quiet-school children failed the screening. A total of 262 subjects (142 from noise schools and 120 from quiet schools) remained in the initial study. Ninety noise-school children and 75 quiet-school children remained in the third grade replication. Individual analyses, however, sometimes contain fewer subjects because of missing data.

Data compiled from the parent questionnaire allowed us to determine the degree of similarity of the prematched noise and quiet samples. In the initial study, there were no differences between the samples on the various social class factors. These data are presented in Table 5. For the parent education scale, "some high school" was scaled as 3 and "high school graduate" was scaled as 4. In the initial sample, quiet and noise groups did not differ on father or mother's education or number of children per family. However, the racial distribution did differ significantly, $\chi^2(3) = 10.5, p<.01$, with the noise group containing more blacks (32% vs. 18%) and quiet group more Chicanos (50% vs. 33%). Noise and quiet samples had nearly equal percentages of whites (32% and 29%, respectively) and of unidentifiable or mixed-race children (3% in each sample). The two samples also differed on mobility, with children in the quiet sample having lived in their homes longer (a mean of 49.6 months vs. 41.4 months) and having attended their schools for longer periods (a mean of 43.2 months vs. 36.0 months) than noise children, $F(1, 270)=4.8, p<.03$, and $F(1, 270)=12.9, p<.001$, respectively. Length of school enrollment was not related to father's education, mother's education, or the number of children in the family.

Table 4. Mean Values for Aggregate Factors Used to Match Noise and Quiet Schools

| | Race | | | Socioeconomic index (state percentile) | AFDC | Bilingual | Pupil mobility | Census tract education | | | | | |
	BL	WH	SP					6th Grade or less	7–11	High school grad	1–3 College	College grad	Greater
Noise Schools	40%	18%	39%	13.75	40.78%	37.30%	52.98%	8%	33%	37%	16%	4%	2%
Quiet Schools	39%	28%	29%	5.67	41.77%	34.53%	51.80%	9%	36%	33%	16%	4%	2%

*Table 5. Noise and Quiet Sample Comparability as Indicated
by Parent Questionnaire Data*

	Original study		Replication study	
	Noise-school children	Quiet-school children	Noise-school children	Quiet-school children
No. of children in family	3.54	3.88	3.26	3.57
Father's education (3 = some high school, 4 = high school graduate)	3.75	3.41	4.14	3.81
Mother's education	3.64	3.35	4.10	3.88
Percentage				
black	32%	18%	54%	44%
white	32%	29%	22%	24%
Chicano	33%	50%	24%	32%
unidentifiable or mixed race	3%	3%		
No. of months enrolled in school	41.4	49.6	29.52	34.13
No. of months living at present address	36.0	43.2		

Moreover, the noise and quiet samples were relatively equal on these various social class factors across all durations of exposure. This finding suggests that the decision to continue living in the noise-impacted area was not determined by the parents' socioeconomic status. There were, however, more blacks and whites in the noise group with less than 2 years exposure than there were in the equivalent quiet group, $\chi^2(4)=12.04$, $p<.02$. There were no differences in racial distribution for other exposure durations.

In the replication sample, the only factor on which noise and quiet groups differed was mobility. As in the original sample, quiet-school children were enrolled in their schools (34.13 months) longer than noise-school children (29.52 months), $F(1, 16)=4.03$, $p<.05$. These data are also presented in Table 5. As discussed later, both studies employ statistical controls for number of children in the child's family, grade in school, months enrolled in school, and race.

Air Pollution

A possible alternative explanation for differences between the noise and quiet samples is that the noise group was exposed to higher levels of air pollution than the quiet group. Such an alternative is very unlikely. Sulfur dioxide was minimal at all the school sites, never exceeding the California standard (South Coast Air Quality Management District, 1977; State of California, 1976). Ozone and nitrogen dioxide standards were exceeded, but maximum levels were slightly higher at the control schools than at the airport schools. The maximum 1-hour rates in any school area for ozone (.21 parts per million) and NO (.60 ppm) were below levels that generally show any effects on human behavior or health (Morrow, 1975; National Academy of Sciences, 1977c). Maximum carbon monoxide was slightly higher in the airport schools (30 vs. 27, 22 ppm), but average values were identical (6 ppm). The differences in maximum values of 8 ppm are negligible, and human effects from CO concentrations of less than 40 ppm are extremely rare (National Air Pollution Control Administration, 1970). Note that we have used maximum values in arguing against an air pollution alternative, thus presenting a very conservative counterargument. Average values in all cases were considerably below established standards.

Noise Measures

The intensity of sound is commonly expressed in decibels (abbreviated dB). The decibel scale is logarithmic with an increment of 3 dB reflecting a doubling of the intensity (energy) of the sound. An increment of about 10 dB is required, however, for a perceived doubling of sound. For comparative purposes, California law states that noise abatement must occur when the noise level inside the classroom exceeds 50 dB(A) and federal rules suggest a maximum of L_{10} (noise level exceeded 10% of the time) = 55 dB(A).

Testing Session 1. Interior sound levels (without children) were measured inside each classroom with Tracoustics Sound Level Meters (SLM S2A). Peak decibel level (A scale) was recorded during 1-hour sessions in both the morning and afternoon.

Testing Session 2 and Replication Study. Sound levels (again without children) were measured inside each classroom for 1 hour during the morning and 1 hour during the afternoon with Digital Acoustics (DA605), B and K (4426), and General Radio (1945) noise-level analyzers. The machines were calibrated to a pure tone source every other day and were periodically calibrated against one another to ensure intermachine reliability. Microphones were placed approximately 3 feet (.9 m) from the ground in the center of the room. Data available from all machines included peak decibel level (A scale), the decibel level that exceeded 33% of the time (L_{33}), and the noise level averaged on an energy basis over each hour period (L_{EQ}).

Sound-level data from the longitudinal study are presented in Table 6. Data from the replication study are presented in Table 7. It is important to note that due to limitations in the equipment and duration of the measurement, these measures are presented only to establish relative differences between the sound levels of various types of classrooms, not as evidence for sound-level criterion or threshold levels of effects. Under optimal conditions, time-integrated noise readings would have been conducted for the entire school day at several times of the year. In other words, we can make conclusions about the differences between noisy and relatively quieter classrooms but do not have the quality or quantity of noise measurement required for establishing dose-response relationships.

A number of analyses reported in the following chapters examine the rela-

Table 6. Mean Classroom Noise Levels at Time of
Second Testing Session for Initial Study

	Noise measure		
	L_{EQ}	L_{33}	Peak dB(A)
Noisy[a]	70.29	55.82	91.50
Abated	62.82	49.27	71.27
(Noisy & abated)	65.13	51.71	81.05
Quiet	62.05	48.63	74.42

[a]There were 7 noisy classrooms, 16 abated classrooms, and 19 quiet classrooms in the second testing session.

tionship between *home* noise levels and the various outcomes under consideration. Data on home noise levels were extracted from a noise-contour map calculated by computer extrapolation of noise levels measured at monitoring stations around the airport. These data are reported in terms of Community Noise Equivalency Levels (CNEL). CNEL is a measure of community noise giving more weight to noise occuring between 1,900 and 2,200 hours and the most weight to noise occuring between 2,200 and 0,700 hours (cf. Peterson & Gross, 1972). Because home noise levels are only available for noisy schoolchildren, these analyses include only the noise samples.

Statistical Analyses

A regression technique was used in all the analyses reported in this book to allow additional control over the effects of socioeconomic and demographic factors (cf. Cohen & Cohen, 1975). All data analyses include controls for the number of children, the child's family, grade in school, months enrolled in school, and race. These control factors are forced into the regression first, followed by noise, and

Table 7. Mean Classroom Noise Levels for
Replication Study

	Noise Measure		
	L_{EQ}	L_{33}	Peak dB(A)
Noisy[a]	66.10	54.40	85.60
Abated	63.80	47.60	72.80
(Noisy & Abated)	65.33	52.13	81.33
Quiet	57.01	49.30	75.15

[a]There were 10 noisy classrooms, 5 abated classrooms, and 12 quiet classrooms in the replication study.

then the noise by months enrolled in school interaction. Additional controls are used in the analyses of blood pressure (height and a ponderal index defined as weight/ height^3), school achievement (cognitive aptitude test) and distractibility (performance under ambient conditions). The primary helplessness analyses include factors for success/failure (those who solved and those who did not solve the success treatment puzzle are treated as separate groups) and the interaction between success/failure and noise. Analyses of longitudinal data also include a repeated measure factor (Testing Session 1, Testing Session 2). School achievement analyses were performed with classrooms (nested in noise), rather than individual children as the unit of analysis. A more detailed description of the form of each analysis is provided in Cohen *et al.* (1980).

The various measures were analyzed in predetermined multivariate clusters created on the basis of theoretical consideration. This form of analysis helps to decrease the probability of chance findings that occur when a large number of analyses are necessary (cf. Bock, 1975).

Interpreting Longitudinal Analyses: Sample Attrition Bias. An effort was made to retest all students who were attending school during the longitudinal follow-up of the initial study. Sixty-two percent (163: 83 noise and 80 quiet) of the original sample (262: 142 noise and 120 quiet) were retested. Although a slightly higher proportion of quiet (67%) than noise- (58%) schoolchildren were retested, this difference was not statistically significant, $\chi^2(1) = 1.99$, $p<.16$.

All data analyses that include data from the second testing session (these are all repeated measures designs) were based on the 163 retested students—the attrition sample. Sample attrition (not being retested) may be attributable to either migration or absenteeism. It is our purpose at this point to describe the nature of any self-selection bias in the retest sample; thus, these causes of attrition are not separated.

One purpose of the attrition bias analyses was to determine whether remaining in the study (being retested) was correlated with one or more of the criterion variables, for example blood pressure. A second purpose was to determine whether attrition was correlated with one or more of the criterion variables in one of the study's conditions (noise or quiet) but not in the other. For example, noise-school children who were not retested had higher blood pressures than those who were retested; whereas being retested was unrelated to blood pressure for quiet-school children. This particular attrition bias resulted in a deflated mean blood pressure for noise-school children when analyzing data based only on those present at both Times 1 and 2 (the attrition sample). To determine whether any such attrition biases occurred, data from the first testing session (all of the original 262 subjects) were analyzed in relation to whether or not a student was retested (yes/no), and a Retest × Noise interaction added to the standard analysis (see Table 3). Note that these analyses are not presented in an attempt to make any conclusions about those who were retested versus those who were not but are being used only to provide information about the nature of the attrition bias that may be useful in interpreting analyses presented later in this volume. For this reason, a rather liberal alpha level (.10) was employed, and multivariate analysis is not reported.

There were no significant relationships between any criterion variable and attrition (main effect of retest yes/no variable). Retest biases varying in noise and quiet groups (Retest × Noise interaction) occurred in the case of three variables. On all of these variables, those in the noise condition who showed the greatest strain during T1 were not present at T2. No such relationship (or in some cases a slight reversal) existed in the quiet group. The variables with Noise × Retest interactions suggesting this pattern were the child's perception that noise made it difficult to hear his or her teacher, $F(1, 241) = 3.46$, $p<.06$, and systolic, $F(1, 233) = 8.65$, $p<.004$, and diastolic, $F(1, 233) = 3.39$, $p<.07$, blood pressure. The implications of these biases for interpretation of longitudinal analyses will be addressed in appropriate chapters.

Los Angeles Noise Project Design Criteria. As noted before, the noise project studies do not perfectly fulfill all of our proposed criteria. They do, however, provide an example of both cross-sectional and longitudinal research that addresses both the practical and methodological issues raised in the first section of this chapter.

Personal Control and Environmental Stress

Many early human behaviors, such as visual exploration, grasping, crawling, and attention toward novel stimuli, share a common biological basis of competence, striving for effective interaction between the human infant and the environment. Since White's original arguments on the validity of effectance motivation as a model of human behavior, a large, complex literature has evolved on the importance of personal control in human behavior. The major thrust of this work has argued that human beings have a strong need for self-efficacy (Averill, 1973; Lefcourt, 1973). Furthermore, it is generally more adaptive and healthier to exercise control over the environment when the opportunity to do so is available (Glass & Singer, 1972; Rotter, 1966, 1975). Finally, when confronted with situations that are uncontrollable, people may suffer negative consequences including negative affect and reduced motivation to establish subsequent instrumental behaviors when they have the freedom to do so (Seligman, 1975).

Personal Control

Control will be considered here in terms of the coping taxonomy developed in Chapter 1. When attempts to cope with nonoptimal, environmental conditions fail, important effects occur because of individual perceptions of uncontrollability and/or their inability to cope with the demands of the specific stressor. Conversely, the provision of different types of control or amounts of control can influence feelings of self-efficacy, mastery, and so on. Control over a stressor may also reduce or modify the stressor itself. Our focus in this chapter is on (a) the effects of varying control over stressors; and (b) the influence of losses of control over aversive events.

Types of Controls

One reason why the stress and control literature is so complex is because of the myriad of usages and definitions of control. Reviews of personal control have emphasized that the stress-reducing properties of control depend upon the meaning of the control responses to the person (Averill, 1973; Miller, 1979; Thompson, 1981). An important implication of this principle is that control serves multiple functions with varying degrees of positive or negative utility. Averill, in his 1973 review, distinguished between three kinds of control: *behavioral, cognitive,* and *decisional.*

Behavioral Control. Behavioral control is the availability of a response to the individual that directly influences the objective characteristics of a stressor. Typical examples of behavioral control are regulated administration and stimulus modification paradigms. In regulated administration procedures, individuals prefer to self-administer aversive stimuli in comparison to experiencing externally controlled, aversive events. One important problem with the self-administration paradigm is that it confounds predictability with control. Individuals know when the aversive stimulus is going to occur as well as when to administer the stressor. We discuss this issue further under the subsection, "Functions of Control."

The evidence for the stress-reducing properties of regulated administration is mixed (Averill, 1973; Miller, 1979). During the time period immediately prior to the onset of an aversive event (anticipatory period), reduced anxiety and physiological arousal occurs for self-administered stressors. During the actual impact period arousal seems to be less, but self-report data are equivocal.

In a stimulus modification paradigm, the subject can substantially modify or prevent entirely the onset of a negative event according to his or her response. Note that in this paradigm, unlike regulated administration, the subject is instructed that coping behaviors can be instrumental in affecting the environment. In the regulated administration paradigm, one can control when a negative event occurs, but one cannot change the fact that the event will occur. In both perceived and actual stimulus modification paradigms, the subject believes that she or he can influence the stressor. The difference is that in the former design the subject never actually uses the instrumental response available to her or him. Whereas, in instrumental control studies, the participant does modify the aversive event by his or her coping behaviors.

The effects of instrumental or behavioral control are quite complex depending upon several factors, including (a) temporal factors and (b) feedback following coping responses. During the anticipatory period, most studies reveal reduced arousal and less anxiety if subjects believe they have a behavioral response available that can influence the nature of the aversive event. During the impact period, behavioral control has a very mixed influence on stress. Self-report data are generally equivocal, whereas physiological indexes are contradictory (Miller, 1979; Thompson, 1981). The number of coping responses one must make and the clarity of feedback about one's coping utility are crucial in determining whether behavioral control is helpful or not. If feedback is inconsistent or takes a long time, then behavioral control may exacerbate the stressful impacts of an aversive event (Averill, 1973; Weiss, 1972). As discussed in Chapter 1, Weiss found the highest ulceration rates in animals who attempted to cope instrumentally with shock but received inaccurate feedback on their responses. Animals who passively accepted the aversive stimulation had less ulcers than the active coping, inappropriate feedback groups. Animals that coped effectively (instrumental response that avoided or reduced shock) exhibited the least amount of physiological stress.

Furthermore, work by Glass and Singer (1972) and others (see Cohen, 1980) on perceived control and stressors (e.g., noise) clearly shows that aftereffects of stress are dramatically reduced if the individual perceives that she or he has control over termination of the stressor. This effect holds even though the individual actually does not implement instrumental control. Just believing that one has control over an aversive event has powerful consequences during the aftereffect period following a stressful encounter.

Cognitive Control. Cognitive control is the ways in which events can be interpreted, appraised, or incorporated into some cognitive plan that influences the net stressful impacts of the event. Two major ways in which researchers have analyzed cognitive control are to manipulate the amount of information about an aversive event (e.g., warning signal) or to analyze the various cognitive strategies people use to alter the nature of the stressor's impact on them (e.g., denial, distraction) (Averill, 1973; Thompson, 1981).

As in the case for behavioral control, the data on the efficacy of cognitive control are varied. Informational control has mixed effects depending primarily on the utility of the information. If information about an aversive event provides the person with sufficient time to respond so as to either modify the event itself or to give him or her some time to rest during intermittent periods, then information reduces anxiety and physiological arousal (Averill, 1973; Thompson, 1981). Greater stress occurs, however, if information control is provided without opportunities for response or rest. If one is warned about an aversive event, this allows for the possibility of instrumental coping or at least raises the expectation that it may be possible. However if the situation is structured so that this expectation cannot be met, more negative consequences occur than if no expectation for control had been raised (see Chapter 1).

Information can also be provided about the causes or sensations associated with a negative event (e.g., what the pain may feel like). This type of information is different than a warning signal in that no instrumental coping responses are possible. Information can help reduce stress, however, particularly if individuals are provided some outlet to handle the heightened arousal produced by the preparatory message (Janis, 1983; Thompson, 1981).

Two types of cognitive strategies used for regulating stress are *sensitization* and *avoidance* (*denial*) strategies. Sensitization means focusing on the effects and feelings associated with an aversive event. Moderate sensitization appears to enhance anxiety and arousal during the anticipatory period but reduces negative consequences during the impact and poststressor periods. Extreme sensitization, however, can produce fear and possibly panic. Denial and other avoidance strategies such as distraction essentially produce the opposite pattern of results (Janis, 1958). One reason why information about the causes or consequences of aversive events may facilitate effective cognitive control is because this type of information reduces the complexity and ambiguity of an already negative situation. As discussed more thoroughly in Chapter 5, uncertainty may greatly increase the information overload produced by stressors because more environmental monitoring is required (Cohen, 1978).

Recent analyses of preparatory information for surgery or other aversive events suggests that cognitive control is most effective when it enhances knowledge of impending procedures, provides information and practice with coping skills (e.g., relaxation training for pain), and promotes balanced optimism about the anticipated gains from the experience (e.g., enhanced physical vigor from a coronary bypass). Adaptive, cognitive control induces hope and self-confidence that the procedure will work and the patient will meet the challenge of pain or other stressors associated with the aversive event (Janis, 1983).

Decisional Control. Averill's (1973) third control category, *decisional control,* refers to opportunities to choose among alternative forms of action. As Thompson (1981) notes, however, very little conclusive evidence is available about decisional control because volunteer participants in research, by definition, have decisional control. Averill raises the interesting point that having too much decisional control can prove stressful because one does not know what to do or how to act, particularly if in a novel or complex situation.

Retrospective Control. A final type of control is retrospective control. Retrospective control is beliefs about the cause of an event after it has happened (Thompson, 1981). People need to perceive an orderly and meaningful world. In fact most people consistently overestimate the amount of control they have over environmental events (Weisz, 1983). Thus coping with past, negative events such as disasters or personal accidents is often associated with self-blame. In this way one is able to maintain the pretense of personal mastery over environmental events (Thompson, 1981). This belief also allows us to feel that similar, aversive events can be prevented in the future. Analogously, when a negative event happens to someone, many people tend to blame the victim rather than accept that his or her misfortune was due to fate or an "unjust" world (Lerner, 1970; Walster, 1966).

Alternatively, retrospective control can be interpreted as changing one's beliefs in order to accomodate an event(s) where one really has no control. Rothbaum, Weisz, and Snyder (1982) argue that secondary control, changing the self to fit the situation, is frequently manifested by strong, persistent needs to explain uncontrollable events. This view of control differs from Thompson's in that retrospective control is not seen as an attempt to maintain a sense of mastery or primary control over the environment. Instead, under secondary control the motivation is to accomodate, to fit in with uncontrollable environmental forces.

Comparisons among Types of Control. A few studies have directly examined different types of control over the same set of aversive conditions. Sherrod, Hage, Halpern, and Moore (1977) compared perceived control over the initiation, termination or both types of controls of an aversive event. In the initiation control condition, subjects were told the experimenter preferred that they perform tasks with background noise on but that they could leave the noise off if they so desired. Thus, subjects could choose whether or not to turn on the background noise. In the termination control condition, subjects had no choice to turn the noise on but were provided access to a turnoff switch. As in the initiation condition, subjects were encouraged to leave the noise on. In the both-controls condition, subjects had initiation and termination control. Control over initiation and termination of the aversive event had a greater but nonadditive, ameliorative effect than either termination or initiation control. Furthermore, termination control produced stronger stress reduction than initiation control.

Although Sherrod and his colleagues found that the greater the amount of behavioral control over an aversive event, the less stress, more complex findings have been noted when different types of controls are mixed together. For example, Corah and Boffa (1970) manipulated whether subjects could terminate a loud noise (behavioral control) as well as subjects' attitudes about their freedom to respond (decisional control). Subjects with high decisional control were told that it was up to them whether or not they terminated the noise. Behavioral control alone reduced stress, but the combination of behavioral and decisional control produced more stress than either form of control by itself.

Some analogous difficulties with multiple control procedures have been noted by Krantz and his colleagues. Mills and Krantz (1979) examined blood donors' reactions to information control, behavioral control, or both types of controls. Informational control focused on detailed explanations of the donation procedures and sensations the donor could expect to feel. Behavioral control gave the patient choice over which arm would be used to draw blood. Of particular interest, the combination of behavioral and informational controls caused more stress than either form of control by itself. Each single form of control reduced physiological and self-reported stress in comparison to a baseline group.

Both the Corah and Boffa and the Mills and Krantz studies are important because they show that different forms of control are not only nonadditive but can actually produce synergistic, negative effects. One possible reason for the negative effects of combining behavioral control with other forms of controls is that subjects are given too much control, too many decisions, in an aversive situation that is already somewhat uncertain. In such situations, people might prefer to have the more practiced, competent professional in charge (experimenter, nurse, respectively in the preceding two studies) make the appropriate choices.

In a final study on different types of controls, Cornelius and Averill (1980) exposed subjects to electric shock and provided behavioral control in some conditions by allowing participants to advance a counter delivering the shock at the participant's own rate of speed. Cognitive control was implemented by providing some subjects information about the shock effects and sensations. Finally, decisional control was varied by enhancing the importance of the research and individuals' perceptions of their involvement and contribution to the research project. Overall, the results supported an interactive model. Different types of controls had different effects, and their multiple implications were again nonadditive. For example, physiological arousal was higher with behavioral control only under conditions of no cognitive control; whereas with cognitive control, behavioral control reduced arousal.

Psychophysiological research on control and stress also lends support to Averill's contention that control is a complex, multifaceted concept with variable consequences in a stress context. Throughout the taxonomic descriptions of the different kinds of controls as well as the preceding studies where different types of controls are examined together, it is apparent that control may reduce or increase physiological arousal. Several recent studies, as reviewed in Chapter 1, suggest that the degree of effort necessary to exert control is an important factor in accounting for the psychophysiological consequences of control. When efforts to exert control are relatively easy, behavioral control does appear to reduce arousal. However, when control is more difficult to exert, the effort involved may increase sympathetic activity.

Summary

Control has been operationalized in several different ways in research on stress and control. Four major types of controls—behavioral, cognitive, decisional, and retrospective controls—have been discussed. Table 8 provides a summary of these four types of controls. Some important distinctions in the various uses of control that influence its impact on the effects of stressors include:

1. whether control creates the perception that one's own behavior can modify the aversive event;
2. whether information about an aversive event allows some modification of the event or facilitates appraising the situation as nonstressful;

Table 8. Types of Control

Behavioral control—availability of a response to influence the characteristics of a stressor
  1. *Regulated administration*—self-administered exposure to aversive stimuli vs. externally regulated exposure
  2. *Stimulus modification*—individual control over onset or termination of a stressor
Cognitive control—cognitive strategies for interpreting aversive events
  1. *Information*—warning signals about impending events
  2. *Cognitive strategies*—information about causes/consequences of aversive events
    a. *Sensitization*—focus on effects and feelings associated with events
    b. *Avoidance*—repression or denial of effects, distraction
Decisional control—opportunities to choose among alternative forms of action (e.g., volunteer)
Retrospective control—beliefs about the causes of aversive events after they have occurred
  1. *Illusion of control*—overestimation of objective control; self-blame in accidents; blaming victims of fate
  2. *Accomodation*—changing one's beliefs to accomodate an uncontrollable and unavoidable aversive event

Note. See text for more details and references.

3. how many coping responses are required to control the stressor and the clarity of feedback about their utility;
4. how much effort is required to exert control over the aversive event; and
5. to what extent the presence of control engenders feelings of self-efficacy in the individual.

All of the major reviews of control conclude in some form that the meaning of the control response to the individual is critical in determining whether control will reduce stress. In particular, a critical aspect of the meaning of control seems to be feedback to the individual about his or her competency in effectively interacting with the environment. Effectance motivation has remained a central underlying construct in research on stress and control (White, 1959).

Functions of Control

The belief that persons are in control in the sense that their actions potentially can influence their environment seems to be a deep-seated aspect of human functioning (Lefcourt, 1973; White, 1959). In this subsection of the chapter, we overview models of control.

Mastery. According to this perspective, control is helpful to the extent that it provides individuals with evidence that they have mastery and personal competence. The primary evidence for this model of control is data showing the powerful negative effects that transpire when freedom to act is restricted (Brehm, 1966), or when one learns that the outcomes of his or her behavior are independent of his or her responses (Seligman, 1975). This model of control follows from the general effectance model of motivation presented by White (1959). Folkman (1984) also alludes to this function of control, noting that control is especially important for self-identity when the issue or target of control is particularly important to the individual.

When extremely stressful experiences (e.g., cancer diagnosis) occur, people reflect strong desires to enhance their personal control over the event. These efforts include positive social comparisons to others in difficult circumstances and focusing on enhancement of life satisfaction in other domains that accompany the disease (Taylor, 1983). There is also evidence that young children in particular as well as adults overestimate the amount of actual control they have in various circumstances (Weisz, 1983). Presumably, overconfidence in control provides greater optimism and persistence in the face of difficult or aversive circumstances. Depressed individuals are more accurate in their assessments of personal control (Weisz, 1983).

Control and Coping. Control can also be understood in terms of its influence on the stress and coping process. Greater instrumental control, for example, is more likely to engender problem-focused coping as opposed to emotion-focused coping. Problem-focused coping includes attempts to manage environmental conditions producing stress, whereas emotional-focused coping refers to regulation of emotions/distress associated with reactions to aversive environmental situations (Folkman & Lazarus, 1980; Lazarus & Launier, 1978).

The match between one's appraisal of controllability and actual control available in a situation can also help explain why control is sometimes helpful and at other times not useful or even harmful (Folkman, 1984; Rothbaum *et al.*, 1982). If a situation is actually controllable and one copes with it instrumentally, one is more likely to have a positive outcome from the exercise of control. On the other hand, if the situation is, in fact, not controllable, persistent instrumental efforts may lead to negative outcomes. Under such uncontrollable environmental circumstances, more palliative coping strategies are likely to be more fruitful (Folkman & Lazarus, 1980). As one example of more adaptive use of palliative coping strategies in the midst of uncontrollable environmental circumstances, Collins *et al.* (1983) found that residents near the Three Mile Island nuclear power plant who coped with more emotion-focused strategies experienced fewer psychological symptoms than nearby residents who tried to cope instrumentally with the Three Mile Island disaster. Similar results have been found in laboratory studies of coping with unavoidable stressors (cf. Monat, Averill, & Lazarus, 1972).

Rothbaum and colleagues suggest that many of the behaviors associated with more adaptive emotion-focused coping in the face of negative, uncontrollable events can be seen as actions to change the self in order to accomodate to an environment that cannot be altered through instrumental efforts. Thus, disappointment at failure to control an uncontrollable situation can be avoided by self-attributions of severely limited abilities and lowered instrumental expectancies. Alternatively, one could attribute outcomes to chance or the power of others in positions of greater authority (Rothbaum *et al.*, 1982). Another type of accomodation may be to search for alternative areas in one's life to assert control (Taylor, 1983).

Folkman's description of control and stress is also helpful in explaining some of the mixed findings on control and stress. For example, several studies report that self-regulated administration of an aversive event seems to help during the anticipatory period, whereas, during the impact period itself, control avails little or no positive effects (Averill, 1973). According to Folkman's model, we could interpret the match between appraisals of control and actual control during the anticipatory period as reasonably congruent. However, during the impact period of a stressor, there is no control over the situation that is possible. As some work suggests (Averill, 1973; Janis, 1958), persons frequently suffer greater negative outcomes if expectations of control remain high during impact periods. Whereas, if expectancies of control are lowered during impact periods, there is subsequently a reasonably good match between actual and appraised controls. Folkman's model is also consistent with the general finding that informational control is helpful only when the person has sufficient time to modify the stressor, or if the information provides signals about intermittent safe periods when rest can occur (Averill, 1973).

Combination effects of mixed types of controls over stressors are also interpretable within Folkman's relational control framework. For example, Mills and Krantz (1979) reported that cognitive control in conjunction with behavioral control during a blood donation procedure caused greater negative responses than either behavioral or cognitive control alone. Expectations or appraisals of control were probably heightened by the combination of cognitive and behavioral controls. These heightened expectancies could have led to a greater mismatch with available control over the aversive event itself (giving blood).

Predictability. Many theorists have noted the confounding of control and predictability or reduction of uncertainty in several studies of control and stress. Berlyne (1960) argued that human beings have a predisposition for prediction, to structure certainty out of the complex, uncertain array of environmental stimuli. Furthermore, as discussed earlier, predictability is likely to reduce cognitive overload because environmental monitoring requirements are lightened, leaving more remaining attentional capacity to deal with task and stressor demands (Cohen, 1978).

Predictability may also reduce stress because it enhances knowledge about when a noxious stimulus is terminated or when it will be delivered (Weiss, 1971, 1972). Seligman (1968) suggested that accurate cues could provide the organism with a safety signal indicating when the organism was safe from the noxious impact period. Animals without reliable warnings or effective behavioral control cannot relax or recover during intermittent nonstress or safe periods because they cannot accurately predict when they are in a safe period.

One difficulty with the prediction model of control and stress is that studies that hold predictability constant and manipulate control still evidence changes in stress outcome measures (Miller, 1979, 1980; Thompson, 1981). For example Reim, Glass, and Singer (1971) found that perceived control over noise caused less physiological stress than did predictable noise with no perceived control. Geer and Maisel (1972) report similar results. On the other hand, performance aftereffects were equivalent for control and prediction groups in the Reim, Glass, and Singer study. Thus, although it is clear that predictability of stressors is an important variable in its own right, all of the effects of control over stressors cannot be accounted for simply by the enhanced predictability provided by control.

Control as Boundary on Aversiveness. Several models of controls suggest that control reduces stress by providing information about the boundaries or extent of negative impact of aversive events (Miller, 1979, 1980; Thompson, 1981). Miller describes one variation of this model as the *internality hypothesis.* The internality hypothesis emanated from critiques of the yoked control paradigm wherein one organism receives aversive stimulation under its own volition while the yoked partner gets the exact same pattern of aversive stimulation, except without volitional control. Critics of this paradigm (cf. Church, 1964) noted that the yoked subject may subjectively experience more aversiveness because it would receive some aversive stimuli during periods of higher vulnerability. The partner with control can regulate adverse stimulation so that negative stimuli are only allowed during periods of low vulnerability or maximum readiness. In other words, control might facilitate confidence that one could handle the situation (Lefcourt, 1973).

A more formal statement of control and boundary processes—the minimax hypothesis—states that when individuals are in an adverse situation they try to minimize their maximum risk. Control sets an upper boundary, minimizing the duration, intensity, or frequency of negative events (Miller, 1979, 1980).

One particular interesting feature of Miller's minimax hypothesis is that personal control is important to the extent that it provides a stable indication of the limits of endurance for the situation at hand. Thus, personal control should be helpful only if it provides a better estimate of the maximum aversiveness of an event or greater assurance that one will not exceed that maximum (Miller, 1979, 1980). Miller's minimax hypothesis also predicts that under circumstances in which personal control is a less stable indicator of maximum aversiveness, individuals will abrogate control. This relinquishment of control is consistent with research demonstrating that under situations of uncertainty (Averill, 1973), when no meaningful choices can be differentiated (Rodin, Rennert, & Solomon, 1980), when too many choices are available (Averill, 1973), or finally, when control has been attributed to highly unstable sources (Schulz & Hanusa, 1978), the provision of behavioral control heightens stress. Rotter (1966, 1975) advocated a similar position with respect to the utility of internal locus of control. Only when meaningful, alternative behavioral options are available will internal locus of control be generally more adaptive. To the extent that behavior options are constrained, internal locus of control becomes increasingly irrelevant or even disquieting.

Furthermore, as Averill (1973) wisely noted, personal control typically entails increased responsibility for the outcomes of exercising autonomy. In some life situations such as medical settings, unrealistic or inappropriately high expectations for personal control and responsibility may lead to negative outcomes (Shapiro & Shapiro, 1979). Furthermore, high responsibility without control may lead to particularly pathogenic outcomes. Workers, for example, who have major job responsibilities but little decision latitude are at highest risk for psychological and physical maladies (Karasek, 1979). When no choices, bad choices, or too many options are available, the gains in responsibility may offset the positive effects of control. After all, competence means *effective* control over one's environment.

Summary

Several functions of control have been posited (see Table 9). Similar to the taxonomies of control overviewed earlier, the meaning of the control response to the individual plus the importance of the target of control is central to understanding how control mediates the relationship between an environmental stressor and individual responses.

A central function of control is feedback about self with regard to environmental mastery or competence. Control also provides opportunities for more problem-focused coping. One key in understanding when control is or is not helpful in coping with environmental stress is the degree of fit between appraisal of control and actual control available over the stressor. When the match is poor and the target of control highly salient, negative outcomes are more likely.

Several theories of control emphasize the importance of its predictive function.

Table 9. Functions of Control

Mastery—control enhances sense of personal competence and self-efficacy
Coping—control facilitates instrumental coping; when behavioral control is not possible, palliative
       coping may be more adaptive
Predictability—control enhances predictability of the presence and absence (safety period) of aversive
       events
Aversiveness boundary—control provides opportunity to minimize maximum risk from negative
       events when it indicates a stable estimate of maximum aversiveness

---

Several types of control provide information about when and/or how a stressor is to be experienced. One type of predictability that appears important is feedback about when the stressor will *not* be present, thus providing opportunities for rest and recovery. Control also may provide feedback on the upper boundary of aversiveness one has to endure.

Some synthesis of the various models of controls may be possible. When a good match occurs between individual ability to predict and regulate maximum aversiveness, self-attributions of competence are likely. Control will be given up when one perceives that he or she is not the best estimator of the control/situation match. When some other person or factor is a better predictor of maximum aversiveness, control appraisal will be low. When a mismatch occurs whereby situational control is low but control expectancy is high, threats to competence and more negative consequences are likely.

Developmental Issues and Control

Surprisingly, there is a marked paucity of research on children and control over stressors. Some researchers have speculated that children are more susceptible to the effects of environmental stressors such as crowding because they have fewer coping options available to counteract the negative effects of stressors (Chapter 1; Evans, 1978). In this section of the chapter we review research on control and children.

Considerable research suggests that locus of control is related to achievement-related behaviors, including greater tolerance for delay of gratification in elementary-school children who are more internal (Lefcourt, 1982). Locus of control describes individual tendencies to perceive the causes of events in relatively internal or external terms. Internal attributions of causality include personal ability or individual effort; whereas external attributions encompass fate or the actions of other persons or institutions. The degree of correspondence between internality and achievement is particularly strong when children believe that their efforts can influence the outcomes of problems rather than ability and/or luck being the major factor affecting outcomes (Weiner, 1974).

Given the fact that locus of control has been linked to achievement, it is not surprising that programs have been developed to alter students' behaviors in the direction of internality. De Charms (1972) developed a curriculum for inner-city minority children to develop stronger self-image of children as the origin of their actions on the environment, rather than as the objects or pawns of environmental forces. De Charms trained elementary-school teachers to promote origin-affirming behaviors through exercises that encouraged realistic goal setting, higher achievement motivation, and greater instrumental self-concept.

Origin scores increased in comparison to control groups, and furthermore, the length of origin training was highly associated with the degree of enhanced origin self-concept. Children in the training program improved their school grades, had fewer absences from school, and evidenced greater overall achievement activity in the classroom. Similar results in enhanced self-esteem have been found for programs using contingency awareness training (Reimanis, 1971) and with wilderness experiences, emphasizing self-reliance and exploration (Kaplan & Kaplan, 1982).

Research on the origins of social learning indicates that children from warm encouraging families that foster independence and exploration are more likely to be stronger internals (Lefcourt, 1982). In one of the most thorough studies of the familial origins of locus of control in children, Crandall (1973) measured maternal behavior during the first 10 years, interviewed the child as an adolescent, and measured locus of control as a young adult. Parents who were warm and supportive while discouraging dependency and encouraging exploration had children who were more internal. Not surprisingly, there is a moderate, positive correlation between parental and offspring locus of control scores (Lefcourt, 1982).

The only study to our knowledge to manipulate control over an aversive stimulus in young children was conducted by Gunnar-Vongnechten (1978). In a control condition, 1-year-olds could turn on a frightening mechanical toy that operated for 3 seconds. Yoked no-control children were exposed to the same amounts of toy operation. Boys in the no-control condition showed strong fear, cried, and fussed more than their control counterparts. Girls in both conditions exhibited little overt fear but appeared very wary, looking frequently at their mothers and exhibiting fewer overall initiation responses to turn the toy on. Both boys and girls also smiled more in the control conditions.

Control and Environmental Stressors

Control has been viewed as one mechanism to explain some of the aversive effects of environmental stressors such as crowding, noise, heat, and air pollution. This section of the chapter briefly overviews theoretical perspectives on control and environmental stressors. Then research on environmental stress and control is reviewed. A later section examines learned helplessness and its relationship to environmental stressors.

Theories of Control and Environmental Stress

Environmental stressors that are unpredictable or uncontrollable produce negative effects including aftereffects of poorer cognitive performance and fewer socially positive behaviors (e.g., altruism) (Cohen, 1980). Cohen argues that uncontrollable stressors place a greater load on information-processing capacities. The amount of effort to monitor an environmental event is an increasing function of the event's uncertainty. With greater cognitive effort comes more fatigue that leads to stronger negative aftereffects. These aftereffects reflect the system's inability to continue monitoring an unpredictable or uncontrollable stimulus. Thus control is an important aspect of environmental stressors because of its relationship to attentional demands and subsequent pressure on information-processing capacities (see Chapter 5 for a more in-depth discussion of Cohen's overload model of stress and performance).

Some authors have suggested that crowding and perhaps noise interfere with certain behavioral alternatives, reducing the range of behavioral options available to people (Proshansky, Ittelson, & Rivlin, 1970; Stokols, 1972a), or blocking individual development and aspirations because too many people in a setting leads to reduced involvement and participation (Wicker, 1973).

An important feature of crowding is whether one attributes the loss of control in a high-density setting to other people in the setting or to physical conditions of the setting itself (Baron & Rodin, 1978; Loo, 1973). Increased social density (increasing group size while holding physical area constant) is more likely to arouse threats to personal control because other people's intentions and behaviors are more variable in comparison to changes in the setting (e.g., reduction in room size). Altman's (1975) model of proxemic behavior also emphasizes the role of social density in thwarting control over interpersonal interactions. As behavioral options for social interaction are reduced, privacy is threatened and negative reactions occur.

Crowding can also be viewed as an evaluative response to uncontrollable situations. Schmidt and Keating (1979) indicate that high density can foster behavioral constraints or information overload. Either of these states in turn leads to less environmental control and heightened stress because of threats to personal autonomy. Individuals under such circumstances make attributions about the cause of their discomfort to various factors present in the environment. The overly close presence of other people or insufficient space to perform various functions may lead to an attribution of crowding.

Control has also been a central factor in discussion of territoriality and related proxemic concepts. Human territories, for example, are largely defined as areas of space where personal control is maximized. Stokols (1976) has suggested that one reason why environmental stressors such as crowding are more stressful in residences than in work settings is because crowding in primary settings threatens control in the very areas where people expect to maintain maximum autonomy. Thus, it is not surprising that density measures of dwelling space (e.g., people per room) consistently yield stronger associations with health outcomes than measures of external density (e.g., houses per acre) (Cohen, Glass, & Phillips, 1977; Cohen & Sherrod, 1978). Thus, situational aspects of control need to be considered in analyzing environmental stressors.

Parallel findings exist in the community noise annoyance literature. People are most annoyed with noise that threatens their safety at home or interferes with important home activities such as sleeping, watching television, or talking on the telephone. Interference with activities outside the home are perceived as considerably less annoying or disturbing (Cohen & Weinstein, 1982).

In addition to the interference properties of noise in its disruption of activities, control is also implicated in noise restriction activities. Studies of high-rise housing (Michelson, 1970), open classrooms, (Ahrentzen, Jue, Skorpanich, & Evans, 1982) and open plan office buildings (Wineman, 1982) all find that people restrict their own or their children's activities in order to minimize disturbing neighbors or adjacent activity areas.

Finally, residential exposures to crowding and noise are consistently more deleterious for certain vulnerable populations, including the very young, the elderly, and persons of lower socioeconomic status (Cohen & Sherrod, 1978; Cohen *et al.*, 1977; Chapter 1). Each of these vulnerable populations is noted by their general lack of control over their environments in comparison to other members of the population at large.

Summary

Thus control has been posited as one mechanism to explain some of the effects of environmental stressors on human health and behavior. Environmental stressors place demands upon information-processing systems. Many chronic environmental sources of stress are largely uncontrollable and unpredictable, thereby increasing information-monitoring demands on the system. Stressors like noise and crowding also directly interfere with various activities. Depending upon the centrality of the location of that interference, the primacy of goals interfered with, and the availability of attributional targets to explain the interference, the reduction of personal autonomy from environmental stressors can lead to negative health outcomes.

In the next section of this chapter, we briefly review research on control and environmental stressors, focusing primarily on crowding and noise but also discussing work on how control moderates heat or pollution. This section is followed by research on institutional settings and control, with discussions of nursing homes for the elderly, hospital settings, and the classroom environment.

Personal Control and Environmental Stressors

Noise. In a seminal series of studies, Glass and Singer (1972) found that when individuals were exposed to loud, uncontrollable bursts of noise, negative physiological effects, distressed emotions, and performance decrements were evident. Of particular interest is the fact that when people perceived they could control the noise, these negative outcomes essentially were eliminated. When subjects were told they could extinguish the noise by pushing a button, subjects had significantly fewer negative effects from the noise than subjects who were not told they could turn off the noise bursts. The actual amount of noise exposure was identical because most subjects did not actually terminate the noise.

Since then, studies of noise and several other stressors find consistently that uncontrollable stressors produce more negative aftereffects including less tolerance for frustration and poorer proofreading (Cohen, 1980). Social aftereffects from uncontrollable noise have also been noted, including reduced altruistic behaviors (Sherrod & Downs, 1974) and increased aggression (Donnerstein & Wilson, 1976). Concurrent social behavior is also affected by control in noise. Geer (1978) reported that aggressive behaviors increased relative to controllable noise conditions in uncontrollable noise situations. Self-reports of affect similarly indicate greater disturbance and anxiety when control over noise is absent (Cohen, 1978, 1980). Furthermore, negative effects in anticipation of exposure to stressors can also be abated by control (Spacapan & Cohen, 1983).

As suggested earlier, the effects of control on physiological measures are complex. Several studies by Frankenhauser and her colleagues reveal that control over noise reduces catecholamine output, particularly adrenaline (Frankenhauser, 1980). However, some research shows, as reviewed earlier, that if efforts for control over a stressor are difficult to assert, then control can increase physiological indexes of sympathetic nervous system activity (cf. Solomon *et al.,* 1980).

Crowding. If control is a principal mechanism involved in human crowding, then it follows that manipulations of control in high-density settings should influence measures of health and well-being. In an early study of control and crowding, Sherrod (1974) told some subjects under high-density conditions that if they wanted to leave the crowded room they were free to do so. Subsequent measures of task performance indicated that these subjects were less stressed than subjects under similar density conditions who were not given permission to leave the room. It should be noted that subjects did not actually leave the room; it was their perceived freedom to do so that was manipulated in this experiment. Control also causes less anxiety and nervousness during exposure to high-density settings (Langer & Saegert, 1977; Rodin, Solomon, & Metcalf, 1978; Wener & Kaminoff, 1983) as well as reduced task decrements (Langer & Saegert, 1977; Wener & Kaminoff, 1983).

Conversely one would expect the negative effects of crowding to be exacerbated when control options are reduced in high-density settings. Several studies have shown, for example, that when goals are blocked or physical interference with tasks is increased, that people perceive high-density settings as more crowded and react more negatively to them (Sundstrom, 1978).

In accord with theories of control and environmental stressors reviewed previously, subjects provided with either behavioral control (Rodin *et al.,* 1978) or cognitive control (Langer & Saegert, 1977; Wener & Kaminoff, 1983) in high-density settings also perceived the settings as less crowded than did subjects in the same settings without control. Additional research indicates that the kind of control provided in a high-density setting may be important. For example, in one crowding study, some participants were assigned the role of coordinator or terminator, respectively. The coordinator initiated and facilitated group tasks. The terminator's role was to decide when the group should finish a task and proceed to the next one.

Remaining subjects did not have any specific leadership roles. Subjects with either type of control felt less crowded, but the terminator role reduced crowding more than the coordinator role did (Rodin *et al.*, 1978). Baum, Fisher, & Solomon (1981) provided subjects with information either about internal events (somatic reactions) or about the objective situation they would face in a crowded space (e.g., interference with tasks). Students unfamiliar with the setting benefited from information about how density might interfere with certain activities.

Research on locus of control has also identified some differences in reactions to crowding in internal and external individuals. Experimental manipulations of density interacted with locus of control with internals' feeling less crowded and less anxious than externals in a high-density laboratory setting (McCallum, Rusbult, Hong, Walden, & Schopler, 1979). Baron, Mandel, Adams, and Griffen (1976), however, found that locus of control of college students did not interact with their reactions to high and low dormitory settings. One possible reason for this discrepancy is that locus of control is useful in predicting behaviors only in situations where people perceive they have some latitude to alter the circumstances. Possibly the short-term experience of crowding in the laboratory was viewed as more malleable than crowding in dormitories.

Several crowding studies have also shown evidence for heightened physiological arousal (Evans, 1979; Epstein & Karlin, 1975). Singer, Lundberg, and Frankenhauser (1978) in an interesting study of crowding on passenger trains concluded that control over high density reduced the aversive physiological effects of crowding on commuter trains. Passengers boarding the trains earlier on the daily route had greater seat selection opportunities and had lower catecholamine levels than others who chose seats later in the route when fewer choices were available.

Research on traffic congestion also indicates that control can mediate the effects of stress on commuters. Novaco, Stokols, Campbell, and Stokols (1979) found that individuals with internal locus of control were more adversely effected by traffic congestion. For example, commuters with highly congested daily routes who were internal had higher blood pressure than externals. The authors explain these data by focusing on the internal's higher expectation for control over his or her daily experiences. These expectations are in marked contrast to the external who believes that most events in his or her life are out of his or her personal control. This interpretation is interesting to consider in light of Folkman's (1985) analysis of control and coping. The internal's stronger expectation of personal control creates a mismatch when traffic congestion is high. Whereas for the external, low expectations of personal control are simply verified by bad traffic conditions.

Lefcourt (1982) also reported mixed findings for locus of control and life stress. For example, several studies have found that internals suffer fewer negative health consequences in response to stressful life events. On the other hand, some research reports no effects of locus of control on reaction to life events. It seems reasonable to suspect that internals might be more likely to use instrumental, problem-focused coping more often than externals. Thus, one prediction that would follow from

Folkman's (1985) model of control is that internals will cope more successfully with stressors where instrumental solutions are possible. For short periods of time, expectations for control may lead to challenge and determine more direct problem-solving coping strategies. On the other hand, with continual struggle and no relief, persons with high expectancies for control may be more susceptible to negative outcomes because their expectations have been directly disconfirmed (Lazarus, 1977). These predictions are also consistent with work reviewed earlier indicating that control efforts without relevant feedback can exacerbate stress.

Air Pollution and Heat. Little research has been completed on heat and air pollution and control. Both Bell and Greene (1982) and Evans and Jacobs (1982) speculated that reduced personal control over heat or air pollution may heighten adverse reactions to these discomforts. Bell and Greene, for example, argued that some performance decrements in heat may reflect individual's giving up in the face of an uncontrollable, adverse situation. Greene and Bell (1980) reported that subjects feel more dominated by hot environments than temperate ones.

Most people feel they cannot do anything to control air pollution and suggest that other, more powerful institutions (e.g., government) need to exert greater power in curbing polluters (Evans & Jacobs, 1982). Some control-relevant research has focused on individual differences in people's opinions about pollution-abatement activities. In one study, Creer, Gray, and Treshow (1970) found that people working for a major pollution source believed that effective air quality control efforts were in operation. These workers in comparison to other local citizens also were more optimistic about the ability of industry to control air pollution.

Another factor influencing coping modes with air pollution is human adaptation. Evans, Jacobs, and Frager (1982) found that people who had adapted to chronic exposure to poor air quality were much less likely to use instrumental coping modes in dealing with bad air pollution in a new residential environment. Groups of people who had just moved to a setting with bad air quality from relatively clear previous residential locations took more direct actions to combat air pollution. The latter groups felt more able to do something about smog (e.g., use mass transit), sought out information about it, reported seeing smog more readily in a photograph, and were more concerned about its possible negative health effects. Furthermore, recent migrants from previously low air pollution areas who were also internal locus of control were more likely to restrict outdoor activities during serious periods of pollution (smog alerts). External, recent residents from low-pollution areas and all recent residents from previously high-pollution areas largely ignored warnings regarding curtailment of outdoor activities during smog alerts.

Only one study has attempted to directly manipulate control over air quality. Rotton (1983) found that perceived control over malodor reduced decrements on task performance and ameliorated frustration tolerance aftereffects. Control was varied by informing subjects they could replace the lid on a foul-smelling jar. As in the Glass and Singer (1972) noise studies, perceived control was manipulated because subjects were told the experimenter preferred they not exercise their control option. Subjects in the odor-with-no-perceived-control condition were not informed they could close the jar lid if it became necessary.

Summary. Several studies of environmental stress and control reveal that control can ameliorate the negative effects of stressors on aftereffects, concurrent complex task performance, and affect. Physiological results are more complex, suggesting that when control can be asserted without much effort, it reduces physiological arousal from stressors. When greater effort is required to exercise control over a stressor, physiological arousal may increase relative to no-control stressor conditions.

At least three areas of further research on control and stress are readily apparent from the review outlined here. First, the cognitive effort demanded by different tasks is an important variable in understanding how stressors influence human performance. Second, nearly all of the control and environmental stress literature has manipulated behavioral control over responses. If you push this button (leave this room, close this lid, etc.), you can avoid the aversive event. With a few exceptions (Epstein, 1973; Glass, Singer, & Friedman, 1969), it is much less clear how some other forms of control (e.g., cognitive control) will modify human responses to noise, crowding, and other environmental stressors. The importance of relevant feedback about stressor modification during control efforts also begs further study.

Third, as noted earlier, very little attention has been given to the implications of control over stressors among children. Some evidence suggests that children and younger organisms, more generally, are more susceptible to the effects of crowding. One possible mechanism for the greater vulnerability of children to environmental stressors may be their relative inability to control exposure to stressors.

Stress and Control in Institutional Settings

Research on control and environmental stressors has typically focused on specific, discrete features of environments. This section of the chapter reviews research on control in the broader overall contexts of institutional settings. Institutional policies on evaluation, incentive, skill building, and decision latitude may influence individual modes of adaptation (Mechanic, 1977).

Nursing Homes and the Elderly. Many of the negative effects of institutionalization stem in part from the minimization of personal decision making and autonomy typically found in such homes. This process may be a prime cause of the diminishing realm of personal competencies evidenced by many elderly residents of nursing homes. Furthermore, with old age, physical capacities decrease, one no longer earns an income through working, and the child-rearing role is diminished if not actually reversed.

Schulz (1976) varied the type of visitation schedule during a 2-month period for elderly residents of a nursing home. One group of residents had complete control over visitations, receiving college student visitors on demand, controlling both the frequency and duration of visits. Two comparison groups received either random visits or predictable visits of matching duration and frequency. For predictable visits, residents were informed ahead of time when a student would visit but had no choice in the schedule. A baseline comparison group received no visits. The prediction and control visitation residents showed equal, marked improvement in affect, enhanced multiple health measures, and greater participation in nursing home activities.

Langer and Rodin (1976) enhanced nursing home resident's control by inducing individual decision making and providing greater personal choice in the operational routine of the institution. Elderly residents in the treatment group were given lectures by the nursing home management staff emphasizing the resident's own responsibility for their care and participation in the community. They were also given a choice of schedules to view films each week and offered a houseplant to attend to. The comparison group was given a plant that was cared for by the staff, assigned a movie-viewing schedule, and given a lecture emphasizing the staff's responsibility for patient care and well-being. The results showed marked positive effects of the control-enhancing intervention including more positive life outlook, better affect, better physical health, and greater participation in community activities. Similar results were found for an orientation program emphasizing predictability and individual autonomy for elderly residents preparing to move to a long-term care facility (Krantz & Schulz, 1980). Moos (1980) has also demonstrated that institutionalized settings for the elderly that provide greater roles for residents in policymaking and daily decisions have higher overall health profiles for residents and greater social cohesiveness. Training in coping skills (e.g., positive rather than negative self-statements when problems occur) also enhance mastery and competence among institutionalized elderly (Rodin, Bohm, & Wack, 1983).

Architects and planners interested in gerontology have also begun to incorporate design features into elderly housing that aim to enhance resident's autonomy and ability to competently interact with their living environments. Among some of the features incorporated are spaces providing hierarchy of social interaction (i.e., from intimacy to public exchange), facilities for maintenance of personal competencies (e.g., cooking, crafts), orientation aides, and small-group living spaces where daily support and care are the mutual responsibility of residents rather than staff. For more information see Lawton (1980).

Rodin and Langer (1977) completed an 18-month follow-up noting that the advantages of their control enhancement program remained. In addition, they also found a slight reduction in mortality rates for those who had participated in the control treatment. The continued impressive results of the personal control intervention were probably due to the feelings of enhanced self-mastery and expanded competence in day-to-day decision making engendered by the intervention.

An important difference between the two research programs is that Schulz's control manipulation was dependent on the behavior of external actors (visiting college students), whereas Langer and Rodin's study focused more on generalized feelings of mastery. Thus, it may not be surprising that in an extended follow-up study of 24, 30, and 42 months, Schulz and Hanusa (1978) found that the control and prediction groups actually did worse in comparison to the random and baseline controls. Health status and zest for life declined more in the previously successful groups. Mortality also showed a slight relative incline.

In order to investigate the differences in long-term effects of the control-enhancing treatments of the two respective programs, Schulz and Hanusa (1979) conducted a series of studies manipulating both competence induction and control strategies in a long-term care facility for the elderly. The results of these studies generally support the attributional analysis stated before. That is, individuals provided control in terms of self-mastery and competence responded more favorably over time than did residents temporarily provided control dependent upon the behavior of others.

Medical Hospitals. Unfortunately, a pervasive aspect of most hospital environments is the lack of individuality and personal control for most patients. Part of this lack of control undoubtedly derives from the patient's ill health status that creates dependency on others plus the fact that she or he has been thrust into an aversive, unfamiliar setting. The patient is typically removed from her or his normal social world where social networks function and detached from areas where he or she previously exercised authority and competence (e.g., on the job, within the family). The hospital setting is also characterized by a strong hierarchical organization with formal arrangements designed to facilitate the delivery of efficient medical services. Furthermore, patients are frequently depersonalized as clinical entities where the disorder, not the patient is the target of medical care (Shumaker & Reizenstein, 1982).

The "good" hospital patient is one who follows orders, asks few questions, and generally plays the obedient "sick role" of passive compliance and sublimation to hospital staff personnel (Taylor, 1979). The "bad" patient, on the other hand, reacts to threats to personal control, asserting individuality, perhaps questioning staff about their disorders and/or treatment regimen. Taylor (1969) notes that many characteristics of good patient behavior may lead to helplessness and thus interfere with recovery whereas, some of the bad patient's behavior are, in fact, signs of positive assertion of control in a negative environment. Recently, some research has directly examined these issues by manipulating forms of control in medical settings.

Two general types of control intervention strategies have been evaluated in medical hospitals. The first type of research has examined various types of cognitive controls in hospital patients. Janis and his colleagues (1958, 1983) in a series of studies showed that when patients are provided with preoperation information about what they are going to experience plus given some guidance on possible coping mechanisms (e.g., encouragement of optimistic reappraisal of the stressor), they experience less postoperative adjustment than patients with no information or guidance. Measures included self-reports, medication levels, and discharge latency. Other examples of cognitive control interventions in medical settings are discussed by Taylor (1979) and Kendall (1983). It is important to note that preparatory information about surgery or other medical procedures is not uniformaly effective.

For example, Langer, Janis, and Wolfer (1975) found that providing information alone on preoperative procedures and postoperative physical symptoms had little influence on postoperative recovery.

The other major approach to control interventions in medical settings has been changes in the designed environment to provide better opportunities for patient efficacy (for a general review of control and the designed environment, see Sherrod & Cohen, 1978). Zimring, Weitzer, and Knight (1982) renovated the residential living facilities of a large state institution for the developmentally disabled. Large open wards were divided into more personal, private livingroom spaces, and huge day rooms were redesigned to facilitate social activities and provide for small-group activities and/or solitude and privacy. These changes produced significantly fewer staff intrusions into the personal areas of patients, higher use of personal spaces by patients, less socially withdrawn behavior in patients, and improved communication among residents and between staff and residents.

Some of the specific design features the authors recommend to enhance control in hospitals include more hierarchical arrangements of spaces with clearly demarcated personal, semipublic, and public areas. Opportunities for personalization are also seen as important design features to improve patient autonomy along with symbolic indicators of ownership and personal identity. The environment itself needs to be malleable with features such as movable furniture, personal light switches and climate controls, and doors and windows that can be used. Design and placement of "signage" and other graphics to facilitate orientation and wayfinding in hospitals is also needed (Shumaker & Reizenstein, 1982; Zimring *et al.*, 1982).

Schools. Control over classroom activities and individual behaviors in classrooms has long been a central concern in educational philosophy and research. The open classroom movement in particular reflects a belief in the benefits of child-directed learning (Ahrentzen, Jue, Skorpanich, & Evans, 1982; Weinstein, 1979). Stipek and Weisz (1981) have reviewed major models and evidence on personal control and academic achievement. Locus of control was originally believed to be related to academic achievement because internal children believe that by personal effort they can influence scholastic outcomes, particularly if those outcomes are valued (Rotter, 1966). There is strong evidence in support of this formulation. More-internal children evidence greater academic achievement than externals do, engage in more achievement-related tasks, persist more on difficult tasks, and appear more able to delay gratification (Lefcourt, 1982; Stipek & Weisz, 1981).

Research on attribution theory has also found that when children attribute achievement, particularly on difficult tasks, to their own effort or ability, they are more likely to attempt other academic tasks. Children who attribute their success to luck or some other external contingency do not respond to success with continued task persistence (Weiner, 1979).

Bringham (1979) reports on a series of studies showing that self-selection of work, self-pacing, self-monitoring of work, and self-scheduling of work periods improve math performance. For example, in one study with fifth graders, when students set goals for a 40-minute math period, they performed better than when teachers set goals for the period. The goal levels set were equivalent when designated by either students or their teachers.

Research on classroom architecture and student behavior also indicates that student personal control is an important factor. For example, several studies show that students strongly prefer private areas in classrooms for studying (Ahrentzen *et al.*, 1982; Weinstein, 1979). Children also like personal territories (e.g., personal desk, storage compartments) where they can control access.

A topic of considerable discussion in the open-school literature is the issue of boundary control mechanisms (see Bechtel, 1977 for an interesting discussion of this problem in open-plan office design). A common criticism of open-plan schools is the high degree of distraction and spillover among different activities. Several studies have found that either programmatic rules or uses of spatial markers and boundaries to separate or close off different spaces are frequently implemented to cope with perceived distraction (Evans & Lovell, 1979; Ahrentzen, 1981).

Summary. It is apparent that personal control can have some powerful effects in the real-world settings of nursing homes, hospitals, and schools. Insufficient research has been completed on control in such settings to draw definitive conclusions on how control influences such processes as health or academic performance. Unfortunately, some of the important distinctions among different types of controls developed in the theoretical work have largely gone untested in real-world settings.

There is evidence that control enhancement, particularly if attributed to mastery or self-efficacy, can positively influence the well-being of elderly nursing home residents. Hospital settings strongly encourage ''good patient'' behaviors that often include docility, compliance, and dependency. This role may facilitate the efficient delivery of medical care but can perhaps impede patient needs for personal autonomy and flexibility in coping with aversive, life-threatening medical events. Providing patients with information and suggestions about appropriate coping avenues appears to ease recovery. Finally, there is evidence that children who perceive themselves as more self-efficacious in the learning environment will achieve more academic success.

Learned Helplessness

This section of the chapter focuses on learned helplessness theory. The original model of learned helplessness was developed by Martin Seligman (see Seligman, 1975, for a review). *Learned helplessness* was defined in Chapter 1 as a failure in coping due to perceptions of uncontrollability. This uncontrollability is induced by exposure to situations where an individual learns that the outcomes of his or her responses are independent of those responses. Learned helplessness occurs when the probability of an outcome is constant no matter what response the organism emits.

Several aspects of the original learned helplessness model are important for subsequent revisions of the model when applied to human beings. First, the organism's response must be *voluntary*. If a response is not perceived as under the individual's volition, learned helplessness does not occur. Second, the organism must be *capable of learning* that outcomes are noncontingent of responses. This second principle has important and largely unresearched developmental consequences.

Learning that the outcomes of voluntary responses are noncontingent has four consequences. First, motivational deficits occur as evidenced by reduced initiation of voluntary responses. Second, emotional reactions including anxiety and depressed affect are apparent. Third, cognitive deficits are seen in helpless organisms. They have difficulty learning that responses produce outcomes. Fourth, helplessness is associated with reduced physiological arousal. Furthermore, some research suggests depletion of circulating neurotransmitters, particularly norepinephrine, in the brain as a possible mechanism for learned helplessness effects.

Human Models of Learned Helplessness

Several complexities found in human studies of learned helplessness have led to reformulations of the original model that had been built upon animal research with uncontrollable aversive stimuli. In particular three major sets of issues have been raised about

1. individual differences in human responses to noncontingency;
2. variability in the generalizability and persistence of learned helplessness across different contexts and time periods; and
3. facilitation as opposed to deficiencies in responding following noncontingency.

Individual Differences. Not all participants in learned helplessness studies evidence the pattern of motivational, emotional, and cognitive deficits associated with learned helplessness. A critical factor in determining whether learned helplessness will occur are the attributions people make to account for their inability to control outcomes. Individual expectations of noncontingency between voluntary responses and outcomes is a critical aspect of this attribution process. If more internal (personal) attributions of failure to effect outcomes are made, learned helplessness is much more likely to occur. Conversely, if one attributes one's inability to control outcomes to some more external cause, then learned helplessness is less likely (Abramson, *et al.,* 1978; Abramson, Garber, & Seligman, 1980; Miller & Norman, 1979).

Several studies have demonstrated, for example, that when attributions of failure are made to task difficulty, learned helplessness is less likely to occur than if failures to effect outcomes are attributed to personal ability or competence (Abramson *et al.,* 1978; 1979; Miller & Norman, 1979). Furthermore, work with locus of control indicates that externals are more susceptible to learned helplessness than internals are (Cohen *et al.,* 1976; Hiroto, 1974).

A closely related attribution of failure is whether the cause is stable or unstable. For example, when failure to control an outcome is attributed to effort (unstable factor), learned helplessness is less likely to occur than if noncontingency is blamed on ability or some other more stable factor (Abramson *et al.*, 1978, 1980; Miller & Norman, 1979). Experiments where attributions for failure have been manipulated or measured post hoc also support the importance of the stability dimension in attributions of uncontrollability. Research on skilled versus chance tasks is also consistent with this factor. When subjects are led to believe that failures to control response outcomes have happened on skill tasks, they are more likely to become helpless than if they believe they have been working on a chance task.

   Generalizations across Contexts and Time. Attributions made to more global as opposed to situationally specific factors create more generalization of helplessness (Pasahow, 1980). The processes by which more global attributions lead to greater generalizations are not well understood. It appears that more prolonged exposure to noncontingency causes greater generalization of learned helplessness. Several studies have also revealed that when the context across different situations is similar, or when the pretreatment and test-phase tasks are similar, learned helplessness is more likely to occur (Abramson *et al.*, 1980; Miller & Norman, 1979; Roth, 1980). Furthermore, when stable versus unstable factors are blamed for noncontingency, generalization is more likely.

   There is also evidence that factors effecting the generalization of learned helplessness in human beings may interact with one another. For example, Cohen *et al.* (1976) found that externals but not internals generalized helplessness from a problem-solving pretreatment induction procedure to a nonproblem-solving test task. Other research has also shown greater generalization of helplessness when the specificity of failure attributions was manipulated from specific to global causes (Pasahow, West, & Boroto, 1982).

   The hypothetical link between learned helplessness and depression has also been incorporated in reformulated learned helplessness theory for human beings. Depressed individuals tend to make internal, stable, and global attributions for uncontrollable, aversive events. Nondepressed individuals, on the other hand, generally attribute such events to external, unstable, specific factors (Seligman, Abramson, Semmel, & von Baeyer, 1979).

   Facilitation versus Inhibition of Responses. A third difficulty with human learned helplessness studies is that not only does learned helplessness sometimes not happen following exposure to uncontrollable outcomes, but under some circumstances facilitation of responses occurs (Roth, 1980; Wortman & Brehm, 1975). When individuals are threatened with loss of freedom, efforts to reassert control (reactance) may occur. Reactance is most likely to happen when the importance of the outcomes is high. Furthermore, the emotional state accompanying reactance is typically hostility and aggression in contrast to depression and passivity as noted by Seligman (1975) in many learned helplessness studies.

   Empirical research is sparse on this topic, but a few studies do exist that are pertinent. First, several studies have shown that short periods of helplessness induction lead to reactance, whereas continuation of the same induction procedures leads to helplessness (Roth, 1980). Some studies have also found negative affect, including hostility with short-term exposure to uncontrollable, aversive stimuli (Roth, 1980).

The threat of loss of control also may interact with length of exposure to noncontingent outcomes. Krantz, Glass, & Snyder (1974) and Glass (1977) have shown that Type A men respond differently than Type B men to noncontingent treatments when the salience of performance is high. Type A men, who in part can be characterized as in high need of environmental control, initially respond to noncontingent reinforcement with greater attempts to reassert control than do Type B men. However, after prolonged exposure to noncontingent response outcomes, Type A men are more likely to suffer from learned helplessness. More recently, Brunson and Matthews (1981) linked the Type A pattern and the reformulated model of human helplessness. They had Type A and Type B male adults think out loud while working on soluble and unsoluble concept formation problems. Performance salience was also manipulated by having half of each group record their data on each trial. Of particular interest to the reformulated learned helplessness model, the high performance salience, Type A men stated and used ineffectual problem-solving strategies and attributed their difficulties to lack of ability. At the conclusion of the experiment, they also said they had not tried hard enough. Type B men, on the other hand, had focused more on task difficulty, chance, or the experimenter as causes of failure. Further, they did not shift as much to inappropriate problem-solving strategies.

Learned helplessness theory has not gone unchallenged. Several criticisms in the animal literature have been made (see Maier & Seligman, 1976). Two major alternatives have been offered to explain learned helplessness in human beings in addition to the reformulated theories discussed previously. In the first alternative, Coyne and his colleagues suggest that the deficits underlying performance following learned helplessness induction procedures can be explained more parsimoniously as cognitive interference associated with anxiety (Coyne, *et al.*, 1980; Lavelle, *et al.*, 1979).

A second alternative explanation of helplessness is that subjects perform more poorly following failure in order to preserve self-esteem (Frankel & Snyder, 1978; Snyder *et al.*, 1981). According to this model, when people believe they did not solve problems that were possible to solve, they will feel they have failed. This failure can either be attributed to their own inabilities or to lack of effort. By not trying in the test phase of the helplessness paradigm, subjects can more readily attribute their previous failure to lack of effort. They avoid confirmation that their previous failure was due to lack of ability.

It is not clear whether these two alternative explanations of learned helplessness are inconsistent with the reformulated model offered by Abramson and colleagues (Abramson *et al.*, 1978, 1980). The reformulated attributional model of learned helplessness states that when the individual's failure to control response outcomes can be attributed to external causes as opposed to internal ones, learned helplessness is less likely. Snyder and colleagues' position could be subsumed under this model because task difficulty and an environmental distraction are both external factors that could explain failure. Snyder's model does raise the important question of whether it is a threat to self-esteem that is causing helplessness reactions or the perception of noncontingency between responses and outcomes.

Coyne and colleagues' model of anxiety and attentional deficits may also be subsumed under the attributional model. High-anxious individuals are likely to make internal attributions for failure, whereas low-anxious persons blame external factors. Furthermore, it seems reasonable to argue that high-anxious subjects do not discriminate as well between situation specific components of the testing situation and thus are more likely to make global attributions of threat.

Summary. In response to difficulties with the original learned helplessness model, recent theories of learned helplessness in human beings have been formulated. Three major sets of issues, individual differences in responsiveness to noncontingency, extent of generalizability of helplessness across tasks and experimental contexts, and evidence of facilitation of responses following exposure to uncontrollable events have led to the reformulation of the learned helplessness model. These new models argue that learned helplessness is most likely to occur in humans when the causes of failure to control outcomes are attributed to internal (personal) causes that are relatively stable. Furthermore, the more global or universal the attribution of uncontrollability, the more likely learned helplessness will persist over time and/or across situational contexts.

Reactance may occur initially following brief exposures to noncontingency that reflects efforts to reassert control. Greater efforts to reestablish control are most likely when the threat value of lacking control is high to the person (e.g., Type A personality) or when the value of successful performance is high.

The learned helplessness model has been applied to theories of human depression with some success, although insufficient data exist to draw any definitive conclusions at this time. Two alternative explanations of human performance deficits following exposure to noncontingent response outcomes have been offered. One model suggests that failure-induced anxiety interferes with cognitive processing because of attention deployment directed toward self-monitoring. A second alternative model indicates that efforts to establish contingency following a helplessness pretreatment are reduced in order to preserve self-esteem. Less effort facilitates the self-explanation that the previous failure was due to lack of effort rather than inability.

The Development of Learned Helplessness

Both animal and human research indicate that the development of competence is a critical prerequisite for healthy social and physical development. Very young children seek out environmental arrays that provide stimulation, challenge, and opportunities for management and control (White, 1959). This section of the chapter examines research on young organisms' reactions to noncontingencies, how attributions about noncontingency may also apply to children, and the relationship between reactions to noncontingent outcomes and achievement-related behaviors.

Early Development and Noncontingency. Strong evidence exists for the importance of instrumental experience with the physical environment early in life. Held (1965), for example, showed that young kittens need to have self-directed motoric interaction with their surroundings during certain critical periods of development for normal maturation of the visual cortex. Held reared kittens in darkness for 12 weeks. Then for 3 hours a day, for 10 days, the kittens were exposed to visual stimulation in the lab. One kitten received active exposure as a function of its voluntary movement around the environment whereas a yoked partner was passively exposed to the same stimulation. The active member of the pair moved around an enclosed chamber while the passive partner rode in a gondala. Thus only the active member of the pair received synchronous motor-visual feedback. At the end of the 10-day exposure period, the passive kitten exhibited clear visual-motor coordination deficits. Unlike the active partners, the passive kittens did not react to frontal approaches by objects, nor did they stretch out their paws before impending collisions or to avoid steep surfaces. Active, self-directed interaction with the physical environment during certain critical periods of development is a necessary prerequisite for normal cognitive development. Such self-directed, instrumental activity allows the organism to experience contingent feedback between movement and visual stimulation changes (see also Denenberg, 1972).

The parent is often a primary partner in the development of organism mastery of the environment. The infant experiments and awaits responses from the parent (Seligman, 1975). Perhaps early maternal separation is debilitating in animals because of the removal of prime opportunities early on for the exercise of competence. Harlow's (1962) research on maternal deprivation in primates and Spitz's (1946) observations of institutionalized human infants find symptoms quite similar to some reactions to learned helplessness. For example, Spitz noted that children reared from birth in institutions with little interaction with other children or adults often were passive, withdrawn, generally depressed, and frequently seemed to have retarded cognitive development. Harlow's maternally deprived monkeys exhibited abnormal social interactions with peers or adult monkeys and would even withdraw from positive, affectionate physical contact.

Some studies have examined how young organisms respond to noncontingent response-outcome relationships. Watson and Ramey (1972) developed an intriguing paradigm to measure whether young infants can learn environmental contingencies. This is an important question. If we return to the original learned helplessness model, recall that learned helplessness occurs when the organism *learns* that outcomes are independent of voluntary responses. The organism must be able to comprehend that no matter what response it emits, the probability of a specific set of outcomes remains the same.

Watson and Ramey electrically connected an overhead mobile to the pillow of 8-week-old human infants. For the contingent condition, when the infant moved her or his head, the mobile rotated. In the noncontingent condition, mobile movement occurred independently of head turning. A third comparison group had a fixed mobile overhead. The authors exposed the infants to the mobile for 10 minutes daily during a 2-week period. Of particular interest here, they found substantial increases in head turning in the contingent group but no changes in either the noncontigent or comparison, fixed mobile groups, respectively. Children as young as 2 months old seem able to learn whether outcomes of instrumental responses are related to those responses. Similar findings exist for young rodents (Joffe, Rawson, & Mullick, 1973).

Kindergartners exposed to either contingent or noncontingent problem-solving situations also exhibit the classic learned helplessness pattern. Children were exposed to the problem of choosing one of two objects for a reward (candy). Children rapidly learn the correct problem-solving strategy that if the first choice gets rewarded, keep choosing the same object. If the first object does not get rewarded, take the other object and stay with that one on subsequent trials. In noncontingent groups, candy is given randomly. For the test situation, children are given a new set of soluble problems. Children who had been in the contingent group learned the criteria for solution of the new problems much sooner than children who had been in the noncontingent group (O'Brien, 1967).

Although the preceding studies suggest that young organisms and children can learn that their behaviors can influence outcomes, it does not necessarily follow that young subjects perceive the contingency (Seligman, 1975; Weisz, 1983). Developmental research shows that not until fifth or sixth grade are children able to distinguish between events that are and are not contingent on personal behavior. Moreover, it takes several more years before children can comprehend some of the logical consequences of contingency (e.g., chance means that the correlation between effort or intelligence and outcomes is expected to be zero) (Weisz, 1983). Thus if learned helplessness is contingent upon learning that voluntary responses are independent of outcome behaviors, then very young children might be less susceptible to the induction of learned helplessness because of their lower cognitive abilities. Parsons and Ruble (1977) evaluated 3 ½ to 5, 6 to 8, and 9 to 11-year-olds' reactions to success and failure feedback on a series of ambiguous tasks. Preschoolers' expectancies for success on future trials were not affected by previous outcome feedback, whereas both groups of elementary-school children were more pessimistic about subsequent task performance following failure feedback. Similar developmental trends have been found for retarded children's susceptibility to learned helplessness. Another reason why younger children may be less susceptible to helplessness is because their estimates of personal competence far exceed actual opportunities for mastery, whereas older children's estimates of personal control are more accurate (Weisz, 1983).

Children's Attributions and Learned Helplessness. Carol Dweck and her associates have developed an attribution model of helplessness to help account for individual differences in children's reactions to failure (Dweck & Elliott, 1983; Dweck & Goetz, 1978; Dweck & Licht, 1980). In two early studies, Dweck and colleagues found similar learned helplessness reactions in fourth and sixth graders as reported earlier by O'Brien (1967). Elementary-school children exposed to insoluble problems exhibited greater deficits on subsequent soluble problems than did children previously exposed to soluble problems. A proportion of children exposed to noncontingent conditions, however, did not exhibit learned helplessness. Dweck has termed these children "mastery-oriented" and found that they are generally more internal in locus of control. Furthermore, these children are more likely to attribute failure in uncontrollable (insoluble) situations to more unstable factors (e.g., insufficient efforts). On the other hand, children who are "helpless" are more external and significantly more likely to attribute failure to invariant factors like lack of ability following exposure to noncontingent reinforcement.

The mastery-oriented and helpless children also use different problem-solving strategies. Performance on soluble problems is equivalent for both groups of children. However, during insoluble problems, the mastery children continue to use sophisticated problem-solving strategies, whereas the helpless children quickly regress to simplistic strategies. Furthermore, when asked to think aloud, the mastery-oriented children make positive self-statements. Helpless children make negative self-statements and numerous comments about causes of failure (Dweck & Goetz, 1978).

Dweck and Licht (1980) replicated these differences between mastery-oriented and helpless children and also found that immediately after the initial and successful problem-solving sequence, helpless children underestimated the number of problems they had solved, anticipated relatively poor future problem-solving performance on an impending sequence of problems, and believed that other children would do better on the upcoming problems. After failing on the insoluble problems, helpless children overestimated the number of problems they had not solved and felt they would have trouble solving the previously soluble problems if given a second opportunity to do so. The mastery-oriented children had essentially the opposite responses, maintaining an optimistic outlook about future task performance and not viewing themselves as having failed.

Helplessness and Achievement. A principal reason why research on helplessness and human development is important is because children who become helpless may exhibit motivational and cognitive deficits that interfere with scholastic achievement. Weiner (1974, 1971) has assembled large amounts of data indicating that failure in children that is attributed to internal, stable factors such as ability is strongly linked with diminished expectancies for future success, reduced persistence on difficult tasks, and lower need achievement.

There is some evidence that children's vulnerability to learned helplessness is related to the development of achievement motivations. A major transition in learning motivation occurs when children enter school. Both developing cognitive abilities and shifting environmental demands from the school setting cause a transition from mastery orientation to achievement motivation (Dweck & Elliot, 1983). Mastery in the young child comes about through accomodation or reequilibration between individual goals or needs and environmental resources. Achievement in the school setting becomes socially defined in terms of normative, comparative perfor-mance. Furthermore, increasing cognitive abilities fosters planning behaviors and the recognition of behavior patterns over time. This in turn leads to a perception that ability is a stable, person-centered trait.

The school environment also facilitates a shift from mastery to achievement-oriented behavior. Elementary-age children are confronted with more intellectual tasks requiring sustained effort and concentration. Previous mastery experiences consisted primarily of physical skills acquisition (Dweck & Elliott, 1983). Further-more, children receive grades and face an increasingly competitive social climate once they enter school. The shift from mastery-oriented learning to achievement motivation is accompanied by increasingly extrinsic reinforcement structures for learning. Thus, the growing child becomes more vulnerable to the effects of failure including the development of learned helplessness.

Causal attributions for failure may shift accordingly with the transition from mastery to achievement motivation. Older children may be more likely to attribute failure to more stable, personal traits like ability. Rholes, Blackwell, Jordan, and Walters (1980) ran children from kindergarten through fifth grade in a standard learned helplessness paradigm using insoluble puzzles. Solubility of pretreatment puzzles did not effect younger children's (kindergarten to third grade) performance on a test-phase soluble puzzle. Only fifth graders exhibited a learned helplessness effect. Furthermore, attributions for failure to solve puzzles during the pretreatment phase were correlated positively with task performance for the fifth graders. For the younger children, however, performance and attributions for failure were unrelated. For fifth graders, helplessness was accompanied by attributions about personal, stable causes of failure (e.g., lack of ability).

Individual differences in school achievement have been directly associated with helplessness and the accompaning attributions about sources of failure or insolubility on problems. Butowsky and Willows (1980) compared good, average, and poor readers (fifth grade) in their responses to soluble and insoluble nonreading-related tasks. Poor readers, in comparison to either average or good readers, had lower initial expectancies for success prior to any task involvement, gave up more quickly in the face of difficulties, and attributed failures to more internal, stable causes (e.g., ability).

Dweck and her associates have begun a series of tests to uncover the etiology of learned helplessness in children following failure experiences in achievement situations. Their work emanates in part from the fact that young boys and girls differ markedly in their responses to failure in academic settings. Following failure feed-back from a teacher or other adult, girls tend to become helpless, attributing their lack of success to inability (Dweck & Goetz, 1978; Dweck & Licht, 1980). Boys, however, are less likely to become helpless, typically blaming lack of effort or some external circumstances (e.g., unfair teacher) for their failures.

Although girls and boys receive the same amounts of negative evaluations, the content and reasons for the feedback differ markedly as a function of the child's gender. Negative evaluation of boys' academic work was typically general and diffuse with little direct comment on intellectual content. Instead, attention was paid to neatness, individual conduct, effort, and other largely nonintellectual aspects of the work. Girls, on the other hand, received more specific feedback about the intellectual content of their work. Girls, for example, received eight times less feedback about lack of effort than boys did.

The preceding pattern of results on teacher feedback to girls and boys could encourage girls to attribute poor performance to ability, whereas boys may believe that poor performance is do to lack of effort or carelessness. Dweck and her colleagues then undertook a series of studies to try to evaluate this hypothetical link more directly. Girls believe that teachers like girls better than boys and view girls as smarter and more hardworking than boys (Dweck & Licht, 1980). Thus, when failures do occur among girls, ability is one of the few possible explanations available, whereas for boys several options are available to account for failure in addition to ability, including effort, conduct, or teacher attitude.

Because girls tend to attribute academic performance to ability that is a stable factor, they should be more negative in their expectancies for success in future performance situations. Because boys are more likely to attribute success or failure to unstable causes like effort or teacher attitude, boys' judgments of future success should be more optimistic. In two studies, Dweck and her colleagues have confirmed this prediction. Girls expect lower grades in the future at the beginning of the school year than boys do. These findings are consistent with earlier research showing that girls consistently underestimate their academic capabilities in comparison to boys (Crandall, 1969). Furthermore, in a laboratory experiment where children were given either a new evaluator, a new task, or both following failure evaluation from an adult on a series of tasks, boys' expectations for recovery of success were greater than were girls' when a new evaluator was present (Dweck & Goetz, 1978).

A more direct test of the gender-related attributions for failure in relationship to feedback from adults was evaluated by manipulating the contents of evaluative feedback to girls and boys. Elementary-school students were presented a series of anagram tasks with mixed success and failure feedback. Half of the children received typical "boy" feedback consisting of a mix of attributions to neatness and correctness of answers. The other half of the children received typical "girl" feedback focusing on the correctness of the answers. After a second phase of initial failures on a different set of tasks, students who had received "boy" feedback blamed their failures on insufficient efforts or on the experimenter. Children who had received "girl" feedback blamed their failures on their own lack of ability (Dweck & Licht, 1980).

Summary. Unfortunately, the processes involved in the development of learned helplessness have not been a major focus of study. Perceptual research on passive exposure to visual stimuli and work on maternal deprivation raise important questions about the role of environmental mastery in development.

Several studies with young organisms indicate that both animals and humans can react to response-outcome contingencies early in life. Further, children's experience with noncontingent response outcomes apparently leads to a similar pattern of learned helplessness as documented in numerous studies with adults. There is strong need for more systematic research on the development of processes that are linked to learned helplessness susceptibility. Because learned helplessness depends upon the comprehension of degree of contingency between voluntary responses and outcomes, it is reasonable that maturation may be related to several aspects of helplessness. For example, the directness and temporal proximity between voluntary responses and outcomes may interact wtth cognitive ability and thus influence resultant helplessness processes.

We also reviewed several studies, relying particularly on Dweck's research, indicating that the attributions children make about causes of failure influence whether or not learned helplessness will occur. These studies, as in the adult research, suggest that when failure or uncontrollability is attributed to more internal stable factors such as ability, helplessness is more likely to occur. Children's attributions about failure, particularly in achievement-related contexts, appear to be related to cognitive development. Younger children are less likely to attribute failure to internal stable factors and thus may be less vulnerable to learned helplessness.

Finally, evidence was reviewed linking helplessness in elementary-school children to school achievement. Experiences of failure attributed to stable factors are likely to lead to reduced school achievement. The performance of children with learning disabilities or of low achievers has been linked to helplessness. Sex differences in achievement and reactions to failure were also explained in terms of the helplessness attribution process. Girls are more prone to blame poor academic performance on personal ability, whereas boys tend to attribute poor performance on insufficient effort, nonintellectual aspects of their performance (e.g., neatness), or on external causes such as teacher attitude toward the student. These gender differences in schoolchildren's reactions to failure may develop because of the typical kinds of feedback teachers give girls and boys, respectively, about the causes of their poor performance.

Learned helplessness is a particularly ripe area of research for developmental psychology. We do not know how cognitive ability influences the formation or longevity of learned helplessness. Another interesting question that has gone largely untested is the potential impacts of early helplessness experiences on later cognitive and emotional development. Although there are a few suggestive findings in the literature, neither of the preceding issues has been thoroughly researched.

Dweck and her associates have proposed a developmental model of achievement motivation and vulnerability to helplessness that warrants further investigation. They suggest that young children are motivated primarily by intrinsically rewarded drives for environmental mastery. During the transition into school, however, both developing cognitive abilities and the school environment encourage the acquisition of more extrinsically determined achievement motivations.

One potential source of noncontingency in childhood is environmental stressors. Recently, some investigators have examined in a preliminary manner whether chronic exposure to uncontrollable, aversive environmental events like noise or crowding leads directly to greater helplessness. Furthermore, some researchers have questioned whether such environmental exposures might predispose some children to greater vulnerability to the effects of helplessness in learning situations. These issues are addressed in the next section of this chapter.

Environmental Stressors and Learned Helplessness

The adaptive cost hypothesis asserts that although human beings can adapt to stressful conditions, there are cumulative costs of adaptation to stressors (Cohen, 1980). Among these cumulative costs are cognitive fatigue, overgeneralization of coping strategies (e.g., tuning out noise), and when coping efforts fail, heightened perceptions of uncontrollability over environmental events (see Chapter 1 on coping taxonomy). This perception of uncontrollability of environmental events may in turn precipitate learned helplessness. Persons with reduced perceptions of control over certain environmental events may have reduced motivation to initiate control when it is possible to do so. This section of the chapter examines research on the relationships between chronic exposure to environmental stressors and learned helplessness. Unfortunately, there is very little research on this topic. Furthermore, few if any of the studies reviewed here were able to isolate noncontingency of behavioral responses to the environmental stressor as the critical variable mediating exposure to chronic environmental stressors and subsequent learned helplessness. Nevertheless, many of these studies have found patterns of behavior that certainly resemble major components of the learned helplessness syndrome.

Three general types of studies have examined helplessness and environmental stress: locus of control and environmental stressors, crowding and noise studies, and residential design research.

Locus of Control. External locus of control has long been associated with poverty and ethnic minority status both in the United States as well as in other societies (Lefcourt, 1982). People in these circumstances have less power and as a result cannot influence what happens to them to the same extent that others with more resources can. Several early studies focused on the impacts of major life events on locus of control. In one study, male undergraduates became more external following the military draft lottery during the Vietnam War. This effect was particularly strong for those who obtained highly favorable lottery numbers (McArthur, 1970). This latter finding is interesting because it shows that important, uncontrollable but positive events can lead to reduced perceptions of self-efficacy. Negative, uncontrollable events have also been found to shift individuals' locus of control in an external direction. Gorman (1968) found that student supporters of Eugene McCarthy's presidential campaign scored more externally the day after the Chicago Democratic Convention than would be expected given the normative levels of externality in the college population at that time. McCarthy lost the presidential nomination to Hubert Humphrey. Furthermore, many McCarthy supporters either witnessed or were themselves the victims of uncontrollable police brutality by the Chicago police force during the convention period.

Although neither of the preceding studies conclusively demonstrates an effect of noncontingent environmental events upon helplessness, they are suggestive. In both cases, college students experiencing an important, uncontrollable event became more external in their general expectancies for control. Greater externality may also predispose individuals toward greater helplessness in the face of uncontrollable events when they do occur. External persons are more susceptible to standard learned helplessness treatments than are internal persons (Hiroto, 1974). Furthermore, learned helplessness is more likely to generalize to different test conditions from the pretreatment phase for externals than it generalizes for internal locus of control individuals (Cohen *et al.*, 1976).

Crowding and Noise Studies. Rodin (1976) reasoned that if crowding in the home led to decreased ability to regulate social interaction as well as reduced behavioral options, children from high-density homes might be more susceptible to learned helplessness. In order to evaluate this hypothesis, Rodin conducted two studies. In the first study, 6- to 9-year-old children who had lived for at least 2 years in a low-income housing project participated in a laboratory experiment. The children were taught an operant learning procedure and then given a choice to select their own reward upon successful completion of a set of responses or to allow the experimenter to choose the reward. Children from higher density apartments were more likely to prefer the experimenter choosing their reward as opposed to self-selecting their own reward.

The second study compared junior high school students' reactions to soluble and insoluble puzzles in a standard learned helplessness paradigm. Children who lived in higher density homes were significantly more likely to suffer from learned helplessness following exposure to noncontingent problem-solving experiences. These children solved fewer soluble puzzles following the helplessness pretraining than did the children from lower density homes.

There is also some preliminary evidence from a field study of airport noise in France that indicates that children exposed to airport noise at school may suffer from some control-relevant maladies. Moch-Sibony (1985) compared two elementary schools in the impact zone of an airport (see Chapter 2 for details on exposure levels). One school was thoroughly sound attentuated but otherwise similar in socioeconomic composition of families and of children's entry-level aptitude scores. Children from the noisy school were less likely to persist on difficult but soluble puzzles than were children from the quieter sound-attentuated school.

As discussed earlier and in Chapter 1, Glass and Singer (1972) reported in a series of studies that exposure to unpredictable, uncontrollable noise produced poststimulation deficits in performance on a number of tasks. Of particular interest to a helplessness explanation for some of these stressor aftereffects is the consistent finding in the noise literature that people previously exposed to uncontrollable noise are less tolerant of frustration as measured on the Feather tolerance-for-frustration task. This measure requires the subject to work on two soluble and two insoluble line puzzles for a fixed amount of time. Subjects previously exposed to uncontrollable noise (Glass & Singer, 1972) or high-density conditions (Evans, 1979; Sherrod,

1974) attempt fewer insoluble puzzles. All of these studies involved short-term exposure to stressors under laboratory conditions. Glass and Singer suggested in their early work that subjects who cannot predict or control a stressor learn that the outcomes of their behaviors are independent of their responses. This learning produces motivational deficits that are manifested by more giving-up responses on subsequent poststressor tasks. As discussed in Chapter 1, an important assumption made here is that helplessness can occur even when performance on the experimental task during the stressor is not instrumental in avoiding or ameliorating the stressor. In the classical helplessness pretreatment, as reviewed earlier in this chapter, the organism believes that it can avoid or minimize exposure to an aversive stimulus by some instrumental action. In the typical aftereffect paradigm used by Glass and Singer and others, the subject has no expectation that he or she can avoid or reduce aversive stimulation through his or her own actions. Noncontingent feedback during a concept formation problem also produces similar frustration tolerance aftereffects as found in the Glass and Singer paradigm (Cohen *et al.*, 1976).

Residential Design Research. Andrew Baum and his colleagues have undertaken the most thorough examination of the relationship between chronic exposure to environmental stress and learned helplessness (Baum & Valins, 1977; Baum, Aiello, & Calesnick, 1978; Baum, Gatchel, Aiello, & Thompson, 1981; Baum & Davis, 1980; Rodin & Baum, 1978). Baum and his colleagues reasoned that exposure to residential living conditions that interfered with the regulation of social interaction and led to unpredictable and/or unwanted personal encounters could lead to increased feelings of helplessness. Furthermore, Baum hypothesized in accord with early theories of crowding (Proshansky *et al.*, 1970; Stokols, 1972a) that increased behavioral constraints from the unwanted social interactions would lead to goal blocking, fewer behavioral options, and increased perceptions of crowding in the living space.

In order to study the relationship between residential design and helplessness, Baum and his colleagues compared college students living in different dormitory settings on two different campuses. On one campus, students were assigned either to suite designs with small common lounges and shared bathrooms for a small group of suitemates, or to long, double-loaded corridor dorms with one lounge and one bathroom per floor. On a second campus, Baum compared residents of a short-corridor dorm to those living in dorms with longer corridors. Because there are several studies and large amounts of data in this program of research, we will focus our discussion only on the data most pertinent to learned helplessness and other control-related findings. For a more complete summary, see Baum and Valins (1977).

Even though each respective dormitory living-condition comparison on each campus was of comparable spatial density, corridor as opposed to suite residents and long-corridor versus short-corridor residents perceived their respective living environments as more crowded. Consistent with the authors' hypotheses, these same residents on each campus also felt less perceived control over common living spaces in their dorms, had fewer friends in the dorm, avoided other residents more often, and reported greater unwanted social interaction in their living settings. Of particular interest here, the authors also found evidence that some of these effects generalized outside of the dormitory setting. Several studies showed that residents of long-corridor dorms in comparison to suite or short-corridor groups avoided social interaction with strangers in a laboratory experiment or, when forced to interact with strangers, felt more uncomfortable.

To test more directly that living conditions that interfered with the regulation of normal social interaction might lead to learned helplessness, the authors conducted several other studies both in and outside of the dormitory environment. In an early laboratory experiment, after a brief waiting period students were given a very vague description of an ensuing experimental procedure. Short-corridor residents were much more likely to ask questions about the impending experiment than were residents of long-corridor dormitories. Furthermore, residents of these short-corridor dorms when directly queried felt less helpless in general and were more likely to believe that they could influence outcomes in their own lives. Long-corridor dormitory residents, on the other hand, asked few if any questions about the ambiguous experimental procedures to follow, felt generally more helpless, and were pessimistic about the utility of changing things in their lives that were in need of modification.

Dormitory residents also played a modified prisoner's-dilemma game in the laboratory. In this task, residents played a bargaining game with an experimental confederate. The interaction of each game player's choice determined a payoff value for each member of the pair. Subjects could play the game competitively, trying to maximize individual gain, cooperatively, maximizing total gain for both players, or withdraw and make choices with little or no strategy, frequently allowing the opponent to win.

The researchers felt that residents of long-corridor dorms would play a more competitive game strategy because their typical social interactions with strangers were unwanted because of the difficulties in social interaction associated with the residential design of their dormitory. When social interaction was permitted between the game players, the hypothesis was confirmed. However, when interaction between the pair was prohibited, long-corridor residents in comparison to residents of short-corridor dormitories were more likely to withdraw.

Although this set of findings on the prisoner's-dilemma game is provocative, its relationship to the helplessness model is not clear. If helplessness is accompanied by reduced ability or motivation to assert control when contingent-based effects are possible, why should residents experiencing greater crowding and problems regulating social interaction in their living space be more competitive in the prisoner's-dilemma game? The withdrawal response of the long-corridor residents when interaction during the game was prohibited is more consistent with learned helplessness studies that often note reduced affect and task disengagement following exposure to noncontingent, aversive stimuli (Seligman, 1975).

In order to examine the process of learned helplessness and dormitory setting conditions more closely, Baum and colleagues have completed several recent studies focusing on the development of helplessness over time in the dormitory setting.

Baum hypothesized that when college students initially arrive on campus, they expect to make friends and have normal social interactions that would include the ability to regulate social interactions much of the time. If these expectations were disconfirmed because of the high degree of interference found in long-corridor dormitories, reactance might occur (Wortman & Brehm, 1975). Initially, residents of long-corridor dormitories might try to reassert control over social interaction. During this reactance period, more control-oriented behaviors including competitiveness might be expected. However, after prolonged repeated exposure to uncontrollable social interactions, helplessness might set in and be manifested by greater withdrawal in social encounters and reduced expectancies for control.

Baum, Aiello, and Calesnick (1978) and Baum, Gatchel, Aiello, and Thompson (1981) compared several measures of helplessness over the period of students' first semester in the dormitory. On the prisoner's dilemma game, long-corridor residents in comparison to short-corridor residents were more competitive at Weeks 1 and 3 but by the seventh week of their initial semester in the dormitories were more likely to play withdrawal strategies in the prisoner's-dilemma game. Short-corridor residents played a mixture of competitive and cooperative strategies and became more cooperative over time. A separate group of residents from the same dormitories were also surveyed during the course of their first semester about their satisfaction with living conditions in their respective dormitories. In parallel with the results of the prisoner's-dilemma game, long-corridor residents prior to their seventh week in residence felt low satisfaction with the dormitory environment and had more negative feelings toward fellow dormitory residents. By the seventh week of the semester, however, these differences were largely absent. Furthermore, long-corridor residents felt increasingly more crowded, were more apt to ignore other residents, and experienced greater unwanted social interactions as the semester wore on. Finally, these same long-corridor residents felt a lack of control over the way things were in their lives in general by the seventh week of the semester; whereas, at Weeks 1 and 3, they did not differ significantly in these feelings of self-efficacy in comparison to their short-corridor dormitory comparison group.

Thus, over the course of the initial semester living in the long-corridor dormitory, control-related problems became more salient and were accompanied by concurrent changes in the strategies typically played in the two-person, prisoner's-dilemma bargaining game. These time-related data trends in helplessness reactions are interesting in light of learned helplessness theory that states that reactance is more likely to occur during short periods of uncontrollable events, particularly when the outcomes of these events are important (Roth, 1980; Wortman & Brehm, 1975). College students at the beginning of the year presumably place a high stake in making friends with fellow dorm residents and getting along with their neighbors. One of the most striking aspects of Baum and colleagues' findings is that the effects of problems with social encounters in the dormitory generalize to the experimental game in a laboratory context with strangers.

If one of the major factors in producing the helplessnesslike phenomenon observed in both self-report data and game strategy behavior is due to the architectural layout of the dorm as argued by Baum and his colleagues, then an architectural modification of an existing long-corridor dormitory ought to change some of the patterns of data noted before. In order to investigate this possibility, Baum and Davis (1980) partitioned a long-corridor dormitory into two short-corridor dormitories. They reasoned that this intervention would reduce the number and frequency of unwanted social interactions by reducing group size and creating more semipublic lounge spaces where one would be more likely to socially interact with friends or known neighbors. As expected, this architectural modification substantially changed the interpersonal dynamics of the previous long-corridor dormitory setting. The intervention reduced perceived crowding and feelings of unwanted social interaction. Furthermore, feelings of self-efficacy were enhanced, and students felt more satisfied with the dormitory and reported getting along with their neighbors better. Furthermore, in a laboratory experiment with strangers, residents whose long-corridor dorm had been partitioned behaved similarly to residents of short-corridor dorms in their interactions with a stranger waiting to be in an experiment. Long-corridor residents, as in the earlier studies, evidenced more social withdrawal, sitting farther from the waiting confederate stranger, making less eye contact, and initiating substantially less conversation.

In a later study, the authors speculated that the attributions dormitory residents make about the causes of their living-condition problems might shift over time in accord with the reactance hypothesis stated before (Baum, Gatchel, Aiello, and Thompson, 1981). Residents of long-corridor dorms as in earlier studies reported more problems in the living conditions of their dormitories than reported by short-corridor residents. Of particular interest is that residents of long-corridor dormitories were more likely to attribute problems initially in the semester to themselves. Over the course of the semester, the explanations for problems in the dormitory became more and more external. This shift in attributions for problems correlated highly with decreasing expectancies for control and more withdrawal strategy in the prisoner's-dilemma game. Short-corridor residents, on the other hand, consistently made self-attributions for the few dormitory problems they were experiencing. Furthermore, the pattern of their attributions for problems in the dormitory did not shift over time.

These investigators noted correctly that internal attributions about lack of control are more likely to lead to learned helplessness and are more likely to generalize to another context (Abramson *et al.*, 1978, 1980). Yet, paradoxically it was not until *external* attributions about problems in the dormitory occurred that the long-corridor residents began to manifest symptoms of helplessness. Baum and colleagues suggested that the students living in the long-corridor dormitories may have become sensitized to noncontingent events and as a result were more likely to prematurely judge that a situation was helpless when personal control was threatened. Thus, when these residents are confronted with a new situation, they may make initial attempts to control the situation but more quickly relinquish control if it is not immediately forthcoming.

To evaluate this hypothesis, the investigators examined the behaviors of long-

and short-corridor residents on each trial of the prisoner's-dilemma game over the course of their first semester in their respective dormitories. Short-corridor residents played a fairly consistent strategy of mixed cooperative/competitive choices over the course of the semester. During the first 5 weeks of the semester, long-corridor residents increased competitive strategies over the course of trials in each game. During this time period, attributions for problems in the dormitory were made to self. At the sixth week and thereafter, a noticeable shift in game strategy occurred. Initial trials of the prisoner's-dilemma game (approximately 6–10 out of 20 trials) were played competitively, but thereafter withdrawal behavior increased with a sharp drop in competitive game strategy. For the first half of each game, residents of long-corridor dormitories appeared to try to control the game by playing competitively. After these initially unsuccessful attempts to control the outcomes of the game, however, these same residents seemed to give up all attempts to gain points and withdrew.

Although not mentioned by the authors, there is a fascinating resemblance between these dormitory helplessness data and the behavior of Type A individuals in control-related circumstances. As you may recall, when Type A individuals are initially confronted with a control-relevant situation, their initial response is highly reactive with vigorous attempts to gain control over the situation. However, after repeated failures to assert control, these same Type A individuals are much more likely to give up even in circumstances where instrumental responses are possible (Glass, 1977). One apparent difference between the two sets of findings is that Type A persons seem to have unrealistically optimistic expectations for control regardless of the objective, control-related conditions of the situation. Residents of long-corridor dormitories, however, seem to acquire a pessimistic expectation that control over social interaction, in particular, will be difficult if not futile in other settings.

To summarize the several findings from this program of research by Baum, residents of dormitories with long corridors and shared public space (e.g., lounge, bathroom) consistently experienced greater problems in their living environment related to difficulties in the regulation of social interaction. These residents in comparison to those living either in suites or in shorter corridor arrangements perceived their living arrangements as more crowded, had fewer friends among their neighbors, and reported frequently avoiding social contact in the dormitory. Initially, residents of long-corridor dormitories attempt to assert control over social interaction but after repeated failures to accomplish regulation of normal social interaction experience helplessness. Evidence from self-report data and game-playing behavior during the prisoner's-dilemma game support this sequence of events. Furthermore, these effects were replicated in several studies. There is also evidence that shifts from internal to external attributions about the causes of dormitory-related problems accompany the change in behavior of long-corridor residents over the course of their initial semester in the dormitory.

Summary. One of the cumulative costs of adaptation to chronic exposure to uncontrollable environmental stressors may be heightened perceptions of uncontrollability over environmental events. This heightened perception of uncontrollability may in turn lead to helplessness. Uncontrollable life events as well as crowded residential living environments may lead to reduced feelings of self-efficacy. Children from higher density homes are more likely to relinquish control to an experimenter and are more susceptible to a standard learned helplessness paradigm employing insoluble puzzles during the pretreatment phase. Children exposed to loud airport noise at school are also less tolerant of frustration and give up on insoluble puzzles sooner than children from quieter schools.

Furthermore, in a major research program on residential architectural arrangements, Baum and his colleagues have shown that college students living in long-corridor dormitories experience greater crowding and difficulties regulating social interactions in the dormitory than other students living, respectively, in short-corridor dormitories or in suite arrangements. Helplessness associated with unsuccessful attempts to regulate social interactions was replicated in several studies of self-report data, game-playing strategies in a bargaining game in the laboratory, and in reactions to experimental waiting room procedures. In addition, Baum and colleagues have been able to document the development of learned helplessness over time. Initially, college students in long-corridor dormitories attribute problems related to social interaction and crowding to themselves and make concerted efforts to overcome these difficulties. Eventually, however, after repeated unsuccessful attempts to regulate social interaction, more external attributions about problems in the dormitory are made. At this same point in time, these residents manifest behaviors associated with learned helplessness.

Several issues have been raised about the relationship between chronic exposure to environmental stressors and control-related problems. First, are people exposed to chronic, uncontrollable environmental stressors more helpless in general, or more susceptible to control-related contingencies, specifically? There is some evidence confirming the former. Rodin found children from higher density homes more likely to relinquish control to the experimenter during operant-learning procedures in which no exposure to noncontingent stimuli occurred. Furthermore, in her second study, children from higher density homes took longer to learn the correct solution pattern for puzzles in the soluble pretreatment condition of the learned helplessness paradigm. Thus, prior to any experimental, helplessness induction, children chronically exposed to high density appeared more helpless. Furthermore, several studies noted reduced perceptions of self-efficacy among residents of more crowded settings. Finally, the prisoner's-dilemma game employed by Baum did not involve exposure to any noncontingent stimuli, yet consistent differences in game-playing strategy were noted by those living in different types of architectural arrangements.

A second question that arises directly from the preceding discussion is whether these various studies of environmental stress have demonstrated that persons exposed to uncontrollable environmental stressors perceive environmental events as less controllable and, as a result, become more susceptible to learned helplessness? With the exception of the second Rodin study on the reactions of junior high-school children from low- and high-density homes to a standard learned helplessness paradigm, none of the studies discussed have clearly demonstrated learned help-lessness as defined by Seligman (1975). There is evidence of relinquishing control to others, reduced feelings of self-efficacy and greater externality, more giving up in the face of difficult or insoluble tasks, and changes in game-playing strategies that are all consistent with the hypothesis that exposure to uncontrollable environmental stressors may produce heightened susceptibility to control-related problems. Nevertheless, only one study has demonstrated a clear contingency-linked learning deficit for individuals exposed to environmental stress.

Furthermore, none of these studies has been able to pinpoint the precise environmental characteristics that precipitate the kinds of outcomes noted. Baum's research comes closest when it demonstrates a clear, parallel set of findings between shifts in game-playing strategies with changes over time in self-reports of problems in the dormitory environment and residents' explanations for those problems. Nevertheless, lack of contingency *per se* over environmental events has not been isolated in any of these field studies on environmental stressors and learned help-lessness. Given that nearly all environmental stressors are uncontrollable and often unpredictable as well, it has not been possible to separately analyze the effects of exposure to controllable and uncontrollable environmental stressors under natu-ralistic conditions. Only experimental studies of controllable and uncontrollable noise and crowding have clearly documented the critical importance of control in the influence of environmental stressors on perceptions of uncontrollability and subsequent task performance deficits (Cohen, 1980).

The Los Angeles Noise Project: Control and Environmental Stress

One of the effects of human exposure to chronic, uncontrollable environmental stressors is reduced perceptions of control over the environment. A potentially deleterious consequence of perceived uncontrollability over environmental stressors may be decreased motivation to initiate instrumental responses. Research on the aftereffects of exposure to uncontrollable environmental stressors including noise, crowding, and air pollution indicates poststressor task impairments. Furthermore, considerable work on learned helplessness demonstrates that when human beings are exposed to noncontingent, aversive events, decrements in motivation, affect, and cognition occur. These behaviors happen because people learn that their instrumental responses to a stressor are independent of the effects of that stressor on them. Furthermore, learned helplessness is most likely to occur when the person believes that the cause of the noncontingency is stable, global, and personal.

In order to investigate the impacts of chronic exposure to loud aircraft noise at school, children from noisy and quiet schools were compared on their problem-solving behaviors on a moderately difficult puzzle following a pretreatment phase with either soluble or insoluble puzzles. In the third grade replication study, children were also given the opportunity to choose a game to play or have an experimenter make that choice. Finally, children's locus of control scores was measured.

Methods, Results, and Discussion

Locus of Control

Prior research with adults has shown that externals are more susceptible to the impacts of learned helplessness induction (Cohen *et al.*, 1976; Hiroto, 1974). In addition, several studies were reviewed indicating that internals and externals react differently to environmental stressors, including crowding (McCallum *et al.*, 1979) and air pollution (Evans *et al.*, 1982). Internals may respond more positively initially to environmental stressors, but if continued instrumental efforts to modify the stressors are unsuccessful, internals may in the long run react more negatively to stressors (Lazarus, 1977; Novaco *et al.*, 1979).

Furthermore, children chronically exposed to uncontrollable noise may become more passive and/or develop greater perceptions of uncontrollability over their environment. Thus, they may become more external in their generalized expectancies for environmental control. As reviewed earlier in this chapter, there is some evidence that adults exposed to uncontrollable circumstances become more external in locus of control. This research, however, has generally focused on more acute, critical life events rather than on more chronic low-level stressors like airport noise.

All children were administered the short form of the Crandall Intellectual Achievement Responsibility Scale (IAR) prior to any experimental tasks. The IAR was administered only at Time 1. The original IAR consists of 34 forced-choice items, describing either a positive or negative achievement experience for elementary-aged schoolchildren (Crandall, Katkovsky, & Crandall, 1965). There are two subscales of the IAR that measure extent of beliefs in internal responsibility for success (I+) in achievement situations as well as belief in internal responsibility for failure (I−). Split half and test–retest reliabilities for the original scale are .69 and .65 for the total IAR and in similar range for the two subscales.

Validity data for the IAR are preliminary but clearly encouraging. The subscale structure has been supported by correlational analyses, and the scales are not highly correlated to IQ or social desirability. Furthermore, the scale is a modest predictor of academic achievement (see Crandall *et al.*, 1965, for more details).

The scale used here is a 14-item subset from the original scale, consisting of 7 items each from the two subscales. These items were the highest loading items from a factor analysis of the original IAR and maintain maximum item to whole, internal consistency. Evidence for the predictive validity of the shortened version as well as test–retest reliability are both comparable to data on the whole test (see Crandall, 1978, for more details).

Helplessness

Performance on a cognitive task preceded by a success or failure experience was employed to examine the effect of noise on response to failure and on persistence on a difficult task. Response to failure is a standard measure of susceptibility to helplessness (cf. Seligman, 1975). Previous research by Rodin (1976) indicated that residential crowding caused greater susceptiblity to helplessness in children. Furthermore, children from noisy schools persisted less on difficult tasks (Moch-Sibony, 1985). Thus, if noise-school children are more susceptible to helplessness, they will show greater effects of a failure experience than their quiet-school counterparts. A lack of persistence ("giving-up" syndrome) is considered a direct manifestation of helplessness.

First Testing Session. Each child was given a treatment puzzle to assemble after the tester demonstrated the task with another puzzle. One-half of the children received an insoluble (failure) puzzle and one-half a soluble (success) puzzle. The soluble puzzle was a circle, and the insoluble puzzle was a triangle. After time (2 ½ minutes) was up on the first puzzle, the child was given a second, moderately difficult puzzle to solve. The second (test) puzzle was the same—a square—for all (success and failure) children. The child was allowed 4 minutes to solve the second puzzle. Whether the puzzle was solved, how long the solution took, and whether the child "gave up" before the 4 minutes had elapsed were used as measures of helplessness.

Second Testing Session. Treatment puzzles were *not* readministered during the second session. Each child was given only the test (square) puzzle to solve. As previously, the child was allowed 4 minutes to solve the test puzzle, with the same measures of helplessness assessed as in the earlier testing session.

During Time 1, a large proportion (34%) of the children assigned to the success condition who received a soluble treatment puzzle failed to solve the treatment puzzle within the 2 ½ minutes allowed. Although the fact that a number of children self-selected themselves into a failure condition makes it difficult to interpret main effects for success/failure and interactions between success/failure and noise, comparisons between the children from noisy and quiet schools, irrespective of (controlling for) their pretreatment, are of primary interest.

Choice Task

Rodin (1976) found that children from crowded residences were more likely to relinquish control over choosing rewards in an operant-learning task to an experimenter than were children from uncrowded homes. In the present study, after all tasks had been completed, children were told that three games were available for them to play. The three games were tick-tac-toe, drawing a picture of an airplane, or a stopwatch game for guessing a 30-second interval. Children were not told any details about the games and were asked: "Do you want to choose the game, or do you want me to choose it for you?"

Data were recorded on whether the child requested information about the games and whether he or she chose the game or asked the experimenter to do so. This task was used in the third grade replication only.

Locus of Control

Direct effects of noise on locus of control scores and the interactive effects of locus of control and noise on helplessness were analyzed. The interaction effects are discussed under the learned helplessness results. The total IAR scores and expectancies for success (I+) as well as failure (I-) in achievement situations were analyzed. The standard set of covariates as explained in Chapter 2 were incorporated into analyses. Because the IAR was only given at the first testing session, there are no longitudinal data on this measure.

Cross-Sectional Results. There were no main effects of noise or months-by-noise interaction effects on any of the locus of control measures. There was, however, a potentially interesting effect of home noise levels on locus of control scores. As can be seen in Table 10, children in noisy homes become more external in their I+ scores over time, whereas those from quiet homes become more internal over time in residence, $F(1, 81) = 2.97$, $p<.09$. The I+ score measures children's perception of internal responsibility for successful intellectual achievement (Crandall *et al.*, 1965). Similar but nonsignificant trends were found for total IAR and I− scores. It should be recalled that home noise measurements were only evaluated for children attending noisy schools.

Discussion. In general, there were few if any effects of environmental noise levels on children's locus of control scores. The marginal changes in I+ scores as a function of exposure to home noise over time are potentially interesting. The data suggest that as children live under noisy conditions they become more external. Children in quiet homes, however, become more internal over time. We might expect that over time children would generally become more internal because with increasing residential experience, competencies and autonomy would increase. Yet exposure to noise is able to counteract the normal progression in internality that occurs with maturity. Crandall, for example, has consistently found that older elementary-school children score more internally on the IAR than do younger

Table 10. I Plus Scores on the Locus of Control Scale: Home Noise by Years in Residence

	Years in residence	
	<3 years	>3 years
	Home noise	
Low	5.56[a]	6.00
High	6.00	5.43

Note. Includes only children attending noise impacted schools.
[a]Lower scores indicate greater externality.

children (Crandall *et al.*, 1965). Our data must be treated cautiously because the effects are marginal and were not statistically significant for the total IAR or I− subscale. In addition, high mobility or some other third variable may have been associated with home noise levels. Nevertheless, the data are interesting enough that follow-up work would be worthwhile.

There are at least two problems with the measure of control used here that may have precluded an adequate test of the hypothesis that prolonged exposure to uncontrollable environmental stressors would lead to diminished feelings of environmental mastery. First, the IAR and its subscales are concerned with control-related efforts in scholastic achievement. Questions probe children's beliefs in their own control of reinforcements in achievement situations such as math and reading work, tests, or teacher–student interactions. Thus, the IAR may not be sufficiently general to register changes in individual attitudes toward environmental mastery and competence.

Second, perhaps exposure to stressors does not produce a general change in expectancies about environmental mastery but instead impacts responses only to the specific stressor. That is, children from noisy settings may not necessarily feel less control over the environmental in general but simply expect little or no control over noise.

Therefore, in addition to measuring generalized expectancies about environmental control, it would have been good to measure children's attitudes about control over noise in their home and school settings. As suggested earlier (see also Folkman, 1985), strong, continuous expectations for instrumental control in the face of uncontrollable events may be maladaptive. To relinquish behavioral control to another more powerful or predictable source in the context of an uncontrollable environmental situation may lead to fewer problems. The difficulty arises when the abandonment of control becomes overgeneralized to other situations in which the exercise of control is possible and likely to ameliorate aversive stimuli. The ability to distinguish between different sources of threats to control and the flexibility to modify coping strategies accordingly may become disrupted by chronic exposure to uncontrollable environmental stressors like airport noise.

As suggested by Baum's important research on dormitory crowding, the development of helplessness under chronic stressful conditions may take time to unfold. More research needs to carefully monitor individual's perceptions of uncontrollability over environmental stressors across extended time periods. These perceptions of uncontrollability should be compared to patterns of reactance and and helplessness that are both situation specific (e.g., in the noisy school) as well as more general (e.g., behavior in a quiet environment).

Learned Helplessness

There are no interactions of locus of control with noise levels or pretreatment conditions in the helplessness procedure (i.e., soluble vs. insoluble first puzzle). Thus, all data reported are collapsed across locus of control scores. One possible reason why locus of control did not interact with noise or the helplessness pretreatment is because the IAR scores were skewed in the external direction. Previous work with locus of control has also found a disproportionate number of externals among poor and minority samples (Lefcourt, 1982) similar to the ones found in this study. Measures of learned helplessness included whether or not the second (test) puzzle was solved, time to solution, and the child's persistence or giving up before the 4-minute time period for solving the soluble puzzle had elapsed. Because treatment puzzles were *not* readministered during the second phase of the experiment, performance was measured 1 year later on a soluble (test) puzzle.

The reader is reminded that a reasonably large proportion (34%) of children assigned to the success pretreatment (soluble puzzle) failed to solve the treatment puzzle during the pretreatment period. Thus, except for the first analysis, which includes only those children who worked on soluble treatment puzzles (success condition), the following analyses also include factors for success/failure (those who solved and those who did not solve the success treatment puzzle are treated as separate groups) and the interaction between success/failure and noise. The control factors were forced into the regression first, followed by success/failure (dummy coded), noise, and the Noise × Success and Noise × Months Enrolled interactions. Because of the difficulty in interpreting success/failure effects, they are not discussed. Moreover, because there were no significant interactions between success/failure and school noise level, the reported results are limited to the overall effects of noise.

Cross-Sectional Results. First, an examination of only those who were assigned to the success treatment condition indicates that children from noise schools were more likely to fail to solve the treatment puzzle (41% failed) than children from quiet schools were (23% failed). This effect, however, was only marginally significant, $F(1, 131) = 3.62$, $p < .07$. Second, there were similar effects of noise on the second puzzle, which occurred irrespective of whether the child received a success (solved or not) or failure treatment. As was the case with the first puzzle, noise-school children were more likely to fail the second puzzle (53% failed) than quiet-school children were (36% failed), $F(1, 246) = 5.99$, $p < .03$, and were more likely to give up, $F(1, 246) = 11.15$, $p < .001$, than their quiet-school counterparts were, multivariate $F(3, 244) = 4.59$, $p < .004$. As is apparent from Figure 2, a marginal interaction between noise and months enrolled in school, $F(1, 243) = 3.27$, $p < .07$, suggests that the longer a child had attended a noise school, the slower he or she was in solving the puzzle. However, the multivariate F for this interaction was not significant.

Although the preceding analyses indicate that children from noise schools are generally less capable of performing a cognitive task (at least puzzle solving) than children from quiet schools are, they provide only suggestive evidence that noise-school children feel or act as if they have less control over their outcomes. The strongest hint that failure on these puzzles on the part of noise-school children is related to helplessness is found in the data indicating that noise-school children were more likely to give up before their allotted time had elapsed than their quiet counter-

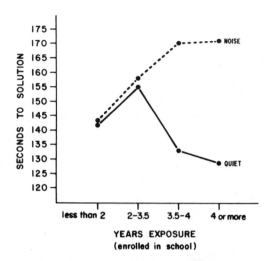

Figure 2. Performance on the second (test) puzzle as a function of school noise level and duration of exposure for Time 1 cross-sectional sample. Each period on the years coordinate represents one-quarter of the sample (from Cohen, S., Evans, G. W., Krantz, D. S., & Stokols, D., 1980. Copyright 1980 by the American Psychological Association. Reprinted by permission).

parts were. It is possible, however, that a constant proportion of children who failed on the second puzzle gave up. It would follow that the amount of giving up in the noise condition was inflated by the fact that there was a greater pool of failures. This interpretation suggests that increased giving up under the noise condition cannot necessarily be viewed as a sign of helplessness. A final analysis addresses this point. This analysis, which includes only those children who failed the second puzzle, indicated that the failures of noise-school children were associated with giving up (31% of those who failed gave up) more often than the failures of quiet-school children were (7% of those who failed gave up), $F(1, 103) = 5.85, p<.025$. Thus, even though all of these children failed to solve the puzzle, noise-school children were less likely to persist than their quiet-school counterparts were.

There were also some effects of noise levels on helplessness for children living in residences of varying sound levels. As discussed in Chapter 2, children attending noisy schools only were grouped according to home noise levels into two categories based on CNEL contours: quiet or noisy homes.

Children from noisy homes were less likely to solve the first helplessness task puzzles than the quiet-sample controls were, $F(1, 132) = 3.04, p<.10$. The longer a child had attended a noisy school, the less likely he or she was to solve either the first puzzle, $F(1, 130) = 4.06, p<.05$, or the second puzzle, $F(1, 240) = 2.07$, $p<.15$. Moreover, children from quiet homes but noisy schools were more likely to fail, $F(1, 244) = 6.20, p<.01$, and to give up, $F(1, 244) = 11.95, p<.001$, on the second puzzle than children from quiet schools were, multivariate $F(3, 244) = 4.71, p<.003$. Further, their failures on the second puzzle were associated with giving up more often than the failures of quiet-school children were, $F(1, 102) =$ 6.27, $p<.025$.

Adaptation to Noise Results. As in the analysis of the entire Time 1 cross-sectional sample, there are effects of noise on test puzzle performance that occur irrespective of whether the child received a success (solved or not) or failure treatment. Noise-school children were more likely to fail the test puzzle (58%), than quiet-school children (40%), $F(1, 133) = 5.37$, $p<.02$, and more likely to take longer solving the puzzle (166.71 seconds), $F(1, 133) = 2.88$, $p<.09$, than quiet-school children (145.76 seconds—multivariate effect for noise, $F(3, 133) = 1.92$, $p<.12$. Although differences between proportion of children failing and time to solution were stable across the two testing sessions, quiet/noise differences in the percentage of children giving up occurred only at T1—Noise × Testing Session interaction, $F(1, 133) = 3.90$, $p<.05$. The multivariate effect for the Noise × Testing Session interaction was not, however, significant, $F(3, 131) = 1.57$, $p<.20$.

As pointed out in the Cohen *et al.* study (1980), an analysis of the proportion of children giving up that includes only those children who failed the test puzzle provides the most direct measure of helplessness. This analysis looks at the degree to which failure is associated with giving up as opposed to unsuccessful persistence. Although data for the entire T1 sample indicated increased giving up on the part of noise-school as opposed to quiet-school children, there was neither a noise nor Noise × Testing Session interaction in the present analysis.

Although not directly relevant to problems posed in this chapter, it is of general interest to examine whether the soluble or insoluble puzzle given at T1 affected performance on the test puzzle administered 1 year later, irrespective of (i.e., controlling for) noise exposure. This comparison provides a rough measure of the duration of the learned helplessness effect. That is, does a failure as opposed to a success pretreatment affect subsequent task performance as much as a year later? As suggested earlier, because of a selection bias created by subjects who were assigned to a soluble puzzle condition who failed to solve their soluble puzzle, there were three levels of the success–failure factor: success group who solved their pretreatment puzzle, success group who failed their pretreatment puzzle, and failure group. At T2, children who had received a success treatment puzzle and solved that puzzle at T1 were more likely to solve, $F(2, 133) = 5.39$, $p<.006$ (84%), and were faster at solving the test puzzle, $F(2, 133) = 3.16$, $p<.05$ (86.62 seconds), than both those who failed to solve the success treatment puzzle (68%, 117.85 seconds) and those who received a failure treatment puzzle (73%, 117.59 seconds). There were no differences between these conditions on the proportion of children giving up; multivariate for success/failure, $F(6, 262) = 2.16$, $p<.05$. These data suggest the possibility of a helplessness effect persisting over a 1-year period, but they are difficult to interpret because of the self-selection problem.

Noise-Abatement Results. Do noise-abatement interventions (and their resulting reduction in classroom noise level) decrease or ameliorate the effects of noise in impacted classrooms? Both cross-sectional data collected during the first testing session and longitudinal data looking at changes in the responses of children who moved from noisy to quiet classrooms are relevant to this question. As in the previous section, longitudinal data are based on the attrition (163) sample and, thus, are subject to the attrition bias. The cross-sectional data reported in this section are based on the entire T1 sample (262).

Cross-Sectional Analyses. Several of the classrooms in noise-impacted schools had been treated with noise-reducing materials several years before the first testing session. Because they were still relatively noisier than quiet comparison classrooms and because of the presumption that the high-noise levels in the homes and play areas of noise-school children were as important as the actual classroom level, these treated classrooms were not separated from other noise-school classrooms in the previous article (Cohen *et al.*, 1980). To evaluate the effectiveness of this treatment and assess the relative impact of a somewhat quieter classroom on the criterion variables, data from the first testing session were reanalyzed, with classrooms categorized as noisy (97 children), abated (45), and quiet (120). The regression analyses on criterion variables are identical to those described previously except that the noise variable had the three levels described previously instead of two.

The percentage of failure on the second helplessness puzzle was consistent with the expected order between quiet, abated, and noisy classrooms. The noise group was more likely to fail the second helplessness puzzle (57% failed) than either the abated group (47% failed) or the quiet group (35% failed), $F(2, 235) = 4.12$, $p < .02$. There was no difference between noise groups on the time required to solve the second puzzle. Preplanned contrasts comparing proportions of students solving the second puzzle indicate marginal differences between the quiet and abated groups, $F(1, 235) = 3.10$, $p < .08$, and the noise and abated groups, $F(1, 235) = 2.70$ $p < .10$, and a significant difference between noise and quiet, $F(1, 235) = 8.03$, $p < .005$. These data suggest that noise abatement marginally affected puzzle task performance, with children in abated classrooms performing at a higher level than those in nonabated room, but not as well as those in quiet rooms.

Both the noise and the abated group ''gave up'' on the second puzzle (17% for noise, 16% for abated) more often than the quiet group (3%). The multivariate effect for noise did not, however, reach statistical significance. An analysis including only those children who failed the second puzzle indicated that the failure of noise- (29%) and abatement-group (35%) children were associated with giving up more often than were the failures of quiet-group children (7% who failed gave up).

Longitudinal Analyses. These analyses compared children who were in nonabated classrooms during both testing sessions with those children who were in noisy rooms during the first testing session and in abated rooms during the second testing session, 1 year later. Quiet classrooms were not included in these analyses because of the difficulty in evaluating change when initial scores are significantly different.

There were no significant multivariate or univariate interactions of noise conditions and testing session on any of the learned helplessness measures.

New Third Grader Replication Results. As in the first-phase results of the longitudinal study, more children from noisy schools failed to solve the second puzzle. Preplanned contrasts indicated that children from quiet classrooms (40%)

failed the second puzzle significantly less often than those in noisy classrooms, (55%), $F(1, 144) = 3.95$, $p<.05$. Overall effects were marginal comparing quiet and noisy classrooms, $F(1, 146) = 2.24$, $p<.14$, and when comparing quiet, attenuated, and noisy classrooms, $F(2, 144) = 2.20$, $p<.11$. Children from attenuated classrooms did not differ significantly from either noisy or quiet classrooms. There were no interactions of months enrolled in school with noise.

Home-noise-level analyses also closely parallel the findings from the first study. Of the children attending schools in the flight path of the Los Angeles Airport, children from noisy homes in comparison to children from quiet homes were more likely to fail the second puzzle and to give up on it.

Discussion. Both the cross-sectional and longitudinal results show that children from noisy schools failed the soluble puzzle more often than their quiet-school counterparts. In addition, for the cross-sectional results, these failures were associated with giving up on the test puzzle before the total time allotted for solution had elapsed. Furthermore, this effect did not diminish over time. The interaction of months enrolled in school and noise was not significant.

However, the increased giving up on the part of the noisy as opposed to quiet-school children was not found in the longitudinal analysis. This nonsignificant effect was probably due to subject attrition and not adaptation over time. The Time 1 cross-sectional sample did not adapt over time, and the main effect of noise was replicated in the new third grader study. Both cross-sectional and longitudinal analyses suggest little if any adaptation in helplessness over time.

There is also evidence that home noise levels may influence helplessness. Children from the noise-school sample only who lived in homes with louder CNEL contours were more likely to fail the second puzzle. Their failure was more likely to be associated with increased giving up on the soluble test puzzle.

The effects of noise abatement on helplessness were generally nonsignificant. Abatement had a marginal effect on whether the child could solve the second test puzzle but was unrelated to the giving-up measure. The giving-up measure that provides the most direct assessment of helplessness in this study was only affected by the noisy-school versus quiet-school distinction. The longitudinal data replicated the lack of abatement effects on giving up and did not substantiate the marginal effects on failures to solve the second puzzle. The failure to mimic the cross-sectional data may emanate from an attrition bias or to the weakness of the effect itself.

One final set of findings worth mentioning briefly is the helplessness effects themselves. As noted in the results sections preceding, at both Time 1 and Time 2, children who had received a soluble pretreatment were more likely to solve, and were faster to solve, the test puzzle than children who had been in the insoluble pretreatment condition or had been unable to solve the success treatment puzzle. The Time 1 results were replicated in the new third grader study.

Therefore, there is strong evidence that young children when exposed to an insoluble puzzle are less likely to solve a subsequent, soluble puzzle. Furthermore, the effects may be quite long lasting. Children who had received an insoluble pretreatment puzzle or who had failed to solve the success puzzle 1 year before were still less likely to solve the soluble test puzzle at Time 2. These results, although fascinating, are preliminary and clouded by a self-selection problem because, as previously noted, some children (about one-third) who were given the success pretreatment puzzle did not solve it. Nevertheless, the intriguing possibility that some residue of helplessness may have persisted for 1 year is worthy of further investigation.

We reviewed several studies earlier suggesting that other factors influencing the efficacy of control in stressful situations were the amount of effort needed to exert control and the clarity of feedback about the utility of having expended effort to ameliorate a stressor. Future research on human responses to chronic environmental stressors should survey individual's perceptions of these two factors. It might also be possible to manipulate required effort and feedback clarity in experimental tasks using the same stressor. The reactions of individuals chronically exposed or unexposed to the same stressor could then be evaluated under conditions of variable effort and feedback.

Another factor in the relationship between exposure to stressors and helplessness outcomes may be the importance of the activities interfered with by the stressor. Thus, in addition to probing how much effort is required to perform various behaviors (e.g., reading, socializing) during exposure to the stressor, it would be useful to measure personal judgments about the importance of these various activities to the person. The relative importance of activities may have implications for whether reactance or helplessness is likely and might also effect the generalization of the effects. Finally, it seems reasonable to assume that children who felt that important activities were being interfered with might suffer the most in self-image with respect to self-efficacy or environmental mastery.

Choice Task

Only the third grade replication study used this task. Children from noisy schools were less apt to choose which game they wanted to play than were children attending quiet schools, $r(146) = .43$, $p<.01$. So few children asked for further information about the three games (7 out of 147) that no further analyses were conducted with this measure. We had predicted that children from noisy schools would be less likely to ask for more information. There were no effects of home noise levels on the choice task.

Discussion. Data from the choice task complement the learned helplessness findings. Again, we see evidence that children chronically exposed to noise are more prone to problems with control-related behaviors. As in Rodin's earlier study with residential crowding and children (Rodin, 1976), children exposed to a chronic, uncontrollable stressor are more likely to relinquish control to another person in a free choice situation. The strength of the effect closely fits Rodin's data. She found a correlation of .45 between exposure to density and abrogation of control.

We did not replicate Baum's findings where residents of crowded dormitories were less likely to ask for clarification when confronted with ambiguous information about an experimental procedure (Baum, Gatchel, Aiello, & Thompson, 1981). There are marked procedural differences between the two studies that makes this failure to replicate understandable. First, Baum and colleagues research utilized college students not third graders. Second, we presented ambiguous information about games to play that was probably construed by the children as a fun, positive experience. Baum and colleagues, however, were presenting adults with ambiguous information about a laboratory experiment that conceivably engendered some anxiety and concern about what would happen. The experimenters in the Los Angeles Noise study tried to maintain a friendly, gamelike atmosphere for most of the tasks and procedures used. Thus, the children probably had no fear or concern about the next "game" to play. Recall that the choice task came after the other procedures.

Conclusions

There is a large, complex literature on control and learned helplessness. Considerable human data show that the meaning of control to the individual has potent mediation effects between stressors and well-being. Control reduces the negative effects of stressors when the individual clearly believes that his or her ability to alter an aversive situation will directly lead in a relatively brief period of time to reduced negative impacts. This reduction in negative impact typically occurs either through modification of the stressor itself or through the utilization of an effective coping strategy.

Laboratory research has consistently demonstrated that environmental stressors like noise or crowding produce more negative effects on human beings when the stressors are more unpredictable and uncontrollable. Furthermore, field studies suggest that the control and prediction of chronic stressors are the most salient aspects to impacted residents. There is also evidence that architectural and organizational features of institutional settings may have important, control-mediated effects on clients. The provision of greater personal choice and individual autonomy has been found effective in ameliorating some of the negative effects associated with institutionalization.

The consequences of lack of control for human beings are just beginning to be understood. Early work suggesting that lack of control over environmental contingencies directly led to learned helplessness has been challenged. The current view is that people's beliefs about the causes, stability, and generalization of lack of control have critical influence on the nature and extent of helplessness.

In Chapter 1 of this book, we developed a taxonomy of human coping with stressors. This model of coping suggests that there are cumulative costs of adapting to chronic stressors. Although various adaptations may be successful in the sense that they allow one to accomodate to noise exposure, evidence is accumulating that fatigue, overgeneralization of some responses, and diminished feelings of self-efficacy may accompany these adaptations to chronic environmental stressors.

This chapter on control and environmental stressors has focused primarily on what happens when coping efforts fail, when individuals' efforts to modify a chronic environmental stressor are unsuccessful because the stressor is uncontrollable. We have suggested that failed attempts at instrumental coping may lead to heightened feelings of uncontrollability of the stressor that in turn create a growing sense of helplessness in the person.

Several findings from our studies of children's reactions to airport noise converge with previous data on control-related consequences of chronic exposure to uncontrollable environmental stressors. Three sets of measures, locus of control, a standard learned helplessness task, and a choice task were used to evaluate the hypothesis that chronic exposure to uncontrollable noise would lead to greater helplessness in children.

Of the three sets of measures, the weakest data are the locus of control findings. Overall, there was little evidence that children from noisy schools were more external. The total IAR scores as well as I+ and I− scores were unaffected by the school-noise condition in this research. Only one potentially interesting locus of control finding emerged. Children living for longer periods of time in noisy homes became marginally more external (see Table 10).

One reason the locus of control data may have been largely unaffected by chronic noise exposure is that the IAR and its subscales focus on perceptions of control in achievement-related situations (e.g., classroom). Thus the IAR may not be general enough to pick up feelings about control over environmental noise or the environment in general. More research is needed on the relationship between exposure to chronic environmental stressors and generalized expectancies of control. This research should focus on perceptions of control over specific environmental stressors as well as more general assessments of mastery.

Unfortunately, we were unable to measure susceptibility to learned helplessness as directly as we had intended. As you will recall, we employed a standard learned helplessness paradigm adapted for young children. One-half of each noise sample (noisy school/quiet school) worked on either a soluble or insoluble puzzle for a short period of time. After this pretreatment induction phase, all children were then given a moderately difficult but soluble test puzzle. As expected, there were consistent and strong effects of helplessness induction throughout. Children in the insoluble condition were much more likely to fail the test puzzle. Furthermore, their failures were frequently associated with giving up before the elapsed solution time had passed.

However, a sizable proportion (about one-third) of children in the soluble pretreatment condition did not solve the puzzle. Thus, most of our analyses of the learned helplessness measure focused on the main effects of noise on the second test puzzle. Data consistently showed that children from noisy schools were less likely to solve the test puzzle or took longer to solve it. Of particular interest, focusing only on those children who did not solve the test puzzle, substantially more children exposed to noise gave up on the test puzzle before the allotted time period had elapsed. These effects were replicated for school-noise and home-noise levels in several studies. Furthermore, the effects of environmental noise did not adapt over time.

An additional learned helplessness finding worth reiterating was the apparent persistence of the effects of helplessness induction for nearly a year. At both Time 1 and Time 2 of the main study, children who had solved a success treatment puzzle were more likely to solve the test puzzle than those who had not solved the pretreatment puzzle. The test puzzle at Time 2 was given 1 year after. As noted earlier, because of the self-selection problem of children who did not solve the success treatment puzzle, caution is warranted in concluding the helplessness effect persisted for 1 year. Nevertheless, this finding is intriguing and worthy of future exploration. To our knowledge, this is the longest duration of a helplessness-induction effect shown to date.

Our data on the helplessness task and in particular the giving-up measure is consistent with earlier crowding studies by Rodin (1976) showing that children from more crowded residences were more susceptible to learned helplessness. Our data also concur with field research (Moch-Sibony, 1985) and laboratory studies (Cohen, 1980; Glass & Singer, 1972) that have demonstrated that exposure to uncontrollable, unpredictable noise leads to reduced frustration tolerance as measured by persistence on difficult or insoluble puzzles.

Finally, data from the choice measure are also consistent with the hypothesis that chronic exposure to uncontrollable noise would lead to control-related, motivational, or cognitive deficits associated with helplessness. Children from noisy schools were more likely to relinquish choice over which game to play to the experimenter. Children from quiet schools were more apt to make that choice themselves. These findings are quite similar to earlier research by Rodin (1976).

As noted before, we did not find any effect of environmental noise levels on children's tendency to ask for more information about the choice task. Thus, these data did not replicate earlier findings by Baum and colleagues in crowded dormitories. Several procedural differences among the studies, however, preclude any direct comparisons between our findings and his.

An important limitation of our research and previous studies on environmental stressors and control is the lack of precision on the mechanisms involved in the control-related effects. Although we can state with reasonable confidence that continuous exposure to uncontrollable environmental stressors like noise or crowding leads to some reduced ability or motivation to assert control when it is available, we do not know precisely what causes this sequence. We also are uncertain about how robust or persistent the effects are. Research on learned helplessness and control indicates that it is the perception of uncontrollability *per se* and the individual's attributions for that noncontingency that are the critical elements linking exposure to chronic stressors and helplessness.

One aspect of the source of environmental stress in this study, airport noise, warrants some discussion. Throughout this monograph we have noted that most environmental stressors are uncontrollable and unpredictable. Although it is clear that airport noise is uncontrollable for most if not all people, such noise has certain characteristics that make it somewhat predictable. Even though a plane passed over the noisy schools on the average of one per 2 ½ minutes, this does not mean, of course, that planes flew over the schools on a regular, fixed 2 ½-minute interval. In this sense the noise was unpredictable. On the other hand, because of the nature of aircraft noise, peak noise levels are always preceded by a regular, ascending shift in volume. One usually knows beforehand that a plane is approaching and can predict reasonably well when the peak noise level will arrive. Thus, even though the aircraft noise was random in its periodicity, it did not occur without warning.

One issue that the Los Angeles Noise Project touched on only marginally is how the process of coping with uncontrollable stressors unfolds over time. With the notable exception of Baum's important dormitory research, this question has gone largely unresearched. The reformulated learned helplessness model would suggest that individuals' beliefs about the lack of control and the importance of desired outcomes should influence the persistence of helplessness over time. Baum's study and some laboratory work indicate that initial exposure to uncontrollable stressors may lead to reactance, particularly if the goals and activities interfered with by the uncontrollable stressor are important. Over time with continual, unsuccessful attempts to assert control, reactance ceases and learned helplessness may develop.

Developmental questions about control and learned helplessness, although not the focus of our research, are an important topic of future research on environmental stress and coping. Children of different ages may vary in the attributions they make about uncontrollable events. Dweck's research program indicates that characteristic differences in the attributions children make about the causes of their own failures have profound influences on achievement motivation. Furthermore, these differences in attributions seem to be strongly associated with early school experiences, and in particular, teacher feedback on academic work. We know extremely little about the interplay among environmental stressors, control, and helplessness in children. Sufficient research exists, however, to raise concerns about whether important long-term negative consequences may occur from early experiences with chronic uncontrollable environmental stressors in the home and school environments.

Some specific developmental issues warranting further attention are whether children of varying ages are differentially susceptible to exposure to uncontrollable environmental stressors? A related question is whether critical periods exist when environmental mastery experiences are crucial for normal social-cognitive development? Both Dweck and Elliott (1983) and White (1959) have reviewed data suggesting that the early environmental mastery experience is an important component of normal development.

Finally, some methodological issues about the measurement of control and helplessness have been raised in our research. Most studies on environmental stressors and control have not used precise measures of perceived control over stressors or clear-cut indexes of learned helplessness. Are changes in general expectancies for control (e.g., locus of control), tendencies to abrogate control of choice to another, lack of inquisitiveness about ambiguous experimental procedures, or withdrawn, noncompetitive game-playing strategy adequate measures of the hypothesis that chronic exposure to uncontrollable stressors leads to reduced self-efficacy and greater susceptibility to helplessness? More attention must be given to these and other tasks, particularly with respect to their construct validity.

Nevertheless, there are converging findings with different measures and across several laboratory and field studies that indicate that when human beings are chronically exposed to uncontrollable, unpredictable environmental stressors like noise and crowding, some problems in self-initiated instrumental performance on cognitive tasks are likely to occur. Children chronically exposed to uncontrollable noise are more likely to give up on difficult tasks than children from quiet areas. Furthermore, these effects do not seem to habituate over time. Children exposed to noise may also develop a reduced sense of competence or mastery. Children from the noisy schools were more likely to abrogate choice to an experimenter. There is also weaker evidence suggesting a shift in locus of control scores in the external direction with continuous exposure to noise.

4
Environmental Stress and Health

In this chapter, we critically review the health effects of environmental stressors. To keep the scope of this discussion manageable, we limit our domain in several ways. First, although we will consider the health effects of crowding and air pollution, emphasis will be placed on the noise literature. Second, because possible effects on the cardiovascular system have been among the most widely studied health consequences (particularly with respect to noise), these effects will receive the most attention here. Third, the review will be selective, rather than exhaustive, emphasizing major trends in the literature. Intensive reviews of the particular health effects of individual stressors such as noise, crowding, air pollution, and heat can be found in other sources (Cohen & Weinstein, 1982; Evans, 1982; Evans & Cohen, in press; USEPA, 1980, 1981).

Recent interest in the possible health effects of noise and other physical stimuli reflects a view highlighted in this book and introduced in Chapter 1. Specifically, it is recognized that environmental events are more than purely sensory stimuli. Therefore, physiological, biomedical, and epidemiological researchers have examined not only the influence of physical and environmental stimuli on the sensory modalities they directly impact (e.g., auditory-related effects of noise on hearing and sleep disturbance) but also the role of these stimuli in disease processes that involve other organ systems of the body. Thus, for example, a considerable literature has accumulated regarding the extraauditory effects of noise on physical and mental health.

This chapter reviews various physiologic and behavioral processes linking behavior and health. This will be followed by a discussion of the role of stress in cardiovascular disease and reviews of the health effects of noise, crowding, and air pollution. We then consider the effects of stress on children and present health and blood pressure data from the Los Angeles Noise Project.

Psychophysiological Mechanisms in the Study of the Effects of Stress on Health

Stress is a process central to the relationship between behavior and health because it helps to explain how psychologically relevant events translate into health-impairing physiological changes and illness. As suggested in Chapter 1, research has identified a number of direct physiological effects of stress that may influence susceptibility to disease (e.g., immunosuppression and elevated blood pressure). Indirect effects of stress involving changes in behavior that are harmful to health, for example, cigarette smoking, alcohol or drug abuse, and increased symptom reporting, have also been noted. These indirect effects are in many cases mediated by the stress-induced changes such as helplessness and selective attention described in previous chapters. Taken together, these influences provide an interrelated and often complex network of mechanisms by which psychosocial stressors can affect health (see Chapter 1; Baum, Singer, & Baum, 1981; Elliot & Eisdorfer, 1982; Krantz, Grunberg, & Baum, 1985). In this section, we briefly outline physiological and behavioral means by which environmental stressors impact on health.

Pathophysiological Influences

Many of the pathophysiological concomitants of stress are thought to result from responses involving the sympathetic-adrenal-medullary system and the pituitary-adrenal-cortical axis (see Chapter 1). These physiological responses involve neural and endocrine activities that regulate bodily processes including metabolic rate, autonomic nervous system functioning, and immune reactions (Mason, 1971). Short-term stress responses include hormonal and cardiovascular reactions (e.g., increased heart rate and blood pressure) that may precipitate clinical disorders in predisposed individuals. Resulting disorders may include stroke, cardiac arrhythmias and pain syndromes, and psychosomatic symptoms. The assumption that is frequently made—although difficult to document and quantify in humans—is that prolonged or pronounced stimulation from environmental stressors leads to long-term physiologic changes that promote chronic dysfunction in one or more bodily systems. As we will describe later, one must be extremely cautious in making generalizations to chronic disease processes from short-term physiological reactions that are elicited by acute stimuli.

Affective and Behavioral Influences

Psychological or behavioral responses to stress may also affect health via several mechanisms. Psychological effects might manifest themselves as mental health or mood disturbances such as depression and chronic anxiety syndromes. These mood disturbances may also influence physical illness and illness behavior. For example, the individual's motivation for self-care and preventive health behaviors such as exercise and proper diet may be lowered. Furthermore, it has been noted that bodily sensations or changes are experienced more or less continously, but perception and reporting of them vary according to a number of psychological factors. When attention is focused on the body, symptom experience becomes more salient (Pennebaker, 1982). Thus, changes in symptom perception may reflect changes in attentional processes, but an individual might interpret the increased awareness of bodily changes as disease and seek out medical assistance.

Because stressful events and circumstances lead to bodily changes due to physiologic arousal and also alter attentional processes as described before, it is not surprising that stress is associated with increased reporting of physical symptoms (Mechanic, 1968; Pennebaker, 1982) and increased absenteeism and health-care utilization (House, 1975). However, norms for seeking medical care and the interpretation of symptoms are also determined to a considerable degree by cultural background, socioeconomic status, age, and a host of other demographic factors (Mechanic, 1968; Riley & Foner, 1968). It is therefore important for epidemiological studies to control for such demographic factors lest sociocultural differences in symptom reporting and health-care utilization be spuriously attributed to differences in physical environmental factors.

Neurochemical Changes in the Brain Relevant to Mental Health Consequences of Stress

Research with animals by Jay Weiss (1985) suggests that exposure to uncontrollable stressors such as shock can lead to a depletion of norepinephrine levels in certain areas in the brain. In this regard, the uncontrollable shock paradigm used by Weiss, Seligman, and others results in behavioral changes similar to clinical depression in humans. Although animal models provide a means for studying physiological factors underlying depression, considerable additional research is required to evaluate the relevance of such models for humans. At the present time, it is difficult to apply animal models that employ high-intensity aversive stimuli to environmental stressors encountered by humans during everyday life.

Effects on Immune Function

The immune system is a surveillance mechanism responsible for combating disease-causing microorganisms and other foreign agents in the body. Research in the last decade suggests that rather than being an autonomous defense agency, this system is integrated with other physiological processes and is sensitive to changes in central nervous system and endocrine functioning, such as those that accompany psychological stress. Accordingly, a new interdisciplinary area called ''psychoneuroimmunology'' (Ader, 1981) studies interactions between behavioral and immunological processes and is particularly relevant to the study of psychological influences on infectious diseases and cancers. This research area not only encompasses the study of stress and immune function but also such factors as learning and conditioning effects on immune system components.

Laboratory stressors can suppress the responsiveness of the immune system in animals, and stress-responsive hormones, including cortiocosteroids, can directly or indirectly alter components of the immune response (see Ader, 1981, for a review). Animal and human studies demonstrate that laboratory and naturalistic stressors can reduce the number of lymphocytes (cells important in the immune process), lower the level of interferon (a substance that may prevent the spread of cancers), and cause damage in immunologically related tissue (see reviews by Ader, 1981; Jemmott & Locke, 1984; Sklar & Anisman, 1981). Of particular interest for the relevance of the learned helplessness phenomenon for health, in yoked designs in which animals receive aversive stimulation (i.e., shock), those with an ability to terminate the stimuli or escape show slower tumor growth and increased ability to reject tumors compared to animals with no option for control (Sklar & Anisman, 1981; Visintainer, Seligman, & Volpicelli, 1983).

Studies of humans also show evidence of stress-linked changes in immunocompetence. Conditions such as bereavement, surgery, and sleep deprivation have been associated with decreases in immune responsiveness, usually followed by an eventual return to normal levels (see Jemmott & Locke, 1984). These effects have been inferred indirectly from observed relationships between stress and the incidence of infectious diseases. However, several recent studies provide more direct evidence of stress-related changes in the function of lymphocytes—cells important in the body's immune defenses (e.g., Dorian, Keystone, Garfinkel, & Brown, 1982).

Stress and Cardiovascular Disorders

Despite the fact that there has been a decline in cardiovascular mortality in the past 10 years (USDHEW, 1979), the major cardiovascular disorders, including coronary heart disease (CHD) and essential hypertension (EH), still account for one in two deaths in the United States. These disorders are also among the most thoroughly researched topics in the study of psychosocial influences on health, partly because numerous behavioral variables are important in phases of cardiovascular diseases, ranging from etiology to treatment and rehabilitation (see Gentry & Williams, 1979; Krantz, Baum, & Singer, 1983; Ostfeld & Eaker, 1985; Steptoe, 1981, for reviews of psychological contributions to this area). Because we will discuss the role of environmental stress in CHD and EH in some detail, a brief description of these disorders is in order.

Most cases of coronary heart disease are thought to be complications of coronary atherosclerosis, which is a progressive condition characterized by narrowing of the coronary arteries, the blood vessels that nourish the heart. An excess accumulation of cholesterol and related lipids in the blood forms a mound of tissue, or plaque, on the inner wall of the coronary arteries (Hurst, Logue, Schlant, & Wenger, 1978). The formation of these plaques may occur undetected for years, affecting cardiac function only when they cause a degree of obstruction sufficient to diminish blood supply to the heart. Once this occurs, coronary atherosclerosis has evolved into clinical coronary heart disease.

In one form of CHD, instances of inadequate blood supply, or ischemia, cause the individual to experience attacks of chest pain. A more severe consequence of CHD is myocardial infarction, or heart attack, in which a prolonged state of ischemia results in death of a portion of heart tissue. Cardiac arrhythmias, disturbances in the conductive or beat-regulating portions of the heart, are often another clinical manifestation of CHD. Certain types of ventricular arrhythmias may be precursors of sudden death, and arrhythmias may also emerge without a readily identifiable physiological basis. There is evidence that central nervous system influences and increased circulating levels of catecholamines, such as those that accompany psychological stress, can increase vulnerability to arrhythmias (Verrier, DeSilva, & Lown, 1983).

Essential hypertension (EH), another major cardiovascular disorder, is a condition of unclear etiology in which blood pressure shows chronic elevations. Current thought holds that this disorder is "multifactorial," that is, involving a complex pathogenesis (Genest, Kuchel, Hamet, & Cantin, 1983; Weiner, 1977). In addition, among the many etiologic variables that have been proposed (e.g., genetics, salt intake, stress), no single factor seems to apply to all hypertensive individuals. Moreover, the physiological characteristics of earlier stages of EH seem to differ from the established or later stages of the disorder. Thus, it is evident that EH is a heterogeneous disorder.

Epidemiological studies reveal that numerous environmental and behavioral variables are involved in the etiology and pathogenesis of both CHD and EH. For CHD, a set of standard or physical risk factors has been identified, many of which involve elements of life-style or habits of living (e.g., age, hypertension, smoking, excessive intake or dietary fat, lack of exercise, etc.). These risk factors also differ in the consistency and strength with which they have been associated with CHD.

For EH, the two best predictors of the future development of the disorder appear to be family history of hypertension and borderline hypertension (Julius & Schork, 1977). However, these predictors successfully identify only a small subset of future hypertensives and fail to identify many others who later develop this disorder. There are also significant race differences in the incidence and prevalence of EH (at least in the United States), with blacks being significantly more likely to develop the disorder than whites (Gillum, 1979). The "physiological profile" likely to be associated with hypertension differs as well between blacks and whites, particularly with regard to the nature of kidney function in established hypertension. Specifically, black hypertensives, in comparison to white hypertensives, are more likely to have EH characterized by low levels of plasma renin activity and high levels of blood volume (Gillum, 1979). Hypertension among whites, by contrast, tends to involve a more diversified physiological profile. Although such race differences are probably related to genetic factors, race differences in diet (e.g., greater salt intake among blacks) as well as race differences in exposure to or sensitivity to environmental stressors cannot be presently ruled out as contributory factors.

Psychosocial and Other Environmental Risk Factors

Research on psychosocial factors in the development of cardiovascular disorders was fostered, in part, by shortcomings in the ability of the standard risk factors to predict many new cases of coronary disease. As a result, a broadened search for mechanisms and influences contributing to coronary risk has examined a variety of social indicators (e.g., social mobility and socioeconomic status) and psychological factors (e.g., anxiety, psychological stress, and overt patterns of behavior). These investigations have been encouraging, though not definitive (see Jenkins, 1976; Ostfeld & Eaker, 1985). The two most promising psychosocial risk factors for CHD and hypertension to emerge in recent years are psychological stress, and the Type A "coronary-prone" behavior pattern for coronary heart disease.

Stress and Critical Life Events in Cardiovascular Disease

Although the concept of *stress* is widely used and often imprecisely defined, there is evidence that certain chronic conditions can enhance coronary risk. These conditions include excessive work overload or job responsibility coupled with low job autonomy as well as an accumulation of stressful life events (see reviews in the volume edited by Ostfeld & Eaker, 1985). However, the literature on the relationship of critical life events to CHD has produced inconsistent findings (see Kasl, 1983; Wells, 1985), and the relationship of several sociological variables (e.g., occupational status) to CHD seems to depend on the way individuals perceive their life situations. Thus, one must consider mediating factors such as perceived demands and degree of autonomy and social support at home and work in order to understand the stress–heart disease link (Haynes & Feinleib, 1980; Karasek, Theorell, Schwartz, Pieper, & Alfredsson, 1982; Ostfeld & Eaker, 1985).

In addition, a review by Kasl (1983) concluded that although physiologic mechanisms (such as those reviewed in previous sections of this chapter) have been proposed to explain linkages between psychosocial stressors and disease, relatively little research has, within the same design, measured stress, illness outcomes, and the presumed pathophysiological mechanisms involved. To the extent that environmental stressors, such as noise, can be quantified according to physical characteristics, they may provide some advantages over less easily quantified psychosocial stressors in studying mechanisms linking stress and cardiovascular disease. Furthermore, it is worth remembering that ethical imperatives frequently preclude the invasive procedures needed to monitor pathophysiological mechanisms in relatively healthy, functioning adults.

Type A Behavior and Coronary Heart Disease

The Type A or "coronary-prone" behavior pattern has been widely studied as a risk factor for CHD. This behavior pattern, first identified by cardiologists Friedman and Rosenman (1959), is characterized by excessive competitive drive, hostility, impatience, and a rapid, clipped manner of speech. A contrasting Type B behavior pattern consists of a relative absence of these characteristics and a somewhat different style of coping. Type A behaviors are thought to be activated by situations that are stressful or challenging. Thus, Type A is seen to be the product of a person–situation interaction that can emerge in various degrees depending on the environment the individual encounters. Developmental research suggests that Type A behavior can be observed in children as young as 6 to 7 years of age (Matthews, 1982).

Epidemiologic studies have found a relationship between Type A behavior and risk of subsequent CHD in men and women, comparable to the effect of risk factors such as smoking and hypertension (Haynes, Feinleib, & Kannel, 1980; Belgian-French Pooling Project, 1984; Rosenman *et al.*, 1975; Review Panel, 1981). However, several recent studies, such as the Multiple Risk Factor Intervention Trial (MRFIT) and studies of patients awaiting cardiac catheterization, have reported a weakened relationship between Type A and coronary disease, particularly among high-risk subjects (see Matthews, 1985, for review). The precise reasons for recent disconfirmatory findings for Type A are currently not well understood (see Ostfeld & Eaker, 1985). However, Type A component behaviors, such as hostility, anger, and vigorous speech, still remain related to CHD in recent studies (Dembroski, MacDougall, Williams, Haney, & Blumenthal, 1985; Matthews, 1985).

Psychosocial Stress and Hypertension: Is There a Relationship?

Stress deriving from psychosocial causes is a factor that has been implicated by some investigators in the etiology and maintenance of high blood pressure. Psychological stimuli that threaten the organism may result in cardiovascular and endocrine responses that can play an important role in the development of hypertension (see Guttman & Benson, 1971; Henry & Cassel, 1969). The brain and central nervous system, which are involved in determining whether situations are seen as threatening or harmful, thus play a role in physiological mechanisms mediating the impact of noxious stimuli. On a societal level, there is some evidence, albeit controversial (see later), that blood pressure elevations occur under conditions of rapid cultural changes and socioeconomic mobility. For example, there are several studies in which less economically developed, geographically isolated populations living in small cohesive societies were found to have low blood pressure that did *not* increase with age. When members of these societies migrated and were suddenly exposed to Western culture, they were found to have high levels of blood pressure that increased with age. This suggested that over the life span, the new living conditions had a cumulative effect that resulted in higher blood pressure levels.

The notion of a link between psychosocial stress and the development of hypertension has resulted in some controversy in the literature. Reviewers such as Syme and Torfs (1978) have noted that there are a sizable number of studies that fail to confirm an association between cultural change or social stress and the incidence or prevalence of hypertension in population groups. For example, whereas a study of Japanese migrants to Hawaii and California found increased CHD incidence among men experiencing life changes associated with migration, no increase in blood pressure levels was noted in these men (Marmot & Syme, 1976; Syme, 1984).

One possible source of inconsistencies in the stress and hypertension literature is the failure to consider the likely fact that in humans, sustained blood pressure elevations are produced by an interaction of genetic and environmental factors. Thus, chronic stress may elicit blood pressure elevations only among individuals genetically predisposed to develop hypertension (see Falkner, 1984; Obrist, 1981, for reviews). Well-designed epidemiological studies can examine genetic-environment interactions and attempt to rule out confounding factors (e.g., diet) by using carefully matched control groups and statistical control techniques. However, there are inherent limits to conclusions that can be reached from correlational research (see Chapter 2 for some details). It is here that experimental techniques for inducing high blood pressure in animals (e.g., Peterson, Augenstein, Tanis, & Augenstein, 1981) and experimental studies of physiological responses to stress in humans (e.g., Falkner, 1984; Julius & Esler, 1975; Krantz & Manuck, 1984; Obrist, 1981) have contributed to recent understanding of the role of behavioral variables (including stress) in the development of hypertension.

Physiological Reactivity as a Mechanism in the Development of Cardiovascular Disease

It is well known that there are wide individual differences in autonomic nervous system responses to stress (Engel & Bickford, 1961). These individual differences are moderated by biological factors (e.g., genetic susceptibility, presence or absence of disease) as well as psychological attributes (e.g., felt ability to cope). Recent research on behavior and cardiovascular disease has suggested that psychophysiological responsiveness (reactivity) to emotional stress may be a marker of processes involved in the development of coronary heart disease and/or essential hypertension (see Krantz & Manuck, 1984, for review). The measurement of reactivity involves the assessment of cardiovascular and/or neuroendocrine *changes* in response to stress, as distinct from the observation of basal or resting levels of physiological variables. An underlying assumption of this research is that changes over resting levels in response to real life or laboratory stressors give an index of the state of the organism during the challenges of everyday life and/or during exposure to environmental stressors.

For both coronary disease and hypertension, biologically plausible mechanisms involving responses to stressful stimuli have been proposed. Some of these mechanisms focus on the potential importance of reactivity in the early etiology of disease, others view it as a precipitating factor for clinical symptomatology, and still others suggest that cardiovascular and endocrine reactivity may have both preclinical and clinical pathogenic effects. A detailed review of these hypothesized mechanisms is beyond the scope of this chapter and can be found in Krantz and Manuck (1984). Suffice it to say that it is possible that individuals can be classified as either high- or low-cardiovascular "reactors" based on their response to environmental stressors and that such reactivity may be a marker of cardiovascular risk. In addition, although there have been preliminary studies suggesting that consistent cardiovascular "reactors" can be identified in school-age children (Matthews, Manuck, & Saab, 1985), there has been, to date, very little developmental research on physiologic reactivity in children.

Psychosocial Stress and Health: Section Summary

The literature reviewed has suggested behavioral and physiological mechanisms implicating aversive psychosocial stimuli in the development of physical disease. As has been noted (cf. Kasl, 1983; Chapter 1), the term *stress* has often been imprecisely defined in research studies. Our emphasis on adverse changes in health status produced by stressors is not meant to imply to the reader that such effects inevitably occur or that humans show little adaptability and hence are overly vulnerable. Instead, we have suggested that particularly for cardiovascular disorders, a sizable research literature has developed implicating several identifiable psychosocial factors in the etiology of these disorders. Such factors include (a) exposure to prolonged or intense psychosocial stressors; (b) certain identifiable behavior patterns such as Type A and related trait components of hostility and anger; and (c) the propensity to show large cardiovascular and endocrine responses to certain stressful or challenging circumstances.

Cardiovascular Risk Factors in Children

Our discussion of coronary heart disease and hypertension has dealt with adult populations. In recent years, attention has also been directed to the study of potential cardiovascular risk factors in children and how these risk factors develop over time. Although CHD, and for the most part EH, are not diseases of childhood, epidemiologists and clinical investigators have studied the developmental course of blood pressure, serum cholesterol, blood lipids, and obesity in children of different races (Berenson, 1980). The aim of this research is to understand the early natural history of atherosclerosis and essential hypertension.

Essential Hypertension

Hypertension in its early stages is not well understood, in large part because of a relative lack of information concerning time-course changes of blood pressure. As Berenson (1980) noted, in children there is a lack of satisfactory criteria for diagnosis of "primary" essential hypertension, that is, hypertension not following from other identifiable physical problems. For example, among children, in contrast to adults, clinical effects and early tissue damage are not detectable.

Epidemiological studies have suggested that primary essential hypertension can begin early in life and can be detected in large epidemiological study samples (Londe, Bourgoignie, Robson, & Goldring, 1971; Voors, Foster, Frerichs, Weber, & Berenson, 1976). This conclusion, however, remains somewhat controversial. There is also evidence of tracking of blood pressure, that is, the child's relative rank of blood pressure among his or her peers remains constant (Berenson, 1980; Coates *et al.*, 1981; Lauer *et al.*, 1975; Voors *et al.*, 1976). A comprehensive review of the pediatric blood pressure literature can be found in Berenson (1980). Suffice it to say here that pediatric essential hypertension is not common but appears to be detectable in studies that examine large samples. In addition, it has been shown that blood pressure levels in children and adolescents increase with age and are highly dependent on height and weight growth patterns of the child. This childhood increase appears to level out at the age when the child reaches adult stature (cf. Berenson, 1980). Major studies reporting blood pressure and other cardiovascular risk factor levels in pediatric and adolescent samples have been summarized in Berenson (1980) and in a detailed report by the National Heart, Lung, and Blood Institute (1978).

Other Coronary Risk Factors and Early Atherosclerosis

Many epidemiologic and clinical studies have established the concept of "coronary risk factor" as a means of predicting ischemic heart disease, and studies of diet, smoking, blood pressure, family history, age, and so forth, have established a constellation of variables predictive of disease for adults (see Kannel, 1979). These adult studies set the stage for directing attention to the study of early prevention of those risk factors that are environmentally determined and also provide a background for studying the early natural history of atherosclerosis in childhood.

Concern about coronary artery disease developing early in life came originally from autopsy studies of young soldiers killed in the Korean and Vietnam wars. The presence of evidence of atherosclerosis in otherwise healthy young soldiers showed that the disease began in a completely asymptomatic form (Enos, Holmes, & Beyer, 1953; McNamara, Molot, & Stremple, 1971).

During the past 15 years, several large-scale community studies of healthy children have been initiated (see NHLBI, 1978). A major objective of these studies with children is to show the interrelationship of hypertension, obesity, high blood lipids, and other metabolic and anthropometric conditions. Although the Type A behavior pattern can be identified in children as early as 6 to 7 hears of age (see Matthews & Siegel, 1982), there are as yet little data regarding the role of other psychosocial risk factors for CHD, such as emotional stress, in preadolescent children.

Having provided in previous sections a general overview of the aggregate of research on cardiovascular risk factors, stress, and health, we narrow our focus somewhat to review the health effects of specific environmental stressors, namely noise, crowding, and air pollution.

Noise and Health

Although noise can impair hearing and interfere with communication (see Kryter, 1970; Miller, 1974, for reviews), the evidence relating noise to serious nonauditory health problems is more equivocal. In sampling the noise literature, we will attempt to clarify the conditions under which noise, especially community noise, results in annoyance and detrimental effects on nonauditory physical health. Attention will also be given to the effects of noise on subjective states, mental health, and related behaviors.

Subjective Annoyance

Contrary to common belief, sound level *per se* is only a minor factor in response to community noise. Although social surveys often report a positive relationship between noise intensity and the average level of felt annoyance, intensity alone seldom explains more than one-quarter of the variance in individual annoyance. Often explaining over half the variance are psychological factors—the respondents' attitudes and beliefs about the noise and the noise source.

In attempting to explain individual differences, researchers have searched for audiometric measures of noise exposure that will correlate more closely with subjective report. Dozens of audiometric measures of physical noise characteristics are now in existence (e.g., LEQ, LDN, peak dB(A), etc.). Although some of these are more highly correlated with annoyance than others (cf. U.S. Environmental Protection Agency, 1974), it is questionable whether significant improvements in predictive accuracy can be produced by futher refinement of such measures. Furthermore, the limited predictive ability of noise indexes cannot be attributed to unreliable annoyance measures. The multiitem scales employed to measure annoyance generally have quite satisfactory internal reliability (e.g., McKennell, 1963), although the temporal stability of annoyance measures is not so clearly established (Griffiths & Delauzun, 1977; Leonard & Borsky, 1973). On the other hand, there may be some deficiences in the characteristics of annoyance measures that interfere with their scaling properties. For example, B. Berglund, U. Berglund, and their co-workers in Sweden have achieved much higher relationships between sound levels and annoyance ratings using calibrated scaling procedures (Berglund, Berglund, & Lindval, 1975; see also Evans & Cohen, in press).

Reviews of this research literature (Borsky, 1969, 1980; Cohen & Weinstein, 1982) suggest that annoyance is heightened when (a) the noise is perceived as unnecessary; (b) those responsible for the noise are perceived as unconcerned about the exposed population's welfare; (c) the respondent dislikes other aspects of the environment; (d) the respondent believes that noise is harmful to health; (e) the noise is associated with fear; and (f) the noise interfers with activities deemed important by the individual (e.g., interference with sleep or speech). This list is abstracted from several social surveys (e.g., Borsky, 1973; McKennell, 1963; McKennell & Hunt, 1966; Tracor, Inc., 1970), and the operative factors affecting annoyance reactions vary from study to study. Nevertheless, psychological factors are consistently more important determinants of individual annoyance than the sound level itself. Although many beliefs are correlated with annoyance, there is no accepted theory of annoyance to bring order to the data. In some cases, "annoyance" may represent fear or simply aversiveness, but many of the correlates of annoyance are predictable if one views annoyance as a mild form of anger. According to cognitive theories of emotion, anger is produced when people believe that they have been harmed and believe that the harm was both avoidable and undeserved (Brown & Herrnstein, 1975; Crosby, 1976). The harm produced in noisy situations may include threats to health and to property values, blocking of behavioral goals, or simply exposure to an aversive stimulus.

Although many noise effects decrease rapidly in the laboratory (Glass & Singer, 1972; Kryter, 1970), community noise research provides little evidence that noise annoyance habituates in residential settings. Surveys find consistently that long-time neighborhood residents are at least as bothered by noise as more recent arrivals. Longitudinal studies, which avoid some of the self-selection problems inherent in cross-sectional surveys also find more disturbance at the end of the study period than at the beginning (Jonsson & Sorensen, 1973; Weinstein, 1978).

It is possible that the data are misleading and that people do habituate to chronic noise exposure. After a while, they may pay less attention to the noise and fall asleep more quickly, but responses to survey items may continue to reflect their original feelings about the noise rather than their current reactions. Nevertheless, it is worth emphasizing that people appear to have much more difficulty in adjusting to noise than is commonly believed. This seems particularly true when people are directly questioned about their attitudes and feelings with respect to noise.

Noise, Sleep Disturbance, and Health

Is noise-induced sleep disturbance detrimental to health? It seems likely that those deprived of sleep would be more irritable and annoyed than others. There is, however, little or no direct quantitative evidence that noise-induced sleep loss causes mental or physical disability (see Kryter, 1970; Lukas, 1975). Although there is laboratory evidence that the autonomic nervous system responds to noise during sleep, Kryter (1970) argues that this system requires less "rest" than the higher nervous systems and skeletal muscles. Moreover, in contrast to autonomic reactions, electrophysiological responses of the central nervous system (e.g., electroencephalogram) tend to decrease during noise exposure (Griefahn, 1980). Thus, Kryter has suggested that unless behavioral awakening occurs, it may be unreasonable to surmise that the beneficial effects of sleep are not realized (Kryter, 1970). There is some evidence that persons high in neuroticism are quite susceptible to noise-induced sleep disruptions (Griefahn, 1980), and it is possible that these persons are therefore more likely to suffer from related health problems. However, normal persons who lose sleep compensate by spending more time in deep sleep, by becoming less responsive to external stimuli, and by napping (Miller, 1974). In short, it may be difficult to deprive a normal person of enough sleep due to noise exposure to produce serious health effects.

On the other hand, people who are deprived of sleep do complain of the loss and feel that it affects their well-being. Possibly, noise may interfere more seriously with the sleep of some people. For example, sleep disturbances might impede the recovery and aggravate the disability of those who are already ill. There is also evidence that susceptibility to being awakened by noise increases with age (Griefahn, 1980; Lukas, 1975). Thus, noise-induced sleep loss may have its greatest impact on population groups susceptible to environmentally produced disease. Also, continuous sleep disturbance by noise might be likely to reduce feelings of well-being and hence be potentially hazardous to health. Young infants who may need sleep for brain development are another potentially susceptible group to noise-induced ill health.

Noise and Mental Health

If noise causes annoyance and frustration, it seems plausible that prolonged exposure could cause or aggravate mental illness. Some studies suggest the possibility of a relationship between noise and symptoms of mental distress, but the evidence is neither very reliable nor abundant.

Industrial surveys report that noise exposure results in increased anxiety and emotional distress. Workers habitually exposed to high-intensity noise (usually 110 decibels or above) show increased incidence of nervous complaints, nausea, headaches, instability, argumentativeness, sexual impotency, changes in general mood, and anxiety (A. Cohen, 1969; Granati, Angelepi, & Lenzi, 1959; Miller, 1974). Jansen (1961) reports that workers in the noisiest parts of a steel factory have a greater frequency of social conflicts both at home and in the plant. These results are difficult to interpret, however, because the same workers are often subject to other work stressors (e.g., task demands and risks) that may precipitate or contribute to the reported symptoms. On the other hand, those workers most bothered by noise may quit or transfer to quieter settings.

Studies of the impact of community noise exposure on self-reported neurotic symptoms provide inconsistent evidence for a link between noise and pathology. Thus, in a review of this literature, McLean and Tarnopolsky (1977) reported that although a number of community surveys indicate an association between noise and various symptoms—being tense and edgy (Office of Population Census and Survey, 1970), irritable (Finke, Guski, Martin, Rohrman, & Schümer, Schümer-Kohrs, 1974), nervous, and having sleep difficulties and headaches (Kokokusha, 1973)—the questions in these studies were worded in a way that seemed to invite one to blame noise for one's ailments. This methodological issue may also be important in community noise annoyance surveys, and perhaps adaptation to neighborhood noise is better assessed by open-ended questions about community problems or environmental dissatisfactions. When people are directly queried about noise annoyance, this may result in biased attribution of dissatisfactions to noise.

Thus, in studies where aircraft noise was not mentioned as a possible cause, there were no correlations between symptoms and exposure (Grandjean, Graf, Lauber, Meler, & Muller, 1973; Tarnopolsky, Watkins, & Hand, 1980). Some similar findings have been noted in the air pollution annoyance literature (Evans & Jacobs, 1982).

It should be noted that one study that surveyed physicians' observations in high- and low-noise areas around Amsterdam's airport (Knipschild & Oudshoorn, 1977) rather than relying solely on residents' responses to self-report surveys, did find a high proportion of psychological and psychosomatic complaints made to physicians in the high-noise area. Differences in social class of the noise and quiet areas, however, provide a possible alternative explanation. In sum, the lack of sophisticated reliable and valid indexes of psychological symptoms and problems has been a notable shortcoming of research on community noise. More attention to such methodological problems would do much to improve our knowledge in this area. (See Chapter 2 for further discussion of methodological problems and suggested remedies in field studies of stress.)

Hospital Admissions

Work on community response to aircraft noise suggests a possible relationship between noise level and admission rates to community mental health facilities (Arhlin & Ohrstrom, 1978; McLean & Tarnopolsky, 1977). Several studies relating sound level and mental-hospital admissions have been conducted in the vicinity of London's Heathrow Airport. Abbey-Wickrama, Herridge, and colleagues (Abbey-Wickrama, a'Brook, Gattoni, & Herridge, 1969; Herridge & Chir, 1972) compared the psychiatric-hospital admission rates of those residing in noisy and less noisy parts of the same borough. Admission rates were higher for the noisy area; persons most at risk were older, single, widowed, or separated women suffering from neurotic or organic mental illness. This pattern of results, with higher risk among those more likely to be alone, suggests that people with low social support might be more likely to be hospitalized for noise-related mental problems. These results have been challenged by Chowns (1970), who argues that the noise index used was inappropriate and that noisy and less noisy (control) neighbors were poorly matched on demographic factors. A replication of this work (Gattoni & Tarnopolsky, 1973), using a different technique of indexing noise and carefully matching demographic variables, found similar but small (nonsignificant) differences between noise and quiet areas. A comparison of noise and control populations around Los Angeles International Airport (Meecham & Smith, 1977) indicates a marginal increase in mental-hospital admissions among those living in maximum noise areas, as compared to a sample of the community at large. Unfortunately, very poor matching of quiet and noise areas on race and socioeconomic status severely limit confidence in these results. However, recent work on mental hospital admissions around Heathrow by Tarnopolsky and his colleagues has employed more sophisticated methodologies and measures of noise exposure (Hand, Tarnopolsky, Barker, & Jenkins,

1980; Watkins, Tarnopolsky, & Jenkins, 1981). This latter work clearly suggests that community noise levels are not associated with mental hospital admissions.

In sum, existing studies provide little convincing evidence for a difference between mental hospital admission rates of quiet neighborhoods and neighborhoods subjected to aircraft noise. Furthermore, studies reporting effects are all retrospective and involve differences between the admission rates of noise and quiet groups that are so small (between .001 and .003%) as to be considered by many as trivial. It should be noted, however, that admissions to community mental hospitals represent only the severest cases of mental distress. It is likely that many sufferers would see their general practitioner rather than visiting a community mental health facility (e.g., Knipschild, 1976). Moreover, what may be considered a serious deviation from normality in one social class or ethnic group may be an acceptable deviation in another (Mechanic, 1968). Thus, depending on the background of the population, the sensitivity of these admission rates to mental distress could vary substantially. Prospective studies of community mental health that include a wider range of psychiatric care modalities would help clarify this issue. Other less catastrophic indexes of psychological distress that ought to be examined include standard psychological symptom profiles (e.g., anxiety, depression); help seeking with friends and family; and use of tranquilizers, alcohol, and other psychoactive drugs. In this regard, Watkins *et al.* (1981) found increased use of psychotropic drugs by people who report being highly annoyed by noise.

Noise and Physical Health

Is high-intensity noise harmful to the human body? Most would argue that outside of the effects of high-intensity sound on hearing (see Kryter, 1970; Miller, 1974), there is little convincing evidence for a causal link between noise and physical disorders. However, noise can alter physiological processes including the functioning of the cardiovascular, endocrine, respiratory, and digestive systems (see review by McLean & Tarnopolsky, 1977). Because such changes—when extreme and persistent—presumably increase the risk of certain diseases, many feel that pathogenic effects of prolonged noise exposure are likely.

The physiological changes produced by noise consist of nonspecific responses typically associated with stress reactions (Glorig, 1971; Selye, 1956). These include increases in electrodermal activity, catecholamine output, vasoconstriction of the peripheral blood vessels, and increases in diastolic and systolic blood pressure. Most of these reactions have been documented in laboratory studies involving short-term exposure to relatively high-sound levels (cf. A. Cohen, 1979).

Do such physiological effects constitute evidence that noise is detrimental to health? The question is difficult to answer. On the one hand, there is mixed evidence that a number of physiological responses do not habituate to repeated exposure (Jansen, 1969; McLean & Tarnopolsky, 1977) and thus could constitute the physiological bases for long-term harmful effects of noise. On the other hand, others report that habituation of these responses occurs after only short exposure to noise (e.g., Glass & Singer, 1972); thus prolonged exposure might not necessarily produce continuous elevation of physiological processes inimical to normal bodily function. Kryter's (1970) conclusion that "the exact course and degree of adaptation of all these responses has not been very thoroughly studied" (p. 491) probably still represents the state of our knowledge in this area. As Kasl (1983) has noted, in order to establish a causal link between stress and disease, further work is needed examining the sequence of three factors together: the stressor, the physiological changes produced by the stressor, and disease endpoints under study.

Industrial Studies

A review of the non-English language industrial noise literature (Welch, 1979) concludes that there is elevated morbidity among people who have been exposed at work to sound of 85 dB(A) or greater for at least 3 to 5 years. Moreover, the morbidity associated with exposure to relatively high intensities of sound increases with advancing age and years of employment for both men and women. Morbidity also tends to be greater under unpredictable, intermittent, and impulse sound than under periodic, continuous, or relatively steady sound; and it affects those whose work involves mental concentration more than those who do mainly manual work. If these trends are reliable, they may reflect "secondary coping effects" alluded to in Chapter 1 or reflect cumulative fatigue of the body's resources.

Welch (1979) argued that the strongest case for industrial noise impacting health derives from research on cardiovascular problems. He interprets the data (over 40 different studies) to indicate that long-term work under high-intensity sound is associated with at least a 60% increase in risk of cardiovascular disease. Impaired regulation of blood pressure (including hypertension) is the best documented of the effects (e.g., Parvizpoor, 1967; Pokroviskii, 1966). Other concomitants of prolonged routine exposure to intense industrial sound include cardiac morbidity (e.g., Capellini & Maroni, 1974; Raytheon Service Company, 1972), poor peripheral circulation (e.g., Jansen, 1959), and elevated cholesterol levels (Khomulo, Rodinova, & Rusinova, 1967).

Although there is an impressive amount of data suggesting that cardiovascular morbidity is greater among workers exposed to high noise levels, there are a number of reasons to be critical (or at least highly cautious) about Welch's conclusions. All of these studies suffer from the methodological problems associated with correlational field research, and as many as half of the existing studies are either poorly designed (inadequate control groups) or do not report sufficient information to allow a critical analysis of the work. Detailed critiques of this research may be found in reports by the Environmental Protection Agency (USEPA, 1980, 1981). Our confidence in the relationship between prolonged exposure to high-intensity noise and cardiovascular problems would be significantly increased if similar effects were found in prospective research. Cardiovascular response could be monitored over time in a new work group entering a particular noisy environment compared with similar workers not so exposed (cf. A. Cohen, 1979). It would also be of interest to analyze subgroups of workers under noise versus quiet conditions who are already at greater risk for heart disease (e.g., individuals with family history of hypertension, Type A individuals, etc.).

Research with suitable animal models can further our understanding of noise–hypertension relationships in humans. Peterson and colleagues (E. A. Peterson *et al.*, 1981) have demonstrated with rhesus monkeys that prolonged (9-month) exposure to a pattern of daily noise resembling the levels and duration of industrial noise resulted in sustained increases in blood pressure. Both systolic and diastolic blood pressure increased by almost 30% in experimental animals without concomitant changes in auditory sensitivity thresholds. Importantly, high blood pressure levels were maintained for a 27-day period (the length of follow-up reported) after cessation of noise exposure. As most animal models of hypertension have not demonstrated a persistence of high blood pressure levels that mimic the human hypertensive condition, the Peterson *et al.* (1981) study provides noteworthy evidence of a causal link between high levels of industrial noise and hypertension.

In addition to cardiovascular changes, reports have linked industrial noise to a number of other health problems. The data in all of these cases are suggestive but as yet inconclusive (cf. Welch, 1979). For example, there are a number of studies reporting increased gastrointestinal complaints for noise-exposed workers. Reported problems include gastrointestinal ulcers (e.g., Kangelari, Abramovich-Polyakov, & Lyubomudrav, 1966) and general digestive problems (Raytheon Service Company, 1975).

Although there has been a lot of speculation about the effect of prolonged exposure to noise on resistance to infectious disease, the small number of existing studies is mixed and inconclusive (cf. Welch, 1979). For example, some research suggests that factory noise is associated with increased reports of sore throats and laryngitis (cf. A. Cohen, 1969) and increased job-related injuries (e.g., A. Cohen, 1973, 1976). It is likely that the former is attributable to workers raising their voices in order to be heard, and the latter, to some degree, to the inability to hear warning signals in noisy settings. However, A. Cohen's (1976) finding that there are fewer accidents among workers wearing hearing protection than those without protection suggests that the masking of warning signals is not entirely responsible for increased injuries. Finally, a small number of studies suggest that prolonged exposure to high-intensity industrial noise is implicated in functional neural change, with the most convincing data suggesting impaired control of circulation (Welch, 1979).

There are also claims that high-intensity sound can affect the reproductive system, but these are virtually unsubstantiated in the industrial noise literature. There is one epidemiological study of community noise by Jones and Tauscher (1978) reporting higher birth-defect rates in high aircraft-noise areas in Los Angeles. Unfortunately, this study suffered from a serious design flaw in that residents living near the airport were compared to all the other county residents, thus leaving possible confounding demographic and socioeconomic variables uncontrolled.

Several industrial surveys failed to find a relationship between noise and ill health. For example, Finkle and Poppen (1948) report that men working in turbojet noise of 120 dB showed complete adaptation to noise. Results for renal function tests, electroencephalography, and hematological examinations were all negative. Glorig (1971) also reports no increase in cardiovascular problems, ulcers, or fatigue for those working in noisy industries.

In sum, although several years of exposure to industrial noise have been associated with a number of specific nonauditory diseases, only cardiovascular problems have received enough attention and consistent support to allow a convincing argument that they are noise induced. This argument is to some extent corroborated by both short-term laboratory studies of humans (see review by McLean & Tarnopolsky, 1977) and animal studies of long-term exposure (cf. Peterson *et al.*, 1981) showing similar effects on the cardiovascular system. More research will be required before firm conclusions can be made about the relation of industrial noise to other diseases and/or the critical noise levels or durations of exposure in different settings that should be set for health-protective efforts in occupational settings.

It should also be noted that noise may not be etiologically specific to any given disease but may enhance susceptibility to disease in general and thus cause a wide variety of symptoms of physical and psychiatric disorders (Cassel, 1974). Such a perspective suggests that studies of the relationship of noise to any specific disease (or even just somatic or just psychiatric diseases) may be insensitive to noise-induced effects and that a broader definition of health that encompasses both mental and physical aspects may be more productive.

Community Studies

Aircraft Noise. A number of recent studies have examined the effects of community noise on cardiovascular problems and physiological risk factors related to cardiovascular disease. In a series of these studies conducted in the neighborhoods adjacent to Schiphol (Amsterdam) Airport, Knipschild (1977) reported that residents in areas with high levels of aircraft noise were more likely to be under medical treatment for heart trouble and hypertension, more likely (especially women) to be taking drugs for cardiovascular problems, and more likely to have high blood pressure and pathologically shaped hearts than an unexposed population. Although these differences could not be explained by age, sex, smoking habits, or obesity, the noisy and quiet areas did differ in socioeconomic status, a confound that limits our ability to attribute group differences to noise exposure *per se*. A final study, not subject to confounds of demographics, followed subjects in a noisy area before and after aircraft overflights were instituted at night in the noisy area (Knipschild, 1980). Findings indicated that use of cardiovascular and antihypertensive drugs increased as night overflights increased. However, a puzzling feature of these findings is that much of the increased drug use appeared to occur in the year prior to the beginning of night overflights.

One aspect of the Knipschild (1977) studies warrants further comment. In one report, he found an apparent dose-response relationship between aircraft noise levels and prevalence of hypertension, with hypertension nearly twice as common in high noise—68 dB(A) Leq[1]—compared to low noise—55 dB(A) Leq—areas. Such a finding is very provocative and unique because it might provide some guidelines for public policymakers regarding the setting of lower thresholds for noise-induced cardiovascular effects. Unfortunately, in this study (Knipschild, 1977), a rather low percentage of community residents approached (42%) agreed to undergo cardiovascular screening. Thus, the possibility of bias in subject selection weakens definitive conclusions that can be drawn from these data. Pending further replications of the dose-response findings, they should be regarded as only suggestive. In contrast, the case for very *high* noise levels being associated with cardiovascular problems finds support in several other studies.

A Russian study (Karagodina, Soldatkina, Vinokur, & Klimukhin, 1969) suggests that children (9 to 13 years old) in noise-impacted areas around nine airports show blood pressure abnormalities, high lability of pulse, cardiac insufficiency and local and general vascular changes. Unfortunately, the report does not provide any information on the nature of the quiet control population or any details of the measurement procedures.

There are, however, studies reporting mixed findings regarding noise-associated health problems. A survey of residences of areas surrounding London's Heathrow Airport (Watkins, *et al.*, 1981) found no relationship between aircraft noise exposure and (a) the use of prescribed drugs; (b) the use of general medical practitioners; or (c) the utilization of out-patient or in-patient community health services. Consumption of nonprescribed drugs did, however, increase with annoyance in high noise but showed no relationship with annoyance in low noise. A study of aircraft noise in Munich over a roughly similar range of noise exposure (von Eiff, Friedrich, & Neus, 1982) failed to confirm Knipschild's (1977) observations of a direct linear increase in hypertension prevalence with increased noise exposure at low-to-moderate noise levels. They did, however, find that highest blood pressure levels were evident in residents exposed to the highest noise levels.

Summarizing community studies of aircraft noise, chronic exposure to very high noise levels has been associated with increased prevalence of hypertension and/or a variety of other cardiovascular disorders in several studies (see USEPA, 1980, 1981). However, with the exception of Knipschild's (1977) study, there is little available data regarding a dose-response relationship between noise intensity levels and measures of disease. An important methodological issue that plagues many environmental (e.g., noise, air pollution) and community health studies is obtaining accurate exposure estimates. It is difficult to establish dose-response curves or health effect thresholds when accurate daily exposure estimates cannot be calculated. For instance, home exposure, community, and occupational conditions may vary considerably. Regarding the measurement of noise exposure levels, there are multiple indexes that have been employed, each of which gives differential weight to such characteristics of sounds as frequency, range, duration of high-

[1]Leq is a measure of noise level computed on the basis of sound energy averaged over time (see Chapter 2).

intensity noise bursts (e.g., intermittent versus continuous), and time of day the noise occurs (e.g., night versus day) (see National Academy of Sciences, 1977a). Another complicating measurement issue is that estimates of outdoor levels of such environmental factors as sound level or air quality do not necessarily reflect indoor exposures, with indoor levels strongly affected by attenuation due to construction materials, household appliances, and other residential sources or buffers (see National Academy of Sciences, 1977a, for discussion of specific noise measurement indexes and Evans & Campbell, 1983, for further consideration of the issue of problems in measuring human exposure to pollutants).

Road Traffic Noise. Studies of the effects of traffic noise on cardiovascular measures are less consistent than those of aircraft noise in finding higher prevalence of cardiovascular disorders in noise-impacted areas. A German study of children in the seventh through tenth grades (Karsdorf & Klappach, 1968) reports high systolic and diastolic pressure for children from noisy schools, whereas a Dutch study (Knipschild & Salle, 1979) found no evidence for increased risk of cardiovascular disease in middle-aged housewives living on streets with high levels of traffic noise as compared with their neighbors living on quieter streets.

Recently, German investigators Neus, von Eiff, and colleagues have reported a series of studies of traffic noise and hypertension (Neus, von Eiff, Rüddel, & Schulte, 1983; Neus, Rüddel, Schulte, & von Eiff, 1983; von Eiff, Fredrich, & Neus, 1982). Initial findings indicated that prevalence of treated hypertension was higher in noisy—66 dB(A) Leq—compared to a quiet area—51 dB(A). Furthermore, hypertension prevalence was positively correlated with length of residence in the noisy area. In a related study (Neus, von Eiff, Rüddel, & Schulte, 1983), subjects not on antihypertensive drugs were treated at consecutive follow-up intervals of 16 months and 22 3/4 months. Results indicated that increase in diastolic blood pressure was significantly higher for residents of noisy areas, compared to residents of the quiet area. Overall, like the industrial studies, the studies of community noise suggest that high noise exposure is associated with an increased incidence of cardiovascular disease and some factors related to the risk of cardiovascular pathology. The data are more consistent for exposure to aircraft noise than for traffic noise, and the evidence suggests that children as well as adults show noise-associated cardiovascular effects (see Cohen *et al.*, 1973). Only recently have longitudinal studies been conducted, and the literature in this area is characterized by a dissimilarity among noise levels and types of noise examined in various studies and rather poor control for possibly confounding demograhic factors (see USEPA, 1980, 1981).

Community Noise and Other Health Problems

There are a number of other studies investigating the impact of aircraft noise on various noncardiovascular health problems. For example, increased noise has been associated with pregnancy complications and decreases in the health and survival of newborn infants (e.g., Ando & Hattori, 1973, 1977; Edmonds, Layde, & Erickson,

1979; Jones & Tauscher, 1978; Knipschild, Meijer, & Salle, n.d.). As noted before, these data are somewhat inconsistent, and, as yet inconclusive (Knipschild *et al.*, n.d.). It has also been reported that increased noise levels due to aircraft overflights are associated with increased death rates due to strokes and cirrhosis of the liver (Meecham & Shaw, 1979), although a reanalysis of the same data, carefully controlling for the confounding effects of age, race, and sex, found no difference in the mortality rates of the airport and control areas (Frerichs, Beeman, & Coulson, 1980). In passing, it is worth noting that the initial study by Meecham and Shaw was covered extensively by the popular press, despite its methodological shortcomings. This emphasizes the importance of carefully scrutinizing the methods of epidemiologic studies before important public policy decisions are made based on such evidence.

Other studies report noise-associated increases in nervous and gastrointestinal diseases (Karagodina *et al.*, 1969), consumption of sleeping pills and visits to doctors (Gradjean *et al.*, 1973), and self-reported incidence of a variety of chronic illnesses (Cameron, Robertson, & Zaks, 1972). In isolation, this work (some of which suffers from serious methodological flaws) can only be viewed as suggestive of possible pathogenic effects of community noise exposure. Furthermore, some of these outcome measures are not clearly linked to hypothesized mechanisms that might mediate associations between noise, stress, health, and risk. Replications of these studies with more careful attempts to control for confounding variables would greatly increase our confidence in the findings. It is currently too early to speculate at what minimal noise levels such health-related risks appear, largely because few well-designed studies have reported health risks with low or moderate noise intensity. However, a convincing case can be made that high-intensity noise is associated with increased prevalence of cardiovascular problems.

Population Density and Health

Crowding is a second environmental stressor purported to have adverse effects on health. Standard epidemiology texts suggest that crowding increases the risk of both infectious and noninfectious disease (Cassel, 1971; Dubos, 1965), and the popular press reports that a broad array of social problems result from crowding. One basis for such assertions is research conducted in the 1940s and 1950s that indicated higher rates of crime, disease, mental disorder, social disorganization, and mortality for urban than rural areas (see Altman, 1975, for review). However, these early studies typically employed inadequate methodological and statistical techniques, thereby making it very difficult to attribute urban pathologies primarily to rural-urban differences in population density. It is entirely possible, for example, that differences in sanitation, health care, and population composition accounted for the greater urban rates of disease and personal and social disorganization.

Another reason for suspecting that crowded living conditions might be detrimental to health is the well-documented finding that acute exposure to high-density laboratory conditions causes increased skin conductance, blood pressure, and heart rate (Aiello, Epstein, & Karlin, 1975; Evans, 1979; Frankenhauser & Lundberg, 1977). Again, however, these acute exposure situations were rather extreme, and by no means can one directly extrapolate from short-term autonomic responses made in artificial settings to chronic disease processes.

Belief in negative effects of density draws sounder support, perhaps, from studies of crowding in animals. The best known of these investigations is Calhoun's 1962 work with Norway rats: crowded animals developed a wide range of pathological behavior, including increased mortality (especially among the young), lowered fertility, neglect of the young by their mothers, overly aggressive and conflict-oriented behavior, withdrawal, hyperactivity, and sexual aberrations. Similar patterns of pathological effects have been reported for mice (e.g., Lloyd & Christian, 1969; Southwick, 1955) and moles (Clarke, 1955).

However, contemporary research on human populations has been less conclusive. Several early reviews of this literature by Freedman (1975), Fischer, Baldassare, and Ofshe (1975), and Lawrence (1974) concluded that human population density is not related to physical pathology, mental disorder, or emotional instability. On the other hand, reviews by Zlutnick and Altman (1972), as well as reviews by Baum and Paulus (in press) and Moos (1976), although carefully avoiding any definitive conclusions, leave the reader with the strong impression that existing evidence does suggest such relationships. The disagreement is not surprising because results from different studies are themselves inconsistent, and there is considerable controversy regarding the adequacy of the research methodologies employed. In addition, because different assessments of the empirical literature reflect different conceptions of density, important differences in the psychological impact of different types of high-density environments are often neglected. For further discussion of definitional issues in human crowding research, see Baum and Paulus (in press).

A partial corrective to this state of affairs can be found in the conceptual distinction between primary and secondary environments. Stokols (1976) suggested that crowding is more stressing in primary environments—those in which persons spend more of their time and in which they relate to others on a personal basis—than in secondary environments—those in which encounters with others are transitory, anonymous, and inconsequential (Stokols, 1976; Zlutnick & Altman, 1972). We have divided the studies discussed later into those dealing with internal density—dwelling space per person (e.g., rooms per person or square feet per person)—and external density—number of persons occupying a residential area (e.g., people per acre, square kilometer, or square mile). Internal density is a measure of density in a primary environment, the home, whereas external density is a measure of a secondary environment, the neighborhood (cf. Cohen, Glass, & Phillips, 1979). These two measures of density may be independent of one another; for example, a luxury high-rise housing project in New York City would have a high external density but a low internal density, whereas a low-income housing project made up of four-story tenements would have a relatively low external density with a high internal density. Our discussion of both types of density will include only those studies that control for factors that often covary with density (e.g., income, education, sanitation) through the use of partial correlation, multiple regression, or stratification techniques.

Studies of Internal Density

Public-health officials have long been concerned with the minimum amount of household space required for the maintenance of physical and mental health. Informed opinion on this lower limit varies widely: the Chombart de Lauwes (cited by Hall, 1966, p. 172) report on the French working-class family showed minimal pathologies with 90 to 140 square feet of interior household space per person; Madge (1968) indicates expert European consensus at a 170 square-foot minimum; whereas the American Public Health Association Committee on Hygiene of Housing (1950) sets the desirable standard at over twice this figure. It is possible (even likely) that available space alone is not the key determinant of pathological effects of internal density. Freedman (1973) suggested that the critical factor is the sheer number of people living in a restricted space, and Galle, Gove, and McPherson (1972), among others, stressed available privacy, as indexed by the number of rooms (or rooms per person) in a dwelling unit.

Several studies have used a variety of these measures and calculated their associations with social and physical pathologies, such as rates of crime, public assistance (e.g., welfare payments), suicides, and mental-hospital admissions. Although not entirely consistent, the evidence suggests that, at least in single-family households, internal density is not an important factor in physical and mental health among normal healthy adults. For example, in Schmitt's (1966) study of 29 Honolulu census tracts, after statistical controls for income and education were employed, persons-per-room was moderately related to only one of nine measures of health and adjustment—rates of juvenile delinquency. Unrelated measures included rates of deaths, infant deaths, suicides, tuberculosis, venereal disease, mental-hospital admissions, illegitimate births, and imprisonment. Similarly, in a study conducted in The Netherlands, Levy and Herzog (1974) found that the number of persons per room had uniformly low and negative associations with nine indicators of mental and physical health. That is, density in the home reduced rather than augmented various pathologies. The authors suggest that the Dutch family may provide a form of protection from external stress, thereby ameliorating the adverse effects of high family density.

Studies in the continental United States and in Canada also reveal dissociation between internal density and pathology. In a Canadian study, Gillis (1974) examined 30 census tracts in Edmonton, Alberta. After controlling for income and ethnic backgrounds, he found that the proportion of dwellings with more than one person per room was marginally related to public assistance rates and unrelated to rates of delinquency. Similarly, Freedman, Heshka, and Levy (1975) analyzed data from 338 New York City health districts. After controlling for ethnicity and social class, they found no relationship between persons per room and rates of mental illness, delinquency, infant death, illegitimacy, and venereal disease.

Some evidence suggests minimal effects of internal density on the general population. Consider, for example, a study of 75 Chicago communities by Galle, Gove, and McPherson (1972); after controlling for income and ethnicity, the number of persons per room was positively related to morality rates, public assistance to persons under 18, and fertility rates. (This last finding is the inverse of the density–fertility relationship reported in many animal studies). In addition, the investigators found that the higher the average number of rooms per housing unit, the fewer the admissions to mental hospitals. Reanalysis of the Galle *et al.* data by Ward (1975) indicated that density did not account for as much of the variance in each of the dependent measu'es as was reported in the original study.

Although internal density may only minimally affect the general population, certain population groups—the very young (Evans, 1978), the old, and those under other forms of stress (cf. Levi & Anderson, 1975)—may be more susceptible. A survey conducted by Booth (1975) illustrates this point. Members of 560 Canadian households were interviewed and given physical examinations. It appears that densities "seldom have any consequences and even when they do they are modest" (Booth, 1975, p. 11). Nevertheless, household crowding did adversely affect child health and physical and intellectual development and had greater negative effects on people under stress caused by low income or other problems. Booth's study is one of the few large-scale investigations of individual rather than aggregate data. Use of the individual instead of demographic-geographic areas provides measures of pathology that are (a) less sensitive to fluctuations of covarying demographic factors (e.g., income and education); and (b) less likely to be affected by distortions in measurement in particular population groups (e.g., underreporting of mental-hospital admissions among the middle class; see Chapter 2).

Like Booth, several other investigators report that children and members of the lower class are particularly affected by internal density. Winsborough (1965) reports that increased number of persons per room is related to increases in infant death rates. Mitchell (1971) found that among low- but not high-income Hong Kong families, square feet per person was related to superficial signs of psychological stress (i.e., self-reports of "worry" and "unhappiness"). Density, however, was not associated with other indexes of stress such as self-reports of psychosomatic symptoms and withdrawal from family and work roles.

To sum up, although weak to moderate relationships between household density and various pathologies are sometimes reported, internal density in family residences has not yet been shown to be an important factor in physical and mental health for adults. Even rates of infectious diseases, presumed to be prevalent under high residential density, are consistently unrelated or negatively related to density. The findings do suggest, however, that household density may aggravate existing stress conditions (e.g., in low-income populations), and when combined with other stressors, internal density may have deleterious effects on mental and physical health. Caution is needed, however, because all of the studies reported here are correlational and hence do not allow easy causal inference. Another very important qualification to consider in interpreting the results of most of these field studies is their reliance on aggregate means, which precludes analysis of individual differences in response to crowding or noise. It is crucial to realize that there are wide individual differences in response to stress (Evans & Cohen, in press; Lazarus & Cohen, 1977), and aggregate analysis does not allow one to examine these differences (see Chapter 2). However, some preliminary studies of children in crowded residences suggest that negative health and behavioral consequences of high-density housing may be more severe for young children, particularly those from low-income families (see also Evans, 1978, and Saegert, 1981, for reviews of crowding and developmental data).

Studies of Internal Density in Institutions. Internal density may not be a major contributor to ill health in households for adults, but several studies in prisons, naval ships, and college dormitories suggest an opposite conclusion. For example, D'Atri (1975) reported that prisoners housed in dormitories have higher systolic and diastolic blood pressure than those housed in single-occupancy cells. Similarly, three females living in college dormitory rooms designed for two report more health problems than those living with one one roommate (Aiello *et al.*, 1975). There was no effect of "tripling" on the health of male students. A study by Baron, Mandel, Adams, and Griffin (1976) also reported no increase in number of visits to the health center for males tripled in double rooms. Stokols, Ohlig, and Resnik (1978) found a significant relationship between self-reported crowding in dormitories and visits to health centers. Increased visits to the dispensary are, however, reported for males crowded aboard naval ships (Dean, Pugh, & Gunderson, 1975).

Important data on the deleterious effects of crowding in prisons comes from recent research by McCain, Paulus, Cox, and colleagues. Their choice of prisons as a research environment reflected an interest in examining situations where crowding was long-term, intense, and inescapable. In a program of several studies (Cox, Paulus, & McCain, 1984), it was found that increases in prison populations (without concomitant increases in housing space) were associated with increased death, suicide, and psychiatric commitment rates among inmates. The doubling up of inmates in cells or cubicles led to increases in illness complaints relative to single cells or cubicles. In addition, in another study (Paulus, McCain, & Cox, 1978), they found a positive association between crowding and blood pressure levels in prison inmates. Analyses of these data indicated that negative effects were associated with increased numbers of persons per space (social density) in the prison, rather than just the total amount of physical space per person (spacial density) in the prison.

Other dormitory research indicates that density can result in interpersonal problems, for example, a desire to withdraw and avoid others (Baum, Harpin, & Valins, 1974; Valins & Baum, 1973), and a dissatisfaction with roommates (Baron *et al.*, 1975). Thus, residential crowding with strangers may be experienced differently from crowding within a family household, indicating that the nature of the social relationships between residents is important in determining the impact of internal density (Cohen, 1978).

Studies of External Density

A rather different approach to the study of population density focuses on the number of persons living on a specified amount of residential land (the neighborhood). Unlike internal density (which is measured in primary environments), external density provides information about the amount of contact one is likely to have with others in a secondary environment, where encounters with others are transitory, anonymous, and inconsequential. The available evidence on the effects of external density is inconsistent. Several studies report clear detrimental effects on health, whereas others report no such associations. Although some discrepancies may be due to differences in the pathologies being measured and how they are defined, nevertheless, studies using similar measures have also led to conflicting conclusions.

Reports of significant density effects include Schmitt's (1966) analysis of data from a large sample of Honolulu census tracks. Positive correlations were obtained between number of persons per acre and seven of nine indexes of health and well-being. Directly related to external density were rates of death, venereal disease, tuberculosis, mental-hospital admissions, illegitimate births, delinquency, and imprisonment. (Unrelated were rates of infant death and suicides.) Similar effects are reported in a study of 125 Dutch geographic regions (Levi & Herzog, 1974). After controlling for various social-class factors (e.g., income), increases in persons per kilometer were associated with increases in rates of death, male (but not female) heart disease, admissions to general and mental hospitals, delinquency, illegitimate births, and divorces. Booth (1976) also found reliable correlates as between external density and several measures of crime (e.g., burglary, rape), life expectancy, and infant mortality. These analyses controlled for crime and education.

Four other investigations suggest no relationship between external density and various pathologies: Galle *et al.* (1972) found no relation with mortality, fertility, public assistance, juvenile delinquency, and admissions to mental hospitals; Gillis (1974) found no relation with rates of public assistance and juvenile delinquency; Freedman *et al.* (1975) found no relation with rates of mental illness, infant death, venereal disease, and illegitimacy (however, a weak relation was found with juvenile delinquency). Booth's (1975) study of Canadian families also found minimal evidence that densely populated neighborhoods affect mental or physical health or family relations. The few weak associations he does report suggest beneficial (e.g., density stimulates neighborhood political activity) rather than detrimental effects.

Finally, Winsborough's (1965) study of 75 Chicago communities provides mixed results. After controlling for socioeconomic factors, he found increased external density was related to increases in infant death rate and decreases in tuberculosis, public assistance to adults, and age-adjusted death rate, and was unrelated to public assistance to persons under 18.

The inconsistency of findings on the relation of external density to health may be due to differences in measurement or accuracy of data, random error, or culturally mediated responses or adaptations to high density. One likely explanation is that reported effects are due not to external density *per se* but rather to other environmental factors that are often associated with large populations existing in restricted spaces. Thus, future research might focus on such factors as the number of housing units per structure, number of structures per acre, types of structures, and availability of stores, services, and indoor and outdoor common spaces for recreation.

Summary of Crowding–Health Review

Although crowding has been shown to have measurable effects on health, the physical and psychological consequences of this stressor appears to depend, to a considerable degree, on individual, social, and contextual factors. Thus, effects of crowding on health indexes appear among individuals in certain environments (e.g., prisons, ships, college dormitories) where social interactions are likely to be uncertain, overloading, negative, or uncontrollable (see Cox *et al.*, 1984).

Air Pollution and Health

By and large, social scientists have paid little attention to air pollution. What little research that has ensued has focused almost exclusively on surveys of attitudes and awareness of pollution. Recently, however, some research has begun on the behavioral and psychological impacts of air pollution. In this section, we summarize the survey work and physical health research on air pollution. Greater attention is given to recent research on behavioral impacts of air pollution, including social behaviors and psychological health.

Air pollution is a ubiquitous problem impacting greater than 50% of the American public and costing greater than $200 million per year in U.S. health costs alone (Lave & Seskin, 1970). Furthermore, tens of billions of dollars are lost yearly on real estate and agricultural impacts in the United States.

The health effects of air pollution include respiratory infection, irritation, and disease. Photochemical oxidants (smog) produce irritation and respiratory discomfort at ambient levels. Some upper respiratory impairment has also been noted, although this finding is more controversial (Rokaw *et al.*, 1980). Ambient ranges of sulfur dioxide irritate upper respiratory passages, reduce mucosal clearance activity, and reduce pulmonary function. Sulfur dioxide has also been associated with upper respiratory infections, bronchitis, and asthma. Oxides of nitrogen reduce pulmonary function and resistance to some diseases, inflame bronchial passages, and interfere with hemoglobin peroxidation. Some respiratory illnesses may be associated with nitrogen oxides as well. Carbon monoxide has been associated with cardiovascular disease, reduced birth weights, and in cases of acute exposure, with headaches, dizziness, and nausea. Research on various particulates including toxic metals like lead and mercury have revealed various human impacts including pulmonary lesions, carcinoma, mesothelial tissue damage and pulmonary cancer for mercury; gastrointestinal disturbances, anemia, retardation, and impaired neural functioning for lead; and neural dysfunction, upper respiratory inflammation, thyroid disturbances, and cancer for asbestos. For reviews of toxicological and epidemiological research on human health and air pollution, see Coffin and Stokinger (1977), Goldsmith and Friberg (1977), and the National Academy of Sciences (1977b). Although there are a multitude of methodological and conceptual shortcomings in much of the research on air pollution and human health (see Evans & Campbell, 1983, for review), the effects listed previously are consistent and have generally been found in several converging programs of research in field and laboratory conditions.

Survey research on air pollution has focused primarily on assessments of citizens' awareness of air pollution. This is quite high among the public, generally exceeding 80% when individuals are directly probed. When more indirect measures are used (e.g., open-ended questions on community/regional problems), awareness levels are considerably lower (Evans & Jacobs, 1982). There is also greater public ignorance about what causes air pollution or how it affects human health.

The major factors effecting awareness of air pollution are the types of pollutants and media publicity about pollution issues. Particulates, soiling of buildings and households objects, and reduced visibility from haze each contribute to greater awareness of air pollution (Barker, 1976). The reliance on visual cues for detection of pollution is noteworthy for several reasons. Several harmful pollutants are invisible. In addition, many individual factors as well as levels of pollutants themselves influence human detection of visual degradation (Wohlwill, 1974).

Mass media publicity campaigns do heighten public awareness of local and regional air pollution problems, although their staying power is questionable (Barker, 1976). Lower education and lower income levels are associated with less awareness of air pollution as well. In addition, people are typically less cognizant of air pollution where they reside in comparison to awareness of pollution in adjacent areas or regional sectors (Evans & Jacobs, 1982). One likely explanation for the latter finding is the reduction of "cognitive dissonance," or the rationalization of inconsistent information because if one is reasonably satisfied with the area in which one lives and there is poor air quality in that place, one has to either shift his or her level of neighborhood satisfaction or his or her awareness of the poor air quality.

A topic of growing scholarly interest is the analysis of physical and psychosocial factors that influence human visibility judgments. Evans, Jacobs, and Frager (1982), for example, have recently shown with signal detection approaches that long-term residents of high air-pollution areas with poor visibility from heavy smog do not suffer in their visual detection sensitivity to visual cues for smog but experience highly significant, conservative shifts in response biases to report smog detection. For a given low level of smog, persons who have adapted to poor air quality are much less likely to report seeing smog than other persons who have not previously adapted to ambient conditions of poor air quality.

Peoples' attitudes about air pollution also vary with the types of questions asked, personal and social factors, and the nature of the pollutants. As in the case of awareness, direct probes elicit stronger negative reactions. Air pollution is considered a more serious problem by members of more economically developed societies as well as by more affluent members of highly industrialized communities. Some have speculated that individuals who have had more critical life necessities such as housing, safety, food, and occupation are more able to turn their attention to secondary issues like environmental quality (Evans & Jacobs, 1982). Other factors associated with greater annoyance with air pollution include female gender, preexisting respiratory disease, unadapted populations (i.e., newcomers to poor air quality), and those who have little or no direct economic ties to major pollution sources. There is also evidence suggesting that air pollution experts, public officials, and citizens differ in both their awareness and levels of concern about air pollution (Barker, 1976).

Visually apparent pollutants like particulates or smog are more apt to elicit negative evaluations as was found in the case of public awareness of air pollution. Similar to research on noise reviewed earlier, persons are more likely to be annoyed by pollution sources that are perceived as unnecessary and/or unconcerned about resulting public health impacts. Finally, although the vast majority of citizens feel that it is important to reduce air pollution, most people believe that little can be accomplished by individuals to reduce air pollution and that control responsibilities lie with industry and government (Barker, 1976; Rankin, 1969). Thus, it would be worthwhile to study further the possible links between chronic exposure to smog and other pollutants and susceptibility to helplessness. For more detailed reviews of the survey work on air pollution, see Barker (1976) and Evans and Jacobs (1982).

The behavioral impacts of air pollution that have been studied include three areas: leisure activities, social and interpersonal behaviors, and mental health. Suggestive but inconclusive evidence indicates that poor air quality may curtail outdoor and recreational activities. The effects are small, however (Chapko & Solomon, 1976; Peterson, 1975), and some data indicate no relationships (Rivlin, 1974). Some factors may function to reduce the chances of noting relationships between outdoor activities and air quality. Most of the preceding studies rely on aggregate levels of analysis and thus are unable to consider either individual differences in susceptibility to air pollution or to adequately measure individual dose-response relationships. To illustrate the potential advantage of individual-based analyses, consider recent research by Evans, Jacobs, and Frager (1982c), who found out that certain subgroups of exposed populations did curtail outdoor activities during high air pollution episodes in the Los Angeles area. Persons more internal in locus of control who were also new migrants to the area from previously low air pollution regions of the country consistently reduced outdoor activities during high-smog periods. More air pollution and human behavior research should focus on potentially vulnerable individuals such as the elderly, young children, or persons with preexisting respiratory disorders.

Given the consistent finding that people are annoyed by air pollution, several recent programs of research have explored whether the negative effects associated with air pollution might lead to more negative interpersonal behaviors. Laboratory studies have shown that malodor or secondary cigarette smoke cause irritation and annoyance (Bleda & Bleda, 1978; Jones, 1978). Further, Rotton and his colleagues have shown that this annoyance does not occur when other persons are also seen as suffering from the same poor air-quality conditions (Rotton, Barry, Frey, & Soler, 1978).

There is also evidence of greater interpersonal hostility and aggression (Jones & Bogat, 1978) as well as reduced altruism (Cunningham, 1979) when people are exposed to air pollution. James Rotton and his colleagues, however, have suggested a more complex relationship between air pollution and hostility. They reasoned that air pollution, like high temperatures (cf. Baron, 1978), might have a linear relationship with negative affect but a curvilinear relationship with aggression. At a certain level of negative affect, aggression will peak and then drop off as the setting becomes increasingly noxious. At this point, individuals' feelings of hostility are overridden by strong desires to minimize exposure to the negative environmental conditions produced by high levels of air pollution. Withdrawal or escape becomes the predominant mode of operation rather than greater aggression that is not viewed as expediting removal from the situation. This type of inverted-U relationship between air pollution and aggression (Rotton, Frey, Barry, Milligan, & Fitzpatrick, 1979) has been found. As noted earlier, similar nonlinear trends between heat and aggression have been noted both in laboratory and field experiments (Baron, 1978; Bell & Greene, 1982).

Given that pollution is related to negative affect as well as some negative interpersonal behaviors, some researchers have investigated potential links between poor air quality and human mental health. Clinical case studies have documented several cases of psychological symptoms from indoor air pollution exposures. These studies have linked depression, irritability, anxiety, and somnolence with indoor air pollution (Randolph, 1970; Weiss, 1983). Furthermore, some aggregate-level epidemiological studies have found significant associations of ambient pollutants with psychiatric admissions to hospitals (Briere, Downes, & Spensley, 1983; Strahilevitz, Strahilevitz, & Miller, 1979). Because of the correlational nature of these field studies, caution is necessary at this time in drawing causal conclusions about air pollution and mental health.

Two more recent studies, with more sophisticated research designs, have also noted some relationships between air pollution and mental health. Rotton and Frey (1982), using a time series design with extensive controls, found a moderate positive relationship between ambient pollution levels and emergency police and fire calls for psychiatric disturbances. Evans and his colleagues have also completed a very large longitudinal and cross-sectional study of individual responses to air pollution. Their work shows that certain subgroups of the population appear susceptible to negative psychological impacts from poor air pollution. In particular, they found that residents exposed to poor air quality who had also had a recent stressful life event were more likely to suffer psychological distress. They found no overall main effects for air pollution levels on mental health; only the vulnerable subgroups were negatively affected (Evans, Jacobs, Dooley, & Catalano, 1982).

Summary. Summarizing, we can see that although there is considerable evidence showing negative respiratory health impacts from air pollution, very little research has examined behavioral impacts of air pollution. What little work that has been done has tended to focus on public opinion and awareness surveys. These studies show clearly that large numbers of people are worried about and are aware of air pollution. Of interest, some of the variables that mediate the relationship between physical levels of pollutants and annoyance are parallel to those studied in more depth by noise researchers (e.g., when the sources are perceived as unnecessary or unconcerned about public health impact). There is also some scant evidence that air pollution causes greater negative interpersonal relationships especially when surrounding people are also seen as victims of the pollutant. Aggression may also increase up to a point with increasing air pollution but then seems to drop off as the setting becomes increasingly noxious. The similarity of the inverted U-shaped function to heat and aggression research was noted. Finally, there are some scattered pieces of data suggesting the possibility of a relationship between air pollution and psychological distress.

Summary of Environmental Stressor Research and a Note on Psychological Mediators

In sum, environmental noise, residential density, and air pollution all appear to have potential effects on physical and mental health. Although direct physiological response to these environmental stressors may be the cause of adverse effects when they occur, there is increasing evidence (primarily from the laboratory) that many adverse effects are mediated by psychological factors. Although we alluded briefly to the evidence in the first section of this chapter, these factors are emphasized in other portions of this book. More specifically, it appears that beliefs about the controllability of a stressful environment (i.e., the degree to which subjects believe they can escape from a stressor) are important determinants of response to stressors in certain situations (e.g., Glass & Singer, 1972; Sherrod, 1974; see Chapter 4). Also relevant to this discussion is the work of Seligman (1975), which suggests that a psychological state of helplessness results when we continually encounter aversive events about which we can do nothing, that is, events that involve a noncontingency between instrumental responses and outcomes.

Several findings in the epidemiological literature on noise and health are amenable to the interpretation that a sense of helplessness affects response. Consider, for example, data cited earlier indicating that those living in "noise slums" were more likely to be admitted to a mental hospital than those living in less noisy areas. Herridge (1974) suggested that the mental distress of those exposed to prolonged noise was due more to feelings of helplessness than to the noise *per se*. A developmental study by Wachs, Uzgiris, and Hunt (1971) also suggests the importance of control in producing noise effects. In that study, noise had a particularly adverse impact on development when the child was unable to leave the noisy environment. Although consistent with the psychological view, interpretation of these data in terms of control is somewhat risky. The report failed to specify the nature of "escapability," which may have also implied less total noise exposure for children who could escape.

It should also be noted that the mediating influences of subjective reactions on noise-related physiological health outcomes may be limited to noise levels that are moderate in intensity. Thus, Neus, Rüddel, and Schulte (1983) found that self-reports of noise intolerability were correlated with treatment for hypertension only within control areas but not within residential areas heavily impacted with traffic noise. In addition, examination of changes in blood pressure over a 2-year period showed positive relationships between rated measures of noise annoyance and blood pressure only within lesser impacted noise areas. Thus, psychological coping mechanisms might not be effective as mediators of long-term physiological response for extreme physical stimuli, whereas they may be evident for noise stimuli of moderate intensity.

Despite the findings described in this section, most research on the impact of environmental noise on physical health has focused on the direct effects of variations of noise parameters (e.g., intensity). Future research that more closely examines the role of an individual's perceptions of the environment may allow us to establish a more precise link between noise exposure and the body's long-term responses. However, in the Los Angeles Noise Project, we were primarily interested in establishing whether health-related cardiovascular effects of noise could be detected in children attending schools under an airport flight path heavily impacted by high noise levels. For this reason, the primary independent variable was physical, rather than perceived intensity of noise.

Los Angeles Noise Project: Blood Pressure

Perhaps the most consistent health-related correlate of chronic exposure to high-intensity noise is elevated blood pressure. Few studies, however, have examined these effects in school-aged children. Moreover, there are no previously existing controlled longitudinal studies on the impact of noise on nonauditory health in children (Mills, 1975). Accordingly, blood pressure was used as the primary health measure. A second factor, school absenteeism, was also examined as a health index in the Los Angeles Noise Project.

Methods, Results, and Discussions

Blood Pressure Measurement. At each testing session, each child's resting blood pressure was taken on a Physiometrics model SR2 automated blood pressure recorder. This instrument is an electronic infrasonic device with established reliability for use with children (Voors, Foster, Frerichs, Weber, & Berenson, 1976). To accustom the children to the blood pressure measurement technique, an initial measurement not recorded as usable data was made at the beginning of the first day of testing. Each child's blood pressure was measured again on the first day and once more on the second day. The blood pressure data are based on the mean systolic and diastolic pressures for these two measurements.

Other Variables Used in Analyses. Each child's weight and height were measured. Absenteeism, gathered from school records, was also used as an indirect measure of health because school absences are often attributable to illness.

Treatment of Data

As with other measures in this study, statistical control factors included family size, grade in school, months enrolled in school, and race. Health measures were separated into two multivariate clusters: "general health" measures and blood pressure. This procedure became necessary because two of the general health measures—height and ponderosity (weight/height^3) were required controls for analyses of blood pressure. We chose the ponderosity index as a measure of obesity because of its high correlation with body fat (cf. Voors *et al.,* 1976).

Time 1 Data. Analyses of the data from Time 1 testing produced a significant multivariate F for the effects of noise on the "general health" factor, $F(3, 235) = 8.04, p < .001$. Although children from noise schools tended to be shorter and weigh less than quiet-school children, neither of these differences approached significance in analyses of univariate effects (p's $> .15$ for both measures). Surprisingly, analyses of attendance data indicated that noise-school children were present a higher percentage of time (97.5% vs. 94.2%) than children from quiet schools, $F(2, 237) = 21.80, p < .001$.

The effects of noise exposure on blood pressure were clear in the analysis of data from Time 1 (see Figure 2). There was a reliable multivariate F for the effects of noise on systolic and diastolic blood pressure $F(2, 244) = 2.98, p < .05$. As is apparent in Figure 3, children from noise schools had higher blood pressures than quiet-school children. Univariate $F(1, 245)$ for systolic $= 4.61, p < .03$, and for diastolic $= 4.86, p < .03$. Unadjusted means for systolic pressure were 89.7 mm Hg for the noise-group and 86.8 mm Hg for quiet-group subjects. Diastolic means were 47.8 mm Hg and 45.2 mm Hg for the noise and quiet groups, respectively.

The pattern of means for systolic and diastolic pressures suggest that the noise–quiet difference was greatest during the first 2 years of exposure, with the differences remaining consistently smaller after that point. However, the statistical

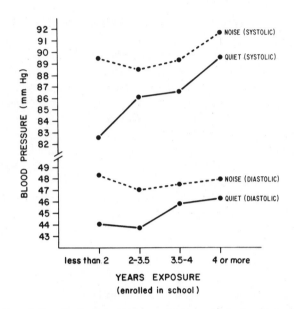

Figure 3. Systolic and diastolic blood pressure as a function of school noise level and duration of exposure for Time 1 sample. Each period on the years exposure coordinate represents one-quarter of the sample. For example, 25% of the sample were enrolled in school less than 2 years (from Cohen, S., Evans, G. W., Krantz, D. S., & Stokols, D., 1980. Copyright 1980 by the American Psychological Association. Reprinted by permission).

interaction between noise and months in school was only marginally significant for systolic blood pressure ($p < .07$) and nonsignificant for diastolic.

It should be noted, however, that although the noise–quiet mean blood pressure differences were statistically reliable, the levels for children attending noise schools do not as a group exceed normative levels for children of similar ages (Voors *et al.,* 1976). In addition, the long-term health consequences, if any, of these blood pressure elevations remain unknown, but the data do provide support for a cardiovascular effect of aircraft noise. However, we should call attention to the epidemiological evidence of blood pressure tracking in children; that is, children measured to be in the higher percentiles of blood pressure for their age tend to evidence similarly high levels when measured several years later (Berenson, 1980; Coates *et al.,* 1981).

To determine whether or not living in a relatively quiet home (at least in terms of aircraft noise) would lessen the impact of school noise, the children living in the 20 quietest homes in the noise sample were isolated. These children were then compared with the remainder of the noise sample and with the quiet sample. Results indicated that the noise-school children from quieter homes *still* had higher systolic blood pressure, $F(1, 240) = 3.59$, $p < .06$, and higher diastolic blood pressure,

$F(1, 240) = 5.32, p < .02$, than children from the quieter school did. Thus, living in a relatively quiet neighborhood did not appear to lessen the impact of noise at school.

Analyses of Habituation versus Attrition. The greater difference during the first few years of school enrollment found in this cross-sectional analysis could be due to noise children habituating to the stressor as the duration of exposure increased. On the other hand, the effect could be due to some kind of subject selection bias; that is, children with noise-induced blood pressure elevations may have quickly moved out of the noise-impacted neighborhood and thus lowered the mean blood pressure for noise-school children who remained exposed for 2 or more years.

Available longitudinal data on *how long* specific noise- and quiet-school children remain enrolled in their schools helped distinguish between these two explanations. A second analysis of blood pressure data from Time 1 was conducted to determine whether families of noise-school children were more likely to move sometime during the 2 years following the original testing session than families of quiet-school children and of noise-school children not showing elevated pressure. In this case, children who were absent at Testing Session 2 but still enrolled were categorized as attending school. For these data analyses, we used a three-level measure of *migration* (not enrolled in school after 1 year/enrolled 2 years later but not after 2 years/still enrolled after 2 years, see Table 3 in Chapter 2). As depicted in Figure 4, noise-school students with the highest blood pressures move out of the

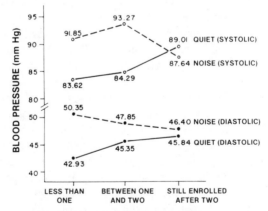

Figure 4. Systolic and diastolic blood pressure measured at Time 1 as a function of school noise level and the number of years enrolled in school following Time 1. Each period on the years exposure coordinate represents number of years students remained enrolled (from Cohen, S., Evans, G. W., Krantz, D. S., Stokols, D., & Kelly, S., 1981. Copyright 1981 by the American Psychological Association. Reprinted by permission).

noise area soon (within 2 years) after the initial testing, and a reverse trend appears for the quiet children. This is supported by a significant Noise and Migration interaction: for systolic, $F(2, 229) = 6.80, p < .001$; and for diastolic, $F(2, 229) = 3.50, p < .03$. Thus, it appears that selective attrition, not adaptation, is responsible for the decrease of the difference between blood pressure of noise- and quiet-school children.

It is important to emphasize that these effects occurred with race and social class statistically partialed out of the analyses. Some possible (and admittedly speculative) explanations for the selective attrition effect among noise children are (a) parents of children with elevated blood pressure were sensitive to their children's experience of stress and as a consequence moved to a less noisy neighborhood; (b) because of a familial bias (either genetic or environmentally determined), parents of children with noise-induced blood pressure elevations experienced similar stress-related reactions that motivated them to move from the neighborhood; (c) the children's elevated blood pressures were a response, not to the noise itself, but to their parents' own noise-induced stress, which motivated the parents to move from the neighborhood; and (d) some unknown third factor is related to mobility, higher blood pressure, and living in a noisy neighborhood. Although we cannot select among these possible explanations, there is recent related evidence that children of hypertensive parents show elevated cardiovascular response to stress (Falkner, Onesti, Angelakos, Fernandes, & Langman, 1979; Obrist, Langer, Grignolo, Light, Hastrup, McCubbin, Koepke, & Pollak, 1983). Thus, parents of children with elevated blood pressure may show elevated responses similar to those of their children. We have no ready explanation for the opposite trend for blood pressure and attrition obtained among the quiet children.

The explanations that suggest that high blood pressure is the *cause* of the migration from noisy neighborhoods assume that the child and/or parent perceive that the child is under stress. It is probable that only those children with blood pressures substantially higher than the group mean would fit into this category. Thus, if elevated blood pressure is responsible for increased migration from the noisy neighborhoods, large proportions of those children leaving the noisy neighborhoods would have relatively high blood pressures. Analyses of the proportion of children moving from their neighborhoods as a function of whether they attend a noise or quiet school and whether they have high (80th percentile or above) or low (below the 80th percentile) blood pressure indicate that the proportion of children with high blood pressure who move from the noisy neighborhoods is higher than the proportion of high blood pressure children who move from quiet neighborhoods (Noise × Blood Pressure interactions: for systolic, $F(1, 246) = 5.42, p < .02;$ for *diastolic*, $F(1, 246) = 5.59, p < .02$. Apparently, a relatively large number of noise-school children who move do have substantially elevated blood pressure.

Longitudinal Analyses. Although the analysis of the complete Time 1 sample indicated higher systolic and diastolic blood pressures for noise-school children, there were no effects on either systolic or diastolic blood pressure in the longitudinal analyses, which include both Time 1 and Time 2 data. *Longitudinal blood pressure effects were not expected, however, because a relatively high proportion of noise-school children with higher blood pressures were lost to attrition (see earlier) and thus not included in the longitudinal analyses.* Because of this attrition effect, our data do not enable us to make a definitive determination about whether the increased blood pressure levels of noise-affected children found in the initial complete cross-sectional sample habituate over time.

Replication Sample. Figure 5 presents cross-sectional blood pressure data from an independent replication sample of third graders. Noise-school children had higher blood pressure levels among those exposed 2 years or less but not among those attending noise schools for longer periods. In sum, the data from both the longitudinal study and the cross-sectional replication clearly indicate that children attending school in the air corridor have elevated blood pressures during their first few years of exposure. Although this elevation does not occur for those who have lived in the neighborhood and attended their schools for longer periods of time, data from the various longitudinal analyses argue that this is due to a bias in who moves out of the neighborhood rather than to habituation to the noise. These data are consistent with previous studies of both adults and children cited earlier, which suggest that those undergoing prolonged noise exposure have persistent elevations in cardiovascular response.

Effects of Noise Abatement on Blood Pressure. Longitudinal analyses revealed no significant effects of noise abatement on blood pressure. Looking at the cross-sectional (T1) data, as is apparent from Table 11, both systolic and diastolic blood pressures appear to vary as a function of noise level, with the highest mean

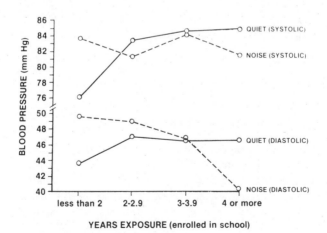

Figure 5. Systolic and diastolic blood pressure as a function of school noise level and duration of exposure for cross-sectional third grade replication study. Each period on the years exposure coordinate represents one-quarter of the sample (copyright in the public domain).

Table 11. Mean Blood Pressure (mm Hg) by
Classroom Noise Abatement for Cross-
Sectional (T1) Data

	Classroom		
Blood pressure	Quiet	Abated	Noisy
Systolic	86.64	88.69	90.09
Diastolic	44.99	46.77	48.46

Note. From "Aircraft noise and children: Longitudinal and cross-sectional evidence on adaptation to noise and the effectiveness of noise abatement" by S. Cohen, G. W. Evans, D. S. Krantz, D. Stokols, and S. Kelly, 1981, *Journal of Personality and Social Psychology, 40*, p. 342. Copyright 1981 by the American Psychological Association. Reprinted by permission.

pressure reported for the noise group, followed by a lower blood pressure in the abated group, and an even lower pressure in the quiet group. Although the analysis of systolic pressure did not indicate a statistically significant impact of noise, there was a main effect of noise for diastolic blood pressure, $F(2, 241) = 3.19, p < .04$. The multivariate analysis was not significant.

Preplanned contrasts between the various blood pressure means indicate that for both systolic, $F(1, 235) = 2.61 \ p < .10$, and diastolic pressure, $F(1, 235) = 5.24, p < .02$, the noise group was different from the quiet group. Comparisons between the quiet group and the abated group indicated marginal differences in both cases—for systolic, $F(1, 235) = 2.21, p < .14$; for diastolic, $F(1, 235) = 3.17, p < .08$. There were no differences between the noise and abated groups for either systolic or diastolic pressure.

Conclusions

Implications of the Noise Project Findings

Results of the Los Angeles Noise Project (LANP) indicate that 7 to 8-year-old children residing and attending school in areas impacted by frequent high-intensity aircraft overflights display higher blood pressures than a sample of children of comparable age, race, and social class living in areas not heavily impacted by aircraft noise. In the study, there was no evidence for habituation; however, it was not possible to definitively rule out the possibility that children physiologically adapt to noise because children with higher blood pressures tended to move out of the noisy areas before longitudinal blood pressure follow-up was conducted.

Results of this study are also consistent with several community surveys (e.g., Knipschild, 1977; Neus, von Eiff, Rüddel, & Schulte, 1983) and with studies from the industrial noise literature (e.g., Welch, 1979) suggesting that chronic exposure to high-intensity noise can have affects on the cardiovascular system. Several cautions are in order in drawing policy-oriented interpretations from these data. First, the noise levels of the impacted schools in this study were extremely high, and we cannot say whether blood pressure differences would have been evident in neighborhoods with lower noise levels. Secondly, the blood pressure levels obtained in the LANP did not exceed normative levels for children of similar age reported in other studies. Comparisons of absolute blood pressure levels across studies are difficult because of differences in samples measurement technique and conditions. In addition, relatively little is known about the developmental course of stress-related blood pressure elevations across the life span from childhood to adulthood, although blood pressure tracking across time in children is widely reported in the epidemiologic literature. For these reasons, the long-term health consequences of any noise-induced blood pressure elevations in children are uncertain.

Regarding the noise-abatement intervention in the schools, we were unable to detect any influences on blood pressure over a 1-year period. It may still well be that because children reside in noisy areas, a reprieve from aircraft noise for several hours during the school day may not have been enough. This interpretation is supported by data we presented indicating that in the initial cross-sectional testing phase, schoolchildren from quieter homes still had blood pressures comparable to other noise-school children; thus, a partial reprieve from noise did not appear to make a difference in blood pressure. It may also be the case that noise-quiet differences in blood pressure would lessen over a longer period than 1 year in sound-attenuated classrooms.

Although the LANP data reported in this chapter have examined the effects of physical intensity, we did find some evidence that after controlling for physical intensity of noise, the child's ratings of noise annoyance seem to be correlated with several dependent measures in the study. For example, diastolic blood pressure is relatively higher among children who rate classroom noise as more bothersome. In addition, noise levels at home and school appear to have an interactive effect, with school noise abatement making less of a difference on blood pressures of children from the noisier homes in the airport community. These results will be discussed more fully in Chapter 6, which discusses home–school interactions in noise–stress levels and the influence of individual perceptions on responses to the aircraft noise stressor.

In retrospect, it would have been informative had the LANP made reliable determinations of which children came from families with a history of high blood pressure in first-order relatives (i.e., parents, siblings). About the time that the study was completed, several reports appeared (e.g., Falkner *et al.*, 1979; Obrist, 1981) that indicated that normotensive offspring of hypertensive parents show elevated response to stressors that require active coping. Because family history of hypertension is a risk factor (albeit imperfect) for high blood pressure, it would have been informative to determine if such children were more likely to show noise-associated elevation in blood pressure.

Conclusion

In this chapter, the evidence for effects of environmental stressors on health has been reviewed, and longitudinal data presented suggesting that chronic exposure to aircraft noise may elevate blood pressure in children. It remains for further research to determine the long-term health effects—if any—of such noise exposure as well as whether certain groups (e.g., children, individuals with family history of hypertension) show greater susceptibility to stress-induced health effects. Because noise levels in areas surrounding the airport were rather high, the LANP does not allow us to determine the lower levels or duration of noise exposure necessary to produce cardiovascular effects. Nevertheless, these data lend further weight to a body of data indicating that environmental stressors can produce physiological changes that may be health relevant.

Few studies in the environmental stress literature have examined the interplay between multiple environmental stressors. This interplay is potentially important because such conditions of high noise, population density, and air pollution often co-occur in naturalistic settings (see Chapter 6). In addition, future studies need to examine the timing of stress measures; some health-relevant stress measures may be associated with short-term physiological and behavioral problems that habituate or improve, whereas others may only be evident after longer term exposure. Our longitudinal analyses of cardiovascular adaptation presumed that a test–retest interval of 1 year offers an adequate test of such hypotheses relating to habituation versus cumulation of negative effects. Unfortunately, these analyses were clouded in our sample by the issue of subject attrition. However, there currently exists in the literature very little conceptual or theoretic basis for determining such parameters as stressor thresholds for effects to occur or the timing of appropriate test–retest measurements.

Environmental Stress and Cognitive Performance

Several general characteristics of research on the effects of stress on cognitive performance are noteworthy. There has been little theoretical development in the area of stress and cognition. We still have no theory of stress and cognition that can adequately account for or predict the circumstances in which stress will impact human task performance. One reason for this shortcoming is that task variables have not been adequately appreciated by researchers studying the effects of stress on cognition. Whether a particular stressor will affect human task performance is strongly influenced by characteristics of the physical stressor, by individual motivation and expectancy, and by task parameters, such as load demands or complexity.

Experimental Studies of Stress and Human Performance

This section of the chapter reviews experimental research on the effects of environmental stressors on cognitive performance. Most of the cognitive research on stress has employed noise as the stressor stimulus.

There are several reviews of environmental stressors on human performance, including articles by Broadbent (1971), Horvath (1959), Poulton (1970), and Wilkinson (1969). Other treatments have been provided by Broadbent (1981, 1983), Loeb (1980), and Hockey (1979). Much of the organizational framework for this section of the chapter is drawn from Hockey's (1979) excellent article. Because these previous reviews on stress and performance already provide detailed treatments of stress and cognition, major findings are presented here in summary form.

Attention

Attention refers to the multiple processes related to the selection of information from the environment. Many early studies of stress on performance were conducted within the ergonomics tradition (human factors engineering) and focused on task efficiency related to attentional demands. Two tasks in particular were used to monitor task efficiency under stress—vigilance tasks and serial reaction time tasks.

Under vigilance conditions, an observer is required to sustain visual attention continuously for periods of 30 minutes or more. A typical vigilance task is similar to detecting a radar blip on an otherwise monotonous background. Serial reaction time tasks require continual, rapid responding to signal cues. A common serial reaction time test consists of a pentagonal array of lights. As each light is illuminated, an observer taps a metal plate adjacent to the light. As soon as the plate is touched, the light is extinguished, and another light comes on.

Noise causes brief periods of inefficiency in this task. When overall accuracy is measured, noise effects are minimal or absent. Momentary deterioration can be compensated for by subsequent spurts of more efficient task performance. One must examine microaspects of task performance to note effects. For example, Woodhead (1964) found very localized effects of intermittent noise on a vigilance task. Errors occurred in noise conditions but only within 30 seconds of the onset of noise bursts. Fisher (1972) reported similar errors in a serial reaction time test, only within shorter time intervals from noise bursts. She also demonstrated that momentary lapses in task efficiency occurred in response to noise onset but not to noise offset. Fisher's research as well as others (Wilkinson, 1969) indicates that intermittent noise functions as a distractor or startle stimulus on attention tasks. Gaps or slightly slower responses occur immediately after noise bursts.

In the vigilance studies, errors that occur are likely to be faster responses that are false alarms (observer reports signal when no signal is present). Vigilance studies, however, mainly use continuous noise, so startlelike responses or distraction is less likely. Errors on vigilance tasks during noise occur because subjects change their response criteria, behaving less conservatively. Subjects in loud continuous noise make more false-positive decisions during vigilance tasks and give fewer medium-level confidence ratings of the certainty of their decisions. Subjects become more certain of their responses under noise (Broadbent & Gregory, 1965; Hockey, 1973; Jones, Smith, & Broadbent, 1979).

Another shift in attention allocation happens under noise when subjects are required to observe multiple signals. In one early study, subjects performed a four-choice reaction time task under two levels of complexity (Boggs & Simon, 1968). In the simple condition, one-half of the subjects pressed one of four switches located directly below one of four target lights. In the complex condition, the four switches did not directly correspond to the position of the lights on the display panel (e.g., leftmost light, switch under second light from right). Subjects were also given a second task—detection of an odd/even sequence of digits to perform simultaneously. This paradigm is called a *dual task*. Loud intermittent noise was presented with the restriction that bursts did not overlap with any secondary task digits. Noise had no significant effects on reaction time but significantly impaired the secondary task. Complexity level also affected the secondary task and interacted with noise levels. Most errors on the secondary task occurred under noise during the complex condition. Finkleman and Glass (1970) also found that noise caused no deficits on a primary task (manual tracking of a visual signal) but negatively affected a secondary task (delayed digit recall). This secondary task required subjects to repeat a previously announced digit upon presentation of the next digit in sequence.

Other work by Hockey (1970a,b) has carefully examined what happens in dual tasks under noise conditions. His data lend strong support to an attention reallocation explanation. Errors only occur on the task that subjects are told is secondary. Thus, when individuals are told to concentrate on tracking a moving target while also (but less importantly) responding as quickly as possible to lights arranged in a half circle in front of the person, only reactions to the lights suffer under noise. The pattern of errors is reversed on the same apparatus when instructions are altered on task priority. Similar results have been found under crowded laboratory conditions (Evans, 1979). The reader should note that there has been a lively debate on the stability of Hockey's dual task findings (see Forster, 1978, Forster & Grierson, 1978, Hartley, 1981a, 1981b, and Hockey, 1978).

Summary. Noise and other stressors have some negative effects on tasks that require rapid detection, sustained attention, or attention to multiple sources of input. Generally, three types of effects have been noted. Table 12 presents a summary of the empirical studies on attention and noise discussed in this section. Under serial reaction time tasks, responses in noise have momentary lapses that appear to be caused by the distraction effect of intermittent noise bursts. Sustained vigilance tasks, where continuous monitoring of relatively infrequent signals is required, are disrupted by noise. Subjects make more rapid, less cautious responses and have more false alarms. When signal rates are low, fewer errors are detected because fewer false alarms occur. Finally, when attention must be shared among several target items, errors in performance during noise occur in the less important, subsidiary cues. These errors are not perceptual but reflect allocation of attention to more important, primary cues at the expense of attention to secondary cues in the task.

Memory

Two basic kinds of memory deficits have been found during stressful conditions. These are explained in more detail than the attentional data were because the memory research is less well known and generally not included in previous reviews of stress and performance. The first memory deficiency that has been noted during stress is in incidental memory. The second deficit is in the processing of more complex information—such as meaning—that requires more time and use of larger amounts of stored material.

Incidental memory tasks measure recall or recognition of information that observers are not explicitly asked to focus on during the learning phase. Thus, incidental memory deficits in stress may occur because of attentional focusing during the presence of the stressor.

Hockey and Hamilton (1970) measured immediate recall of eight-word sequences that were presented on slides. Noise during the learning phase slightly facilitated recall accuracy. Subjects were then asked to recall in what corner of the

Table 12. Effects of Noise on Attention: Tasks That Require the Selection of Information from the Environment

Study	Description of noise	Performance measure	Basic results
Boggs & Simon (1968)	92 dBSPL, RI	Dual task with varying complexity	No effect on primary, deficit on secondary especially under more complex conditions in noise
Broadbent & Gregory (1965)	100 dB, CT	Vigilance detection of slight increment in light brightness, high & low signal rate	Noise causes drop in number of response with intermediate confidence rating at high signal rate, more false alarms at high rate
Finkleman & Glass (1970)	80 dB, I, RI	Dual task	No effect on primary, deficit on secondary with RI
Fisher (1972)	80 dB, RI	5-choice serial response task	Momentary lapse in performance immediately after onset
Hockey (1970a, 1970b)	100 dBA, CT	Dual task	No effect on primary, deficit on secondary task—noise effects reversible by manipulating task importance
Hockey (1973)	100 dBA, CT	Sampling of visual display to detect fault	Increased tendency to sample high-probability source during noise
Jones, Smith, & Broadbent (1979)	80, 85dB(C),CT	Bakan vigilance task (detection of specific digital sequences, e.g., 3 odd digits)	Noise produces impairments
Woodhead (1964)	110 dB, RI	Detection of variable target numbers that change during experiment	No overall effects but errors within 30 sec from noise bursts; subjects fail to detect that target number has been changed

Note. See text for more details. I = intermittent noise bursts; RI = random intermittent noise bursts; and CT = continuous noise.

slide the word sequence had appeared (incidental information). Noise significantly degraded memory for this incidental information. Others have replicated this noise effect (Hockey, 1979) and found similar effects under crowding (Evans, 1979; Saegert, Mackintosh, & West, 1975).

Mathews and Canon (1975) observed that the willingness of pedestrians to assist a stranger (confederate) who dropped several books to the ground was affected by noise levels. The "accident" was staged adjacent to a lawnmower that was either running (noise condition) or was turned off (ambient condition). Under noise conditions, fewer persons assisted the stranger. Of particular interest to a cognitive explanation for this reduced helping effect, the researchers crossed the noise condition with a second variable—presence of an arm cast on the confederate. As expected, the arm cast caused significantly more helping overall. However, this variable interacted with noise in a very interesting manner. Under ambient conditions, there was substantially more helping when the confederate wore a cast. Under noisy conditions, there was no effect of the cast on helping. The incidental cue, "arm cast," was apparently not salient to would-be helpers under noise conditions.

In order to explore the links among noise, attention to incidental cues, and altruism more fully, Cohen and Lezak (1977) devised a memory experiment. In the first phase of the experiment, subjects were shown slides under noisy or quiet conditions. Two types of slides were presented. One type of slide consisted of nonsense syllables. The participants were asked to remember them as best as they could. At the same time, a second group of slides was presented alongside each nonsense slide. The second slide depicted a social scene. The social scenes were of two types. In one type, a central person who was not in any particular distress (e.g., man riding a bicycle) was depicted. In the second type, the same situation was illustrated except the central person was in distress (e.g., man falling off a bicycle). Each subject saw only one version (distress or no distress) of any given social situation. Participants were instructed to pay attention to the nonsense syllables and not worry about the social scenes.

Recognition memory for the social slides was measured by showing subjects either the same slide they had viewed or the other version of that slide (distress or no distress). Consistent with Mathew and Canon's interpretation of their noise and altruism data, Cohen and Lezak found that noise significantly degraded recognition memory for subtle cues of social distress. Noise had no effects, however, on recall of the nonsense syllables.

There is also some fragmentary evidence for a second kind of memory deficit. When individuals learn materials under stress, surface, physical details such as the sequential order of items may be facilitated, but some impairments in comprehension of meaning may occur. Hamilton, Hockey, and Quinn (1972), for example, found that when the order of paired associate items was held constant over learning trials, subjects who learned the pairs in noise actually did better than those who learned them in quiet. When the order of items was changed over each trial, the noise subjects did slightly worse. Furthermore, the probability of a particular word being correclty sequenced when subjects are instructed to arrange a group of words in the order they originally heard them is facilitated by noise (Broadbent, 1981). Memory for order and not for the items themselves may be facilitated by noise and possibly other stressors as well (cf. Dornic, 1975). Nevertheless, these order effects can be manipulated by instructions. For example, when subjects are instructed to be prepared to locate where items occur in the stimulus field, noise does not facilitate order memory (Broadbent, 1981).

When people are asked to recall items under normal learning conditions, free recall will be clustered in categories. For example, if given the words *California, Texas, England, Kansas, Brazil,* and *Chile* to learn, most persons on free recall will cluster items by state and by country (Neisser, 1967). If noise interferes with processing of meaningful types of information, then clustering of free recall material should be disrupted by noise. Disruption of semantic clustering with noise has been found in several studies (Daee & Wilding, 1977; Schwartz, 1975; Smith, 1980). Broadbent (1981) notes that close analysis of the clustering effects indicates that under noise people initially recall items in clusters, and then after exhausting categories they recall a few isolated words. This may show that people quickly move from one category to the next before they have completely finished when organizing information in noise.

More direct evidence for the differential effects of noise on the kinds of information remembered comes from a study by Hockey (1979). Subjects learned a fictional prose passage and then were tested for recall of the names of persons in the passage as well as comprehension of the story's major themes. Noise slightly facilitated verbatim name recall but significantly damaged comprehension of the story.

Better memory for the physical characteristics of stimulus materials during noise coupled with poorer comprehension of complex, meaningful relationships has led to the hypothesis that stressors affect encoding strategies during the learning of material. Specifically, noise may produce shallow encoding of information at the expense of deeper processing. Depth of information processing means the type of organizational encoding strategy employed by a learner (Craik & Lockhart, 1972). Shallow encoding, for example, includes concentration on physical characteristics of materials such as their color, order, type case, and the like, whereas deep processing focuses on the semantic interrelationships among materials.

There is some counterevidence to the hypothesis that noise disrupts deeper, more complex memory as opposed to more surface memory for physical contents. Smith and Broadbent (1981) were unable to confirm this hypothesis in two tasks that should be sensitive to differential interference by noise as a function of processing levels required. In the first task, subjects were shown a list of words for later recall. Depth of processing was explicitly manipulated by classification queries after each word. Shallow processing was encouraged by a query on whether the word rhymed with others on the list. Deeper processing was encouraged by asking whether the word was pleasant, neutral, or unpleasant (one-third of the words were in fact in each category). This manipulation, as expected (Craik & Lockhart, 1972), significantly improved recall over the shallower classification queries. Continuous noise levels during presentation, however, did not interact as expected with depth of processing queries.

Letters that are physically similar (*AA*) can be matched faster than letters that are semantically similar (*Aa*) (Posner, 1969). Smith and Broadbent reasoned that if noise disrupts deeper processing more than shallower processing, then noise should interfere more with name matches than with physical matches. Instead, noise level had no main or interactive effects with type of letter match on reaction time or accuracy. Both of these experiments by Smith and Broadbent raise doubts about a strict depth-of-processing explanation of noise effects on memory. More recent work suggests that noise may interfere with inner speech and thus affect semantic memory under certain circumstances. Strings of words or letters that sound the same or the use of acoustic interference during learning (e.g., repeating the same word aloud), interact with noise levels to affect learning (Wilding & Mohindra, 1983).

There is some evidence suggesting that subjects work faster on speeded processing tasks during noise periods. Hamilton, Hockey, and Rejman (1977) used a running memory paradigm in which the subject is given a long series of items and then asked to recall the last few items in the sequence. The subject does not know when the experimenter will interrupt the sequence and ask for recall. When the list is presented under noise, items most recently present are recalled slightly better, but items farther back in position from the end item are recalled significantly worse than when the list is presented under quiet conditions.

Increase in information-processing speed at the expense of memory storage capabilities during stress may also be reflected in tasks where the individual is required to perform transformations of stimulus materials within short time periods. Hamilton *et al.* (1977) devised a mental transformation task with variable storage requirements to investigate the effects of noise on mental storage capacity. Each subject was presented an input cue—one to four letters in the alphabet—and was required to state as rapidly as possible the next nth letter sequence in position from the cue set. Thus, if set size was 3 and $n = 4$, subjects might be given *FBR* and the proper response would be *JFV*. Set size 4 and $n = 1$ might be DTLY with the correct response, *EUMX*.

The authors found that noise produced deterioration in performance as a function of set size but not of sequence variation (n). When three- or four-letter sets were used, noise produced deterioration in performance. There was also no interaction of sequence and set size on performance. The transformation results are particularly interesting because they demonstrate that task difficulty or complexity *per se* is not the critical variable for determining when noise will negatively impact task performance. In this task, the amount of memory storage required (set size) and not the amount of operation (n) required for transforming the set was sensitive to noise.

Summary. There are two principle memory deficits produced by stress. See Table 13 for a summary of the empirical studies on memory and noise discussed in this section. Memory for items incidental to the major material to be recalled is worse under noise than in quiet conditions. This deficit may be caused by the attentional focusing effects discussed in the previous subsection on attention. Noise also facilitates the fast processing of information in working memory. There are two apparent consequences of the faster throughput of information in working memory. First, certain relatively superficial, concrete details of materials seem to be encoded more accurately. During noise, people acquire better memory for the verbatim

Table 13. Effects of Noise on Memory[a]

Study	Description of noise	Performance measure	Basic results
		Incidental Memory	
Cohen & Lezak (1977)	95 dBA, RI	Verbal recall and incidental recognition of slides	Decrement in incidental recognition
Hockey & Hamilton (1970)	80 dB, CT	Verbal recall and incidental recall of positions of stimuli on screen (which corner)	Slight facilitation of verbal recall, decrement in incidental recall by noise
Matthews & Canon (1975)	87 dBC, CT	Altruism measure of helping person with and without arm cast	Main effects and interaction of noise and arm cast—under noise arm cast did not increase helping
		Complexity and Memory	
Daee & Wilding (1977)	85 dB, CT	Free recall of randomly presented items of different categories	No effect on recall, decline in clustering of related items in noise
Hamilton, Hockey, & Quinn (1972)	85 dBC, CT	Paired associate learning with fixed, changed order over time	Noise facilitated learning when in fixed order but slight decrement when not in fixed order
Hamilton, Hockey, & Rejman (1977)	80 dB, CT	(i) Running memory, unannounced recall of last few items from long list	Noise improved memory for last few items but reduces span—items beyond 4th position from end recalled worse under noise
		(ii) Mental transformation task with varying set (1–4 letters) and varying calculation (1–4th next letters of alphabet)	Noise reduced memory-holding capacity but did not interact with complexity of calculation in the mental transformation
Hockey (1979)	85 dBA, CT	Item recall and comprehension of story	Noise facilitated verbatim item recall but reduced story comprehension
Schwartz (1975)	Not available	Recall of normal sentences and anomolous word strings	Noise disrupted recall of sentences more than anomolous word strings
Smith (1980)	85 dBC, CT	Recall of categorized lists	Disruption of category clustering under noise
Smith & Broadbent (1981)	85 dBC, CT	(i) Word recall with depth of processing manipulated by instructions	No effect of noise
		(ii) Reaction time to make physical matches (AA) and name matches (Aa)	No effect of noise

Note. CT = continuous noise; RI = random, intermittent noise.
[a]Two general types of memory paradigms have been explored: (a) memory for incidental material; (b) memory for information of varying complexity, including meaning, literal content, and spatial position.

contents of prose and the spatial order of lists of materials. Second, some evidence suggests that comprehension of meaningful complex interrelationships among information is disrupted by noise. This may occur because of reduced holding capacity in working memory under noise created by the faster throughput of information. Contextual and semantic or thematic structures may not be held in memory during the faster information processing that happens during noise. These comments on the possible interrelationships between faster processing of information, better encoding of physical details, and poorer comprehension of complex meaning are speculative and in need of further research. It is clear that noise improves learning-of-order information, speeds up processing in working memory, and, under some circumstances, disrupts comprehension of more difficult, complex information such as the meaning of a story.

Aftereffects

One of the most consistent negative effects of environmental stressors on task performance occurs in aftereffects paradigms (Cohen, 1980). Typically, in an aftereffects paradigm, the subject is exposed to a stressor condition while working on some task. After cessation of the stressor, the individual is immediately removed to another place and asked to perform a different task without the stressor present. The most common aftereffect tasks employed are measures of frustration tolerance and proofreading. As noted before, some stressor aftereffects studies have also measured poststressor altruism. The subject's task in the frustration tolerance test is to solve a series of puzzles. The puzzle requires tracing a diagram without lifting the pencil off the paper and without crossing any line twice. Some of the puzzles are impossible to complete. Frustration tolerance is measured by the number of impossible puzzles attempted and/or the amount of time individuals persist on the impossible puzzles. Proofreading tests typically measure accuracy and speed.

Many studies have shown negative effects of noise on poststimulation tasks. For example, in a series of studies, Glass and Singer (1972) found that 25-minute exposure to loud, continuous, and uncontrollable noise caused decrements in frustration tolerance, proofreading, and the Stroop test. Several researchers have replicated negative aftereffects of uncontrollable, continuous noise, particularly on frustration tolerance (Cohen, 1980). Intermittent noise that is unpredictable also has similar negative aftereffects, although the pattern of results is less consistent across studies (Cohen, 1980).

Research on crowding has also uncovered negative aftereffects (Cohen, 1980; Evans, 1978a). Individuals who have previously been in crowded—as opposed to uncrowded—laboratory conditions perform more poorly on the tolerance-for-frustration test (Evans, 1979; Sherrod, 1974).

Summary. Both crowding and noise can produce negative aftereffects in task performance (see Cohen, 1980, for an extensive review of this work). These effects are stronger and more likely to occur if the stressor is not under the subject's control. Aftereffects have been found most typically on the frustration tolerance task and in altruistic behavior. Less frequently, proofreading and the Stroop test have been affected by previous exposure to environmental stressors.

Field Studies of Noise and Performance

The most frequent noise sources found in community studies are from transportation-related activities, particularly airports, train tracks, and highways. There is also literature on adults' occupational exposure to noise (e.g., industrial workers). The conclusions from this body of research are largely equivocal because of poor research methodology (see Broadbent, 1979, and Cohen & Weinstein, 1981, for critical reviews of this literature). Field research on children's cognitive performance under noisy conditions is more varied methodologically, with some researchers taking steps to control for confounding factors in field studies of noise. In addition to the obvious problems of correlations among noise levels with social class, parental education, and the like, problems of subject selection and subject attrition can be serious in field studies (see Chapter 2 for more discussion of some of the methodological issues in field studies of environmental stressors).

This section of the chapter is organized into two subsections: (a) adaptations to chronic noise and (b) school achievement. Whereas most studies of school achievement assess test scores in the noisy environment, several of the studies on problem solving and adaptation to chronic noise test children outside the noisy conditions (e.g., in a soundproof trailer). The distinction between testing children in noise as compared to out of noisy surroundings is important. The former research strategy measures children's immediate reactions to acute noise as well as residual effects that may be present because of their continual environmental exposure to the noise source. The latter strategy of testing children outside of the immediate noise condition attempts to capture the more chronic adaptations noise-exposed children have made to the environmental conditions in which they live or attend school.

Adaptation to Chronic Noise

Several scholars have proposed that when children are reared in noisy settings, eventually they become inattentive to acoustic cues (Cohen, 1980; Deutsch, 1964). Although tuning out the disturbing, distracting effects of noise may be useful, overgeneralization may occur such that the child learns to tune out acoustic information indiscriminately. For example, the child may learn not to distinguish adequately between speech-relevant and irrelevant sounds. This lack of sensitivity to acoustic cues may have negative repercussions for the acquisition of verbal skills, such as reading. The ability to recognize relevant sounds and their symbolic referents, which is presumably a prerequisite for reading mastery, may be disrupted by learning to tune out ambient noise.

A few studies have attempted to examine this tuning-out hypothesis and its implications. Cohen, Glass, and Singer (1973) tested the auditory discrimination abilities of second- to fifth-graders with normal hearing levels under quiet testing conditions. These children lived on various floors of an apartment complex constructed over a freeway. Floor level directly corresponded to noise levels emanating from the highway below. Auditory discrimination was measured with the Wepman test. In this test, observers hear pairs of identical words (e.g., goat–goat) or pairs of similar sounding words (e.g., house–mouse). The subject's task is to indicate whether each pair of words is the same word or two different words.

After controlling for socioeconomic status and parental education levels, floor level significantly correlated with auditory discrimination ability. The greater the noise exposure, the worse the performance on the Wepman. Furthermore, the longer the children had lived in the apartment complex, the stronger the link between floor level and Wepman performance. Moch-Sibony (1985) found similar effects for children attending a noisy school near the Paris airport. Children in a sound-attenuated school had better auditory discrimination scores than children matched on social class who attended a nearby unattenuated school.

Cohen *et al.* (1973) also found that auditory discrimination ability was linked to reading achievement. Children on higher floors of the apartment complex had better reading scores than children residing closer to the highway. Moch-Sibony (1985), however, did not find any differences in reading scores among children in the two schools near the Paris airport. Different measures of reading ability were used in the two studies, and all of Cohen's subjects attended the same school. Some variance in reading achievement is undoubtedly related to individual school programs independent of ambient noise levels.

Several studies have attempted to directly examine tuning out to noise by measuring individuals' responses to distracting sounds. Karsdorf and Klappach (1968) reported that children from schools near busier streets had greater concentration abilities than children attending schools near quiet untraveled streets. Mc-Croskey and Devens (1977) induced a 4-dB increase in the ambient noise levels of several elementary-school classrooms and observed performance under the noise conditions. Children tested in the noisier classrooms had poorer auditory discrimination ability than those tested in classrooms where no additional noise increments were made. Ward and Suedfeld (1973) observed classrooms at a college campus during periods of high ambient noise and during normal ambient conditions. Noise was induced by broadcasting traffic sounds over loudspeakers mounted proximate to the classrooms. During noise periods, there was less classroom participation and reduced attention to classroom activities by students. Students also complained about the noise, had trouble hearing, and felt more tense.

If children chronically exposed to noise learn to tune out acoustic cues, then they should be less susceptible to distraction by noise while performing a task. Heft (1979) found that kindergartners from homes described by their parents as noisy were less affected by an auditory distractor on a visual search task. Each child performed the task under quiet and noisy conditions. In the noisy condition, a male voice read a list of words. The visual search task required the child to select one of several familiar figures from a set to match a standard design. Response latency increased significantly more under distraction for the children from quiet homes than it did for children from noisy homes. One important limitation of this study is that noise levels were not actually measured in the children's homes.

There may be some connection between the tuning-out strategy to chronic noise conditions and the reallocation of attention that occurs under acute exposures to noise. As described earlier in this chapter, many experimental studies of noise find that individuals seem to reallocate attention to the more central, important cues in tasks when they are performed under noise conditions. Perhaps tuning out is related to "tuning in." One way to ignore or tune out distractors might be to focus attention on a specific, narrow aspect of the task at hand.

Heft also used an incidental learning task to determine if chronic noise exposure would also cause decrements in incidental memory as has been found in experimental studies with noise (Cohen & Lezak, 1977; Hockey & Hamilton, 1970). In Heft's study, the child's task was to search for a familiar object from a matrix of cards. After several search trials, the matrix was removed, and the child asked to recognize whether a particular card had been in the matrix. The entire procedure was conducted under quiet conditions. Children from noisy homes had poorer incidental memory than the children from quieter homes.

Summary. A few investigators have examined the effects of chronic noise exposure on cognitive performance (see Table 14 for a summary of the empirical studies on chronic noise exposure and the tuning-out, adaptive-cost hypothesis). Researchers have noted that children living in noisy areas seem to develop a tuning-out strategy to deal with high chronic noise conditions. One possible negative consequence of learning to tune out auditory stimulation is the development of auditory discrimination difficulties. Children living or attending schools in noisy settings may be less able to discriminate among similar speech-relevant sounds. One study has also indicated that children from noisy homes are better at ignoring auditory distractors and may have incidental memory problems similar to those found in laboratory studies of noise and memory.

School Achievement

One of the earliest investigations of individual development and noise is a study that examined 102 infants, ranging from 7 to 22 months of age, on several measures of psychological development (Wachs, Uzgiris, & Hunt, 1971). Significant decrements in psychological development were related to ratings of neighborhood noise, sound levels in the home, and whether the television was on most of the time. Unfortunately, no actual sound levels were recorded in the home or neighborhood. Similar effects have been found for slightly older infants, although these effects may be specific to boys (Wachs, 1978, 1979). The Wachs data are noteworthy because they highlight a potential relationship between home noise and intellectual development in very young children.

High density in the home has also been linked with intellectual development in

Table 14. Field Studies of Noise: Selective Inattention and Chronic Noise Exposure

Study	Description of noise	Performance measure	Basic results
Cohen, Glass, & Singer (1973)	55–66 dBA, CT (corresponding floor level in apartments above freeway)	Wepman auditory discrimination test	Significant positive correlation of noise level with deterioration in Wepman; larger correlation with longer residence
Heft (1979)	Rating scales of home noise level by parents of subjects	(i) Auditory distractor on a visual search task	Children from quieter homes more negatively affected by auditory distractor
		(ii) Incidental learning task	Children from noisier homes have poorer incidental memory
Karsdorf & Klappach (1968)	70 PHON with 84 phon peak, traffic noise	Concentration capacity test	Greater errors and more time to complete concentration performance test under noise
McCroskey & Devans (1977)	4 dBA, CT increase in classroom	Wepman	Poorer auditory discrimination
Moch-Sibony (1985)	22[a], 54[b] airport noise	Massiot–Phillips auditory discrimination test	Poorer auditory discrimination in unattenuated school
Ward & Suedfeld (1973)	63–70 dBA traffic noise	Self-reports and observations in classrooms	Greater discomfort and less classroom participation and attention to activities during noise

Note. CT = continuous noise.
[a]Isolation dBA in unattenuated school (41 in attenuated school).
[b]Speech interference level in unattenuated school (28.5 in attenuated school).

young children. Several studies have noted significant relationships between family size and IQ. High residential density after controlling for family size has also been associated with poor achievement skills (Saegert, 1981). Higher density in the home also is correlated to less maternal contact and child supervision and greater juvenile delinquency and behavior problems in school (Evans, 1978b; Saegert, 1981). These studies have included controls for social class. There is also some evidence that crowding in the home causes restricted social interaction, greater withdrawal, and deficits in motor development, particularly among young boys (Evans, 1978b).

There have been a host of studies on intellectual achievement in children attending school near airports, train tracks, or highways. A common observation in such schools is that teachers and students are bothered by the noise and that teaching adjustments are made in response to noise. For example, in schools near airports and train tracks, respectively, teachers have been observed to momentarily pause whenever a plane or train passed by the school (Bronzaft & McCarthy, 1975; Crook

& Langdon, 1974). This observation raises an interesting question that has not been explored in any depth by noise researchers to date. To what extent are the effects of noise on school achievement mediated by teacher attitudes and responses to the noise conditions in which they teach? For example, studies of open classrooms have found that teachers restrict certain activities because of concerns about disturbing other classes (Ahrentzen et al., 1982). Adults in noisy homes might engage less frequently in conservation, reading aloud, or correcting children's speech (Bronfenbrenner, 1977).

Six studies have examined the effects of noise sources near schools on standardizing reading scores. Five of these studies have found some deficits. The one exception is the Moch-Sibony (1985) study that found no differences between a sound-attenuated and an unattenuated school proximate to the Paris airport.

Close examination of the five studies noting significant detrimental effects of noise on reading tests reveals several patterns. First, negative effects are stronger in the upper grades. Several studies have found slight or no-noise effects on reading ability below the fifth grade. Bronzaft and McCarthy (1975), Maser (1978) and Lukas, DuPree, and Swing (1981) all found significant detrimental effects of noise that interacted with grade level. This age-related noise effect on reading could be because of longer exposure to noise. Recall that Cohen et al. (1973) found that the longer children had lived in the apartment complex constructed over the highway, the stronger the correlation between apartment noise levels and reading deficits. An alternative explanation of the differential effects of noise by grade is that achievement tests for younger children may be less reliable and thus produce lower statistical power. Finally, it is possible that for some reason older children are more susceptible to noise effects on reading achievement.

There is also some preliminary data indicating that low achievers are more susceptible to the detrimental effects of noise on reading scores. Maser (1978) compared standardized reading and math scores of low- , medium- , and high-achiever high school students in noisy and quiet schools. Students with low aptitude attending noisy schools (on an airport flight path) had the clearest noise-associated deficiencies in math and reading.

Home noise levels may also relate to school achievement. Both Cohen et al. (1973) and Green, Pasternack, and Shore (1980) have found significantly lower reading scores in children from homes located near noise sources. Of additional interest, Lukas et al. (1981) report a significant interactive effect of residential neighborhood noise levels and school noise levels on reading achievements. Children who live in noisy neighborhoods and attend elementary schools near noisy freeways have greater reading deficits than children attending the same schools who live in quieter neighborhoods. The interactive effects of home and school noise levels are also discussed in Chapter 6.

The potential interactive effects of home, neighborhood, and school noise levels have implications for the utility of sound-attenuation projects in schools. At least two studies to date have examined the efficacy of sound attenuation for relieving the harmful effects of noise on school achievement. Moch-Sibony (1985) compared reading scores and auditory discrimination abilities in two nearby elementary schools located on the flight path of the Paris airport. One school was sound attenuated, and the other was not. All testing occurred under quiet conditions, and both schools had children of similar socioeconomic backgrounds. Auditory discrimination was poorer in the unattenuated school, but reading scores did not differ. Bronzaft (1981), however, in a follow-up study to Bronzaft and McCarthy (1975) found successful abatement effects on reading scores. Elevated train tracks near an elementary school in New York were restructured with rubber pads, and acoustic materials were placed in the noisy classroom on the side of the school proximate to the train tracks. These procedures substantially reduced deficits in reading ability that had been previously found in those children attending the noisy classrooms.

There is, however, an important difference between these two studies. Children in the Paris study live and play under loud ambient noise conditions near the airport, but only the school setting was attenuated. In the Bronzaft study, children generally live under normal ambient community conditions and are only exposed to the noise while in class. Thus, noise-attenuation procedures in schools may have variable impact dependent upon the extent of noise exposure outside of the school environment. Also, as noted earlier, the reading scores from the Paris schools study may be less sensitive to noise-induced effects because they came from two different schools, whereas the Bronzaft data were all collected in the same school.

Summary. Several studies have shown that noise from airports, trains, and highways near schools can have detrimental effects on school achievement (see Table 15 for a summary of the empirical studies on school achievement and noise exposure). These effects appear to be more pronounced for children in higher grades (above fourth). This effect may be due to greater length of noise exposure, less reliable achievement tests for younger children, or actual developmental differences. Sketchy evidence also suggests that aptitude levels and home neighborhood noise contours may also interact with school noise levels. Children who are lower achievers and children who live in noisy neighborhoods may be more at risk from the detrimental effects of noise in their schools.

Theories of Stress and Human Performance

There have been two major theoretical constructs—arousal and information load—offered to explain the effects of stress on human performance. Alternative approaches to stress and human task performance are motivation-based theories such as learned helplessness. These models focus on changes in human task performance under stress as a function of the individual's reduced willingness or ability to persist on problems. This perspective is discussed in Chapter 3. We acknowledge that these different perspectives on stress and performance interrelate but are separating them in this book for organizational simplicity and brevity and because the two literatures have remained relatively distinct.

Table 15. Field Studies of Noise: School Achievement

Study	Description of noise	Performance measure	Basic results
Bronzaft (1981)	7 dBA sound attenuation in noisy classroom	Reading achievement	Previous reading deficit (Bronzaft & McCarthy, 1975) eliminated
Bronzaft & McCarthy (1975)	89 dBA, peak from trains	Reading achievement	Noise-produced deficits
Cohen, Glass, & Singer (1973)	55–66 dBA, CT (corresponding to floor level in apartments above freeway)	Reading achievement tests	Significant positive correlation of noise with reading deficits, Wepman score also correlated with reading ability
Green, Pasternack, & Shore (1980)	Schools assigned noise exposure according to noise exposure forecast contour	% students reading below grade level	More children in noisiest schools read below grade level
Lukas, Dupree, & Swing (1981)	Schools proximate to freeways	Reading and math achievement	School noise-related reading deficits, particularly in older children (3rd vs. 6th) from noisy home areas
Maser (1978)	Schools proximate to airport, levels not provided	Reading and math achievement	Noise-related deficits, particularly in low-aptitude children in higher grades (7th & 10th vs. 3rd & 5th)
Moch-Sibony (1985)	22[a]; 54[b] airport noise	Reading achievement test	No noise effect
Wachs (1978)	Trained observer ratings of home and neighborhood noise levels	Stanford–Binet intelligence test	Some correlations between noise ratings and test scores
Wachs (1979)	Trained observer ratings of home and neighborhood noise levels	Piagetian measures of psychological development	Some correlations between sound levels and decrements in psychological development
Wachs, Uzgiris, & Hunt (1971)	Trained observer ratings of home and neighborhood noise levels	Piagetian measures of psychological development	Some correlations between sound levels and decrements in psychological development

Note. CT = continuous noise.
[a]Isolation dBA in unattenuated school (41 in attenuated school).
[b]Speech interference level in unattenuated school (28.5 in attenuated school).

Arousal Models

Arousal, or *activation,* as originally conceived, refers to nonspecific changes in brain activity mediated by the brain stem reticular formation and the diffuse thalamic projection area (Hebb, 1972; Lindsley, 1951; Malmo, 1959). Behavioral manifestation of activation is the general waking state of the organism ranging from low activity levels (sleep, extreme drowsiness) to high activity levels. Thus arousal can be viewed as as a general state that energizes behavior (Berlyne, 1960). Stressors, according to the arousal model, change phasic arousal levels, raising or lowering activity level that in turn influences task performance.

The Yerkes–Dodson model of arousal and performance (Kahneman, 1973) posits an inverted U-shaped function between arousal and task performance. Thus persons who are either above (too much activation) or below (too little activation) some moderate optimal arousal range will suffer performance decrements. Furthermore, optimal performance level is hypothesized as a decreasing, monotonic function of task difficulty. Table 16 outlines some of the basic points of arousal theory.

Broadbent (1971) proposed an arousal model for stress and cognition because of the interaction of various stressors on human performance. Stressors that increase sympathetic activity (e.g., noise) have the opposite effect on certain tasks such as vigilance than do stressors that reduce arousal (e.g., sleep derivation). Furthermore, if these two stressors are combined, they cancel out one another's task decrements. In addition, combinations of two stressors with similar arousal properties have additive effects on task performance degradation. Thus, if one combines noise with task incentives, which also heighten arousal (e.g., posting test scores of individuals by name), vigilance performance is worse than under just noise or task incentive alone.

There are several problems with the arousal model of human performance under stress. The most fundamental criticism of the arousal model is the lack of clear evidence of physiological activation. Psychophysiological research has found little support for unidimensional changes in various indexes of sympathetic activity. The norm is a multiplicity of different responses to various stimuli (Jennings, 1985a; Lacey, 1967).

There are also problems related to the inverted U-shaped function between arousal and task performance. The negative effects of suboptimal arousal (e.g., sleep deprivation) on task performance can readily be explained by fatigue. With insufficient mental alertness, one simply fails to detect critical cues, or one is just too tired to respond quickly enough. Why overarousal should degrade performance is more problematic.

Easterbrook (1959) proposed that high levels of arousal narrow the focus of attention so that only the most central cues in a task array are attended to. This attentional-focusing hypothesis can explain the relationship between task complexity, overarousal levels, and task decrements. Complex tasks have more critical cues necessary for successful task completion than are found in simple tasks. A type of task in which stressor effects have frequently been detected are complex tasks that

Table 16. Basic Points of Arousal Theory

1. Arousal has been defined in both physiological and behavioral terms.
    (a) Physiological—nonspecific facilitation of the cortex by the reticular formation
    (b) Behavioral—behavioral continuum ranging from deep sleep to high excitement
2. Yerkes—Dodson construct

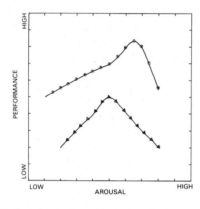

Task complexity has typically been incremented by having an individual attend to two or more signals simultaneously (dual-task paradigm); attend to a high-frequency signal; or attend to a low-probability signal.

3. Stress has been defined in terms of arousal theory by suggesting that either overarousal or underarousal is stressful. Certain environmental stressors may affect arousal level. Noise and crowding, for example, have both been shown to increase arousal. Thus, complex, as opposed to simple, task performance decrements would be expected to occur under crowded or noisy conditions.

4. During overarousal, a narrowing of attention appears to occur where the organism focuses on the most important cues available in the situation. In complex tasks more such cues per unit time are necessary for successful task completion. Thus, when a narrowing of attention occurs in the complex task situation, the organism is much more likely to ignore a cue(s) relevant to successful task completion than in a simple task situation where fewer such important cues are necessary for task success.

involve (a) simultaneous attention to multiple signals (e.g., dual task paradigms); (b) attention to high-frequency signal; (c) attention to a low-probability signal. Thus, as arousal increases and attention narrows, one begins to miss certain cues. Under complex task conditions, some of the cues missed are more likely to be critical for accurate task performance than when a simpler task is used.

There are at least two difficulties with the narrowing-of-attention explanation of overarousal effects on task performance. First, there has been criticism of the nonlinear inverted U-shaped function itself. Logically, it is difficult to critically evaluate this nonlinear relationship between arousal levels and performance. If one does not find deficiencies in task performance during a stressor, the stressor may not be intense enough, the task may not be complex enough (place sufficient demands on information-processing capacity), or both.

Furthermore, some researchers question whether overarousal can ever be detrimental to task performance (Poulton, 1977, 1978, 1979). Heightened arousal counteracts fatigue and improves alertness. Task decrements that do occur under noise are unrelated to the arousal-inducing properties of the stimulus, according to Poulton. What causes task decrements in noise are direct acoustic effects. During intermittent noise, errors can be explained by distraction. Continuous noise causes auditory masking of task cues and/or interferes with auditory processing in echoic memory. Broadbent (1978) disputes this position, however, noting that some task deficits under noise cannot be readily explained by distraction or auditory masking.

Another issue with respect to the attentional-narrowing hypothesis is the meaning of attentional narrowing or filtering. Does overarousal actually reduce the amount of information one can process per unit time, or does it influence individual decisional criterion or response bias in terms of which task cues one should focus on? Broadbent (1971, 1978) argues that the effects of attentional narrowing are decisional, not perceptual. Under conditions of heightened arousal, such as in the presence of noise, subjects are believed to pay more attention to important sources of information. As discussed earlier, the most direct evidence for this is Hockey's work demonstrating that errors occur during noise in those parts of multiple tasks that subjects are told are less important (Hockey, 1970a, 1970b).

The various questions about the arousal model of stress and performance relate to a more profound limitation of the classical arousal construct. The original arousal model of stress and performance says nothing about the role of subjective appraisal of stressors and individual coping responses. Individual responses to stressors are mediated by a host of psychological and situational variables (Appley & Trumball, 1967; Averill, 1973; Lazarus, 1966; Moss, 1973). For example, the attributions people make about the causes of arousal influences behavior.

When physiological arousal changes and there is no immediate explanation for those changes, people will label their somatic state according to their appraisals of the surrounding circumstance. To test this model of arousal labeling, Schacter and Singer (1962) injected subjects with either adrenaline or a placebo (saline). Subjects were informed accurately about the somatic effects of adrenaline (symptoms of sympathetic arousal), misinformed (symptoms of parasympathetic arousal), or not informed about somatic effects. After the injection and the information manipulation, subjects were taken to a waiting room where another "subject" (actually a confederate) acted either angrily or euphorically.

Consistent with their two-factor theory of arousal induction and cognitive labeling, Schacter and Singer found that the type of information about the physiological consequences of the injection was negatively correlated with self-evaluations of affect as well as the subjects' behavioral responses to the confederate. Thus, the strongest feelings of a specific state (anger or euphoria) happened when subjects were misinformed about the effects of the injection.

A two-factor, arousal-attribution model of environmental stress has been applied to research on human spatial behavior. Patterson (1976, 1982) suggested that adjustments in intimacy-relevant behaviors by one individual such as shifts in interpersonal distance change arousal levels in the interacting partner of a dyad. Depending upon the interpersonal context of the dyad, these arousal changes are labeled by the recipient as positive or negative. Positive labeling of an arousal shift leads to reciprocal intimacy behaviors such as moving closer, greater eye contact, or more direct and open bodily postures. Negatively labeled arousal changes cause compensatory reactions such as increased interpersonal distance, less eye contact, or more closed, indirect bodily postures.

The arousal attribution process has also been used to explain the consequences of high density on human behavior. Crowding may occur when heightened arousal produced by high density is labeled negatively. Two specific mechanisms have been suggested as precursors to arousal induction—excessively close interpersonal distance (Evans & Eichelman, 1976; Worchel, 1978) and loss of control (Schmidt & Keating, 1979).

Crowding attributions can be lowered when individuals' attention is directed away from the close presence of surrounding people under high-density conditions. Thus, when people in high-density conditions are given various acoustical or visual distractors they report feeling less crowded (Worchel, 1978). Crowding is also substantially reduced by providing individuals under high-density conditions greater personal control (Sherrod, 1974) (see Evans, 1978a, for further discussion of the arousal model of human crowding).

An alternative formulation of the arousal model suggests that arousal may need to be viewed both in terms of *patterns* of responses as well as *levels* of responses. This idea has been articulated as a physiological model by Naatanen (1972) and Pribram and McGuinness (1975) and linked to cognitive performance by Kahneman (1973) and Hockey (1979). During concentration, that is, thinking, indexes of sympathetic activity increase. Heart rate, skin conductance, and pupil dilation all increase under activities of thought or concentration. Similarly, they all increase with short-term passive exposure to a stressor such as noise as long as no cognitive efforts are required of the subject to maintain task performance. However, when attention or effort to some external stimulation is required, then a different response pattern, directional fractionation, occurs. During a specified anticipatory period, heart rate deaccelerates, whereas skin conductance and pupil dilation are elevated (Jennings, 1985a; Lacey, 1967). Heart rate also deaccelerates immediately following a meaningful event that is unanticipated (Jennings, 1985a). Lacey (1967) suggested that the intention to note and detect external stimuli causes heart rate deacceleration, whereas cognitive efforts to process information further and ignore (filter) immediate surroundings produce cardiac acceleration. Jennings has modified Lacey's theory suggesting that cardiac deacceleration is produced by holding attentional capacity in reserve during input. Cardiac acceleration occurs when further processing of information is required for tasks. Thus, memory storage or the manipulation of information already in memory increases cardiac activity (Jennings, 1985a, 1985b).

Kahneman (1973) and Hockey (1979) suggest that although arousal levels may be useful to describe the continuum between rest and activity, once a person is mentally working on some task, we must consider the individual's attempt to direct his or her cognitive resources and the ensuing influences on performance and physiology. This more complex arousal model proposes that information is selected from the environment more rapidly and perhaps more accurately under high-arousal conditions. In general, a higher level of bodily arousal increases the efficiency of one's attention to external stimuli. High arousal also shifts the balance of normal functioning so that decisional criteria for the allocation of cognitive resources change in response to the interaction of task demands and stressor characteristics. Thus stressors impact both level and pattern of response.

Information Overload Models

The second major model of the effects of stressors on human performance is the information overload model. This view had emphasized the impact of stressors on cognitive capacity and the organism's adaptation to information-processing demands. Milgram's (1970) overload analysis of urban life illustrates this perspective. Large amounts of stimuli in the urban setting cause people to adopt certain coping strategies designed to deflect or eliminate sources of stimulation. As one means of coping with overload, low-priority sources of information are filtered out. The inattention of bystanders to victims in crisis in the city may be explained in terms of this learned-response mode of urbanites to low-priority information. Alternatively, some information is removed prior to reaching the person such as by unlisted phone numbers or using doormen or other devices to block some social inputs.

A more complex overload model has been introduced by Cohen (1978) to explain the effects of environmental stressors on human performance and social behavior. Cohen's model posits that stressors reduce cognitive capacity because they require some allocation of attention to the stressor. As more capacity is allocated to monitor a stressor, less capacity will be available for other information-processing demands, such as task completion. See Table 17 for an overview of this model.

An important feature of this model is that it incorporates subjective appraisal of the stressor and the context in which it occurs. Thus, if a stressor is more threatening, it may be monitored more closely and thus use more capacity. Furthermore, as the uncertainty of a stressor increases (e.g., less subject control over the stress, less predictability of the stressor), greater attention will also have to be allocated for the monitoring of the stressor. Given that stressors exceed cognitive capacity, the organism tries to cope with task demands under these circumstances by allocating available spare capacity to the more central aspects of the task—cues most relevant for successful task completion.

There are several consequences of the hypothesized shrinkage in information-processing capacity with stressor exposure. First, as the duration of task demands increases, greater negative effects of stress on task performance should occur. Horvath (1959), Wilkinson (1969), and Alluisi and Morgan (1982) in their reviews of various sources of environmental stresses conclude that the deleterious effects of

Table 17. Basic Points of Attentional Overload Theory

1. Humans have limited information processing capacity. Information overload occurs at any time that the demand for attention exceeds total available capacity.
2. *Cognitive Fatigue Hypothesis*: Total available capacity is not fixed, but in fact "shrinks" when there are prolonged demands on attention. Thus an individual can attend to fewer inputs after prolonged demands than he can in a rested state. (Cognitive fatigue is a function of *task load* and *task duration*.)
3. The presence of an environmental stressor, because *it requires* an allocation of attention, is likely to create information overload.
   The amount of attention required to monitor an environmental stimulus is an increasing function of the *uncertainty* it arouses concerning its adaptive significance (related to predictability and controllability).
4. The most usual strategy employed to deal with overload is the focusing of available attention on the aspects of the environment most relevant to task performance at the cost of less relevant inputs.

a stressor on task performance increase with prolonged task duration. Furthermore, the negative effects of stressors on tasks of long duration can be largely eliminated by providing participants with intervening rest periods (Mackworth, 1964).

A second implication of the overload model is that cognitive fatigue should increase with increased task load. Typically, task load is incremented by (a) increasing the number of inputs one must attend to; (b) increasing the speed of the input; and (c) reducing signal predictability (Cohen, 1978; Evans, 1978a). All three factors have been shown to reduce task efficiency in humans. Human observers make more errors when monitoring six versus two dials (Poulton, 1958) or in sorting decks of cards according to two versus one criterion (Murdock, 1965). When subjects are asked to respond to two different auditory cues while performing a simultaneous writing task, more errors occur when the digits are presented faster (Kalsbeek, 1965). Finally, considerable evidence shows that when individuals are performing two simultaneous tasks, their performance suffers as the probability of the primary task cue decreases (Bahrick, Noble, & Fitts, 1954; Dimond, 1966).

A final implication of the capacity shrinkage hypothesis is that given sufficient load demands, some residual fatigue effects may occur after the stressor has been terminated, producing aftereffects (Cohen, 1980). These effects are termed *cumulative fatigue effects* in our discussion of coping in Chapter 1. Thus as load on capacity is increased, either by duration of task or varying task load, aftereffects should increase. Cohen and Spacapan (1978), for example, showed that shorter interstimulus intervals, longer task sessions, or greater task load all decreased subjects' performance on a frustration tolerance test administered after cessation of the experimental session. In addition, subjects reported greater mental and physical fatigue with greater task duration or with shorter interstimulus presentation intervals.

On the other hand, at least one researcher has found that negative aftereffects from noise in comparison to a quiet control were not increased by task demands

(Wohlwill, Nasar, De Joy, & Foruzani, 1976). Subjects performed equally poorly after cessation of noise whether or not they had performed a task during the noise.

Comparison of the Two Models

Both models predict greater impact of stress on performance as a function of task complexity. Both models can also explain poststressor effects on performance, although the load model does so more parsimoniously. According to the load model, mental capacity becomes depleted over time with continuous coping with task demands and cognitive fatigue occurs. Reserves are depleted, leaving insufficient capacity to perform subsequent tasks. One prediction that follows from this position is that as the length of time between the cessation of tasks performed under stress and aftereffect measurements increases, some recovery from cognitive fatigue ought to occur. No research has examined this question to date.

The load models can also more readily account for the effects of control and predictability of stressors on aftereffects. Noise that is unpredictable or uncontrollable causes substantially greater negative aftereffects on performance (Cohen, 1980). The load model accounts for this finding in terms of additional capacity demands created by enhanced difficulty of monitoring unpredictable or uncontrollable stressors. Lack of control *per se* may also be a powerful threat stimulus in its own right, thus increasing the need for vigilance. Arousal models can explain the effects of stressor predictability in terms of novelty. Less predictable stimuli are harder to habituate to and impact reticular activation (Sokolov, 1963). Why control should affect arousal levels is less clear.

One problem with the cognitive fatigue model is that the correlation between the degree of physiological adaptation and the amount of negative aftereffects is not very high. Glass and Singer (1972), for example, found little relationship between habituation of phasic skin conductance during noise over time and performance on frustration-tolerance aftereffects tasks. Furthermore, cognitive effort may still be expended on tasks even if adaptation does not occur. Whether coping efforts are successful in maintaining task performance or not may be unrelated to aftereffects. To further complicate matters for arousal explanations of aftereffects, researchers have argued both that tonic arousal falls after attempts to perform tasks when under stress (e.g., Poulton, 1978) and that it rises (Glass & Singer, 1972).

One important phenomenon that load models of stress and performance do not handle as well as the arousal models do are the effects of combining stressors. Some stressors in combination produce *less* performance decrements than either stressor alone (Broadbent, 1971).

Another difficulty with load models of stress is that cognitive capacity is not precisely defined, although some central information-processing capability is implicitly posited. Unfortunately, there is little evidence for a single information-processing capacity. For example, secondary tasks in dual-task paradigms are not interchangeable, nor are they simply additive in their measurement of processing needs. A secondary memory task and a secondary attentional task do not produce equivalent results when combined with the same primary task in a dual-task paradigm (Kerr, 1973).

The arousal and load models make different assumptions about the way humans function. The arousal models posit a largely passive organism, reacting to changes in environmental stimulation. The load models assume an active, information-seeking organism that tries to make sense out of its surroundings. *Meaning* of stimulation rather than *amount* of stimulation is more central to the load models of stress.

An Alternative Perspective on Stress and Cognition

An alternative way to understand stress and human information processing is to view stressors as altering cognitive processing strategies directly. These changes may be adaptive, coping responses to the stressor. This view of stress and performance departs from the arousal and load models in several respects. First, this approach to stress and cognition does not assume that decrements in human performance will occur in response to stressors. Instead, the emphasis is shifted to changes in cognitive functioning. Second, no single or central mechanism is targeted as an explanation for changes in cognition under stress. Human information processing consists of multiple and interactive cognitive activities that are differentially impacted by stressors.

Changes in information-processing activities under stress are functional to the extent they are related to personal attempts to maintain cognitive performance under suboptimal environmental conditions. As discussed earlier in this chapter and in Chapter 1, these efforts to alter cognitive strategies under stress can have physiological and motivational consequences. Furthermore, such efforts can alter habitual information-processing modes with sufficient experience. One may redirect attention during stress toward the most important or salient aspects of the environment. If this strategy persists over time, long-term readjustment of cognitive processing may occur such that information is treated in a different way than it would be dealt with under normal circumstances. Other shifts in cognitive-processing strategy that seem to occur under stress have also been identified. Stressors appear to affect a learning strategy that emphasizes speed of processing at the expense of capacity in working memory. Furthermore, there is some indication that people more accurately encode the position or spatial arrangement of items in memory under stress. See Table 18 for an outline of this alternative model.

Thus, we can speak of task deficits under stress to the extent that a particular task requires particular information-processing strategies. A stressor will influence that task to the extent that the cognitive strategy involved is altered. A stressor will affect cognitive processing if allocation of effort will make a difference. Thus, cognitive processes that are automatic will be unaffected by noise. Because noise seems to cause reallocation of effort to dominant aspects of tasks, other tasks that share some of the same cognitive resources as the primary tasks will decline under noise (Broadbent, 1981, 1983). For example, earlier research has found that seman-

Table 18. An Alternative Perspective on Stress and Cognition

---

1. Stressors cause people to change cognitive strategies.
2. Some of these changes include
   (a) allocation of attention to more dominant aspects of the task;
   (b) faster throughput of information in working memory; and
   (c) better memory for spatial position of information in sequence.
   Both (b) and (c), which may be related, appear to cause poorer comprehension where association of recent input with previously stored materials is required. Working memory works faster but holds less under stress.
3. Task performance will change under stress to the extent that variability in cognitive strategy is possible in the completion of the task. Tasks that only allow one strategy are less likely to be affected by a stressor.
4. Efforts to alter cognitive strategies and maintain optimal performance during stress produces physiological effects that are different than those produced directly by the stressor itself.
5. Persistent habitual efforts to maintain optimal task performance under stress can cause long-term physiological and cognitive effects. Involuntary selective inattention to acoustic cues may occur, for example, in persons who have chronically been exposed to noise.

---

tic clustering of serial lists is disrupted under noise. Recently, people have noted that these effects do not occur if the clustering is very obvious (Broadbent, 1983). Another finding of interest is that noise effects can be changed by instructions to allocate effort to one part of a task versus another (Hockey, 1979) or by instructions to try hard and ignore the noise (Jones, 1984).

Unfortunately, few studies have tested whether subjects put forth more voluntary effort to maintain task performance during stressors. Although there are some preliminary data for physiological changes as a function of task maintenance (cf. Frankenhaeuser & Lundberg, 1977; Lundberg & Frankenhaeuser, 1979; Chapter 1), few investigators have actually asked subjects what they do during the presence of a stressor while performing tasks. Weinstein (1974, 1977) found that subjects are aware that they respond more rapidly during noise while performing a proofreading task. Subjects were unaware, however, of their errors during performance.

This perspective on stress and cognition deemphasizes the quantitative aspects of stressors in terms of amount of stimulation or overload and redirects us to the kinds of information-processing strategies that are invoked by particular tasks. At present, we may tentatively conclude that the following kinds of changes in cognitive strategies occur in the presence of a stressor:

1. a shift in attention toward the most dominant (important) aspects or cues in a task when information must be selected from the external environment;
2. faster processing of information in working memory;
3. reduced holding capacity in working memory, possibly because of the faster throughput of information during stress;
4. better encoding of sequential order of materials in memory if given rehearsal opportunity; and

5. poorer performance after cessation of a stressor on tasks requiring persistence.

After prolonged exposure to stress, some chronic shifts in information processing may occur. Prolonged exposure to noise leads to an attentional strategy characterized by insensitivity to auditory cues. This selective inattention can later be manifested as an auditory discrimination deficit and possibly be related to scholastic skills like reading ability.

The final way in which our view of stress and cognition differs from the arousal and load models is the emphasis here on the relationship between effort, change in cognitive strategy, and physiological effects of stressors. Dramatic or rapid changes in environmental stimuli undoubtedly have some direct, relatively immediate physiological effects. We do not believe that these physiological changes are the primary cause of shifts in information processing. Our approach suggests that effort to change cognitive strategies in order to cope with changed environmental conditions affects physiology.

One important distinction that may operate among the various shifts in cognitive strategies is the amount of *voluntary effort* involved in maintaining cognitive performance. Both overgeneralization and passive coping responses probably do not entail active, attentional efforts, whereas other coping responses may. There are two important implications of the hypothesized distinction between effortful and effortless coping. First, physiological responses and cognitive processes may differ in the presence of a stressor as a function of the amount of voluntary effort summoned. Second, chronic habitual experiences with stressors probably lead to coping responses that do not require voluntary cognitive effort. On the other hand, a stimulus that is highly unpredictable, novel, or difficult to control may demand more voluntary effort for successful allocation of cognitive strategy. The importance of effort in cognitive processing has been discussed by others (James, 1980; Kahneman, 1973; Kaplan & Kaplan, 1982).

Task requirements will influence whether a shift in cognitive strategy impacts performance. In tasks with cues of variable occurrence and/or high-importance, greater sampling of the high-probability and/or high-importance cues may occur under stressful conditions. In tasks where speed or storage is required, speed may be maintained or increased under stress at the expense of memory storage requirements. Tasks that require greater elaboration and use of already encoded material are also prime candidates for reduced performance under stress. Furthermore, certain physical aspects of information such as verbatim encoding or sequential structure may be encoded better under stress. Effort may be linked to some of those changes in cognitive strategies. Greater effort is required for deeper information processing, acquisition of more complex, less accessible information, or wider attentional sampling of both and less important stimulus cues. Thus, effort may be one heuristic to understand choice of cognitive strategy during stress. When under stress, perhaps we choose to process information in the quickest, most simplified, easiest way we can.

Context of a stressor has implications for performance as well. Rotton,

Olszewski, Charleston, and Soler (1978) who replicated the Glass and Singer (1972) noise aftereffect findings found that meaningful speech caused greater negative aftereffects than white noise at the same loudness level. Furthermore, if subjects were told that recall of the text would be required, even greater poststimulus aftereffects occurred.

Summary. It is apparent that the current state of theorizing about stress and cognition is elementary. The two classical models of arousal and information overload do not adequately predict what kinds of task deficits one is likely to see under the presence of a stressor. The alternative perspective suggests a reorientation in theorizing, with a focus on the cognitive strategies that change under stress. Shifts in performance can then be understood by examining the cognitive processes involved in various tasks. Much more work is necessary both to clearly specify which processing strategies change during stress and what kinds of tasks are therefore likely to change. An important task that challenges cognitive psychologists is the development of taxonomies of mental operations utilizing criteria such as processing capacity (cf. Kerr, 1973) and effort (cf. Hasher & Zacks, 1979; Kahneman, 1973). Another area of great interest is the interplay between cognitive processing and physiological changes. Previous models implicitly have posited that physiological changes caused by the stressors were the causes of task decrements. We suggest that the picture is considerably more complex. Efforts to maintain task performance that involve changes in cognitive-processing strategy produce physiological changes. The direct physiological changes produced by stressors may have either no effects on cognition or have interactive effects with the physiological implications of effort expended to maintain task performance. Finally, the role of the meaning of information and the context in which it occurs need to be taken into account by an adequate model of stress and cognition. Given our argument that at least some of the shifts in cognitive strategies during stress are voluntary, then it follows that cognitive appraisals of a situation will be likely to influence cognitive processing during stress.

In this next section of this chapter, we describe data on cognitive performance from the Los Angeles Noise Project. Remember that the data from this project bear only on certain aspects of the issues and questions raised in the various chapters of this book.

Los Angeles Noise Project: Cognitive Processes

Five different measures of cognitive performance were used in our research program. Two of the measures, distraction and signal to noise ratio, were developed to directly test the tuning-out hypothesis discussed earlier in this chapter. We also included a measure of incidental memory because some laboratory research has shown detrimental effects of acute noise exposure on this task. Finally, we included measures of auditory discrimination and school achievement. In this section, we provide methodological details about each measure as well as brief statements of the hypotheses under examination. The next section includes the results and discussion for each measure.

Methods, Results, and Discussion

Distractability

If children chronically exposed to noise become relatively inattentive to acoustical cues (i.e., tuning out), then these children should be less affected by auditory distraction. Thus, distractability was used as a measure of selective inattention in our study. Under both ambient and distracting conditions, children performed a task consisting of crossing out *e*'s in a two-page passage from a sixth-grade reader they were unfamiliar with. Each child was instructed to move from left to right and from top to bottom of the page, as if they were reading. The child was asked to go as fast as he or she could without missing any *e*'s. Each child worked on a short practice paragraph and then on the task for 2 minutes. Two versions consisting of different samples of prose were used.

In the distraction condition, the child worked on one of the versions of the passage while a tape recording of a male voice read a story at a moderate volume. The story was presented over headphones. The story was designed to be comprehensible and interesting to children in Grades 3 to 5. In the no-distraction condition, the alternative form of the task was completed under the same conditions, only no story was presented over the headphones. The distraction and no-distraction tasks were administered on different testing days. Both the order of alternative versions of the task and the experimental conditions were counterbalanced.

The main criterion measure on the task was the percentage of *e*'s found on the distraction task after the scores were adjusted for no-distraction performance. It was expected that the children from noise schools would be less affected by distraction than the children from quiet schools. Because selective inattention is an overgeneralized coping strategy that develops over time, we also predicted that this tuning-out strategy would increase with a longer duration of noise exposure in the school.

Signal-to-Noise Ratio

Another outcome of the selective inattention strategy may be diminished ability to perceive subtle differences in acoustic interferences with speech. In order to test this hypothesis, we developed a signal-to-noise ratio task that required the child to choose a particular auditory channel that he or she could either hear the best or like the best. We expected that children from quiet schools would choose the channel with the highest signal-to-noise ratio, but children from noise schools would not distinguish as well the optimum channel. This task was administered only in the third grade replication study.

Each child was given a box with a switch that could be set in four positions. As the child listened with headphones, he or she was told to make one of two choices. One-half of the children from both the noise and the quiet schools were told to choose the channel that they liked the best. The remaining half of the children were asked to choose the channel where they could hear the story the best. The two conditions were randomly assigned.

We examined both "hear" and "like" best alternatives because we were concerned that children from noise schools might choose a noisier channel because they liked it better because it would be more familiar to them. The hear-best instructions test more precisely the selective inattention strategy hypothesis.

Each box was marked with four switch positions that were randomly paired to four different signal-to-noise ratios. The signal was a male voice reading a story at $\bar{x}$ = 72 dB(A). The story was comprehensible and of interest to third-grade students. The story was different than that used in the distraction task. The four noise background channels were 58, 62, 66, and 70 dBA broad-band continuous white noise. The experimenters were blind to the random switch positions on each box (four different boxes were used.)

Incidental Memory

The incidental memory task was from Heft's (1979) study of home noise. Heft found that children reared in noisier homes (parental self-report) had poorer incidental memory. Acute exposures to noise under laboratory conditions also cause detriments in incidental memory. Unfortunately, procedural errors were made in administering the incidental memory task in our study. Therefore, the results from this task are uninterpretable and will not be reported.

Auditory Discrimination

A primary reason we are concerned about children learning to cope with chronic noise by tuning out is that this strategy may negatively effect auditory discrimination abilities. If auditory discrimination abilities are diminished, reading and math skills may be damaged as well. A child who cannot readily discriminate basic speech sounds faces a difficult task in learning to associate these sounds with their appropriate symbolic referents. Cohen *et al.* (1973), as noted earlier, found that auditory discrimination ability in children living in an apartment complex constructed over a New York City highway was poorer the closer they lived to the highway. They also noted a significant decrement in the reading skills of children from noisier apartments.

Each child in our study was individually administered the Wepman Auditory Discrimination Test (Wepman, 1958). The Wepman test consists of 40 pairs of words, some of which differ from each other in either initial or final sound, for example, *sick-thick* or *map-nap*. The pairs of words are recorded on tape and presented to each child through earphones. The child is instructed to report if the two words in each pair are similar or different. Control word pairs, in which the words are the same, allow for the elimination of children who have problems with same/different judgments or who are not attending to the task. We predicted that children from noise schools would have poorer auditory discrimination than would children from the quiet schools.

School Achievement

Scores on the California Test of Basic Skills (California Assessment Program, 1976) reading and math tests were gathered from school files. These tests are administered in the classroom during the second and third grades by the school system. It is important to note that these are the only cognitive measures administered under ambient noise conditions. All other measures were taken under quiet conditions.

Distractability

Two measures of distractability were analyzed. Separate analyses examined the number of lines completed under distraction and the percentage of e's in the completed lines that were found. No-distraction performance (number of lines in the first analysis and percentage of e's in the second) was added as an additional control variable (covariate) in order to equate the children on their ability to perform the task under quiet conditions.

Cross-Sectional Results. There were no differences between the noise group and the quiet group on the number of lines completed under distraction. The interaction of noise condition and months enrolled in school was also nonsignificant. The main effect for noise condition on percentage of e's found was not significant. There was, however, a significant interaction effect on distractability between noise/quiet and years enrolled in school, $F(1, 237) = 5.05, p < .03$. As is apparent from Figure 6, the children in noise schools did better than those in quiet schools on the distraction task during the first 2 years of exposure and did worse after 4 years. Quiet- and noise-school children who had been enrolled between 2 and 4 years demonstrated equivalent performance on the distraction task.

As an additional analysis, we also compared children who lived in noisy homes with those from quieter residences. As indicated in the methodological chapter, most children from noisy homes also attend noisy schools. Home noise levels were calculated from CNEL contour lines. This information gives a weighted average of noise exposure taking into account the time of day that sound levels occur. It is the standard way in which community noise levels are now measured in the United States. There was a significant effect of home noise level on the percentage of e's found in the distraction task, $F(1, 126) = 3.60, p < .06$. Figure 7 reveals that children from noisier homes were less strongly distracted by the auditory distractor. There was no interaction of home noise level and months in residence.

Adaptation to Noise Results. As is apparent from the lower half of Figure 8, the attrition sample showed a similar T1 pattern found in the cross-sectional analy-

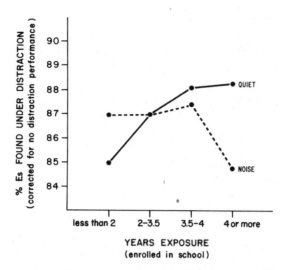

Figure 6. Distraction as a function of school noise level and duration of exposure for Time 1 cross-sectional sample. Each period on the years exposure coordinate represents one-quarter of the sample (from Cohen, S., Evans, G. W., Krantz, D. S., & Stokols, D., 1980. Copyright 1980 by the American Psychological Association. Reprinted by permission).

sis, except that noise-school children who had been enrolled for 2 to 4 years also were less distractable than their quiet-school counterparts. In the cross-sectional analysis, the quiet- and noise-school children enrolled for 2 to 4 years performed equivalently to children enrolled less than 2 or more than 4 years on the distraction task.

The upper half of Figure 8 reveals the same pattern of better performance by the T2 noise group during the earlier years of school enrollment. For this analysis, however, performance of the quiet and noise school groups are comparable after 4 years of enrollment. The noise condition by years enrolled in school interaction is marginally significant, $F(1, 141) = 3.66 \; p < .06$.

Noise-Abatement Results. Analyses of the distraction effects indicated no significant cross-sectional or longitudinal effects of abatement on either distraction measure. Children's ability to resist distraction was apparently unaffected by sound-attenuation procedures.

New Third Grader Replication Results. There were no significant main effects of noise nor noise-by-months-enrolled interactions on distraction in this sample.

Discussion. Findings from both the cross-sectional study and the longitudinal analysis of adaptation effects indicate that children in noise schools have some initial increased ability for tuning out auditory distraction. Furthermore, children from noisier homes are also more resistant to the effects of auditory distraction. Contrary to earlier evidence that auditory discrimination and reading scores decrease as a function of noise exposure (Cohen *et al.*, 1973), the interaction of noise condition with school enrollment shows that as the length of the noise exposure increases, children are more disturbed by auditory distractors (see Figure 6). The similarity of the T1 and T2 data (see Figure 8) indicates the relative stability of this unexpected interaction. On the other hand, the new third grader study did not replicate this interaction effect on distractability.

One reason why children from noise schools at first may be better at ignoring auditory distraction is that they initially cope with aircraft noise somewhat successfully by tuning all noise sources out. Later, however, as they find this strategy is not adequate, they may give it up. This hypothesis is consistent with the helplessness findings discussed earlier in Chapter 3.

Another possible explanation of the crossover interaction is that as the duration of exposure increases, children become more discriminating in the kinds of sound they tune out. Whereas initially all acoustic cues are indiscriminantly gated out, with experience, children in noisy environments may learn to tune out only more noxious acoustic cues such as airplane noise. The distractor stimulus in our task was a man's voice reading a story that might not be screened out indiscriminantly if children were gating out aircraft noise in particular. One way to test this hypothesis would be to utilize a similar task with both aircraft-related distractors and distractors similar to the one we used that have semantic content. Although noise abatement had no apparent effects on auditory distraction, as we shall see later, some important effects were noted on school achievement.

The evidence for the tuning-out hypothesis is generally strong, albeit complex. Children who have attended noise schools as well as children from noisier homes are more resistant to the effects of auditory distraction than are children from quieter settings. With continuous noise exposure at school, however, children become more susceptible to the effects of auditory distraction during performance tasks. Two possible mechanisms for this change in susceptibility to auditory distraction were presented. One explanation centers on reduced motivation to respond because of chronic exposure to an uncontrollable stressor. Alternatively, the data may reflect more fine-tuned inattention to auditory cues with continuous practice in gating out noise. Clearly, at this time both of these explanations are speculative and in need of further investigation.

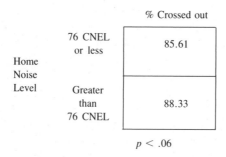

Figure 7. Percentage of E's crossed out under distraction by median home noise level (Airport I).

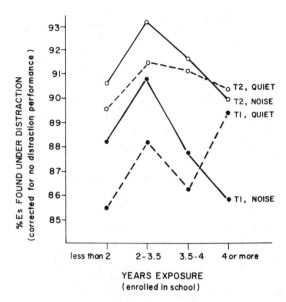

Figure 8. Distractibility at Time 1 and Time 2 as a function of school noise level and duration of exposure for the longitudinal study. Sample includes only those subjects present for both sessions. Each period on the years exposure coordinate represents one-quarter of the sample (from Cohen, S., Evans, G. W., Krantz, D. S., Stokols, D., & Kelly, S., 1981. Copyright 1981 by the American Psychological Association. Reprinted by permission).

Signal-to-Noise Ratio

As noted before, this task was only given in the third grade replication study. Two measures were used. The first one asked children which signal-to-noise channel they *liked* the best. The second task required the child to indicate which channel she or he could *hear* the best. We predicted that if children chronically exposed to noise learn to tune out acoustic cues, then noise-school children should be less able to discriminate among the four signal-to-noise ratios.

For the half of the children who received the preference instructions (which channel do you like the best), there was no significant main effect of noise condition. The interaction of noise condition by months enrolled in school was also not significant.

Inspection of Table 19, however, indicates that children from the noise schools chose a noisier channel than did the quiet-school children when asked to indicate which channel could be heard the best, $F(1, 71) = 2.91$ $p < .09$. Furthermore, there was a marginally significant noise condition by years enrolled in school interaction $F(1, 71) = 2.83, p < .10$. Inspection of the table indicates that the main effect of noise condition is due to the large decrement in signal-to-noise discrimination found in the children enrolled in the noise schools 4 years or more. This result

Table 19. Which Channel Can You Hear Best?

	Years				
	Less than 2	2–2.9	3–3.9	4 or more	Overall
Quiet	13.75	15.00	14.00	14.29	14.17
Noise	16.15	15.00	15.00	30.00	17.14

Note. Signal-to-Noise Ratio: 10 = 72:58; 20 = 72:62; 30 = 72:66; 40 = 72:70.
Noise: $p < 0.09$. Noise × Years: $p < 0.10$.

is important because once again it suggests that children are not adapting to noise effects over a long period of time.

Further research is necessary to examine the tuning-out hypothesis more closely. Both speech-relevant and speech-irrelevant sounds as distractors should be examined. In particular, the relative sensitivity of individuals to distractions that are similar and dissimilar to the ambient noise conditions they face should be studied. Our results from this task and the distraction e's task highlight the complexity of the stress and coping process. Although one cognitive strategy may be initially adopted to cope with exposure to an environmental stressor, with continued exposure that strategy may become modified or replaced by another strategy. More thought is needed on when certain cognitive strategies are adopted and how they shift over time. Presumably shifts in cognitive strategy during stress occur both because of the cognitive demands of the task, individual experiences in similar conditions, and the adaptive demands (costs) the strategies place upon the individual.

Auditory Discrimination

The Wepman scores as well as the achievement test scores included an additional control. To roughly equate the effect of the noise and quiet conditions on the aptitude of the children at the time they entered school, an additional control for the mean cognitive abilities (standardized test administered by school) of the child's class on entering the first grade was included in the auditory discrimination and achievement test scores analyses.

Cross-Sectional Results. Auditory discrimination was unrelated to both noise and the Noise × Months Enrolled in School interaction. No longitudinal data were collected on the Wepman or school achievement tests. The achievement tests for reading and math are administered during the third but not during the fourth grade.

Noise-Abatement Results. There were no noise-abatement effects on auditory discrimination.

New Third Grader Replication Results. As in the main study, auditory discrimination was unaffected by noise condition or its interaction with time enrolled in school.

Discussion. Contrary to our prediction and to findings in other studies of noise and children (Cohen *et al.*, 1973; Moch-Sibony, 1985), we did not find any effects of airport noise on auditory discrimination abilities. We did find some relationships among Wepman scores and measures of school achievement that are discussed later.

We are uncertain why our auditory discrimination results did not replicate earlier findings. One reason why our data could differ from the conclusions in the Moch-Sibony study is because she used a different measure of auditory discrimination, the Massiot–Phillips test, whereas we used the Wepman test. Her study was conducted in France, and we have not been able to uncover information about the Massiot–Phillips test. The Cohen *et al.* (1973) study, however, employed the same Wepman measure that we used. It is possible that differences in noise characteristics between the two studies could explain the different results. Cohen *et al.*'s noisy subjects experienced more continuous traffic noise but at lower average intensities and without high peaks as found in the airport noise schools.

There are at least two important consequences of noise exposure at home versus at school that could affect auditory discrimination abilities. First, noise exposure at school does not occur until children are 5 or 6 years old, whereas home noise exposure can begin for neonates. Second, total time spent in noise may be much greater for home noise than for school noise, especially for children of elementary-school age. Unfortunately, no systematic research has been conducted on auditory discrimination and noise quality, length of noise exposure, and critical development periods of exposure to noise.

School Achievement

Two measures of school achievement, math and reading scores, were analyzed. As noted earlier, children were tested during the second and third grades by the school system. Because children are tested in their classrooms, children in noise schools performed this task in noise.

Cross-Sectional Results. Noise had no significant main effects on either math or reading scores. The noise-by-months-enrolled interaction term was also nonsignificant. Further analyses revealed that children who were better at auditory discrimination were also better at reading, $r = .19, p < .05$, and math, $r = .18, p < .05$. There were, however, no significant correlations between these variables and the selective inattention measure. The same analyses, including only noise-school children, and correlations partialing out control variables for both the entire sample and the noise sample only yielded similar results.

Because noise might affect school achievement by impacting classroom behaviors, a second analysis of the school achievement scores was performed with classroom, rather than individual child, as the unit of analysis. This analysis included the standard control factors and examined school, months enrolled in school, and classrooms (nested in noise) as independent variables. The results of this analysis were the same with no-noise or noise-by-months effects on reading or math scores.

Because reading and math tests are only administered during the second and third grades, no longitudinal analysis of school achievement scores was possible.

There were, however, effects of home noise levels on reading scores. Children from noisier homes had poorer reading scores than did children from quieter homes, $F(1, 116) = 5.23$, $p < .02$, see Figure 9.

Noise-Abatement Results. Because the scores used in the following analysis were recent for third graders (administered at approximately the same time as our own testing), but were 1 year old for fourth graders, we expected that noise abatement would affect the achievement scores of third graders who spent a year in the abated classrooms before (and while) taking the test but not fourth graders. Fourth graders classroom assignment at the time we collected our data was irrelevant to how they performed on a test taken in another classroom 1 year earlier. Unfortunately, we do not know which classrooms fourth graders were assigned to during the year they were administered the achievement tests.

To test the hypothesis that the achievement scores of third but not fourth graders would be affected by abatement, a Grade × Noise interaction term was added to the noise-nested-in-classrooms analysis of the school achievement scores. Although there were no effects for the noise or Noise × Grade interaction on reading scores, there was a marginal Grade × Noise interaction on the math achievement test, $F(2, 32) = 3.06$, $p < .06$; the multivariate for noise was $F(6, 60) = 1.98$, $p < .08$. As is apparent from inspection of Table 20, although grade level did not have a substantial effect on the relative performance of third and fourth graders in quiet schools, third graders in abated classrooms performed substantially better than their counterparts in unabated classrooms. The reverse was true, however, for fourth graders. It is also apparent from Table 20 that there was a similar pattern for reading scores, although the Grade × Noise interaction term was not statistically significant.

New Third Grader Replication Results. The third grader replication results match those of the cross-sectional analysis. There was no statistically significant effect of noise or the Noise-×-Years-Enrolled-in-School interaction term on reading or math scores among the new third grade sample. As in the cross-sectional analy-

		Reading	Math
	76 CNEL or less	49.59	65.14
Home Noise Level	Greater than 76 CNEL	44.53	63.55
		$p < .02$	NS

Figure 9. Raw scores on high school achievement tests by median home noise level (Airport I).

Table 20. Mean (Adjusted) School Achievement
Percentiles for Cross-Sectional (T1) Data as a
Function of Classroom Noise Abatement and Grade

	Reading		Math	
Classroom	3rd grade	4th grade	3rd grade	4th grade
Noisy	30.30	35.96	34.35	39.35
Abated	47.36	37.90	56.24	37.54
Quiet	37.85	39.09	36.96	42.76

Note. From "Aircraft noise and children: Longitudinal and cross-sectional evidence on adaptation to noise and the effectiveness of noise abatement" by S. Cohen, G. W. Evans, D. S. Krantz, D. Stokols, and S. Kelly, 1981, *Journal of Personality and Social Psychology, 40,* p. 342. Copyright 1981 by the American Psychological Association. Reprinted by permission.

sis, Wepman scores were significantly correlated with math and reading achievement. Of additional note, third grader performance on the signal-to-noise task under the hear-best condition was significantly correlated to reading achievement ($r = .35, p < .01$). Children choosing the channel with the best signal-to-noise ratio (i.e., the least amount of noise) had better reading scores. Wepman scores, signal-to-noise performance, and distraction measures on the crossing-out-*e*'s task were all unrelated to one another.

Discussion. In general, the results of our study suggest that aircraft noise exposure in school had little or no effect on school achievement in math or reading. This conclusion does not replicate the findings of many studies discussed earlier where significant school noise effects on reading have been noted (Bronzaft & McCarthy, 1975; Cohen *et al.,* 1973; Green *et al.,* 1980; Lukas *et al.,* 1981; Maser, 1978). One previous study, however, similarly reported no airport noise effects on children's reading scores (Moch-Sibony, 1985).

As we suggested earlier, one reason that some studies do not find noise-related school achievement deficits may be attributable to insensitivity in experimental design. In both the Moch-Sibony and the present study, noise-sample and quiet-sample children attended different schools, were in different classrooms, and had different teachers. In the other noise and achievement studies, students attended the same schools and in some cases (e.g., Cohen *et al.,* 1973) were in the same classrooms. Thus, the lack of noise effects in our study and the Moch-Sibony study may be attributable to extraneous factors that make the detection of a small effect of noise on achievement quite difficult.

Some indirect evidence for this comes from the fact that Wepman scores that moderately relate to reading scores here ($r = .19$) generally corelate more substantially with reading scores (e.g., Cohen *et al.,* 1973, $r = .53$). Extraneous sources of variance in achievement test scores (e.g., different schools) would weaken the generally strong correlation between the Wepman test and reading achievement scores. The fact that achievement scores for third grade children was higher in abated classrooms in noise schools than those of children in quiet schools (see Table 20) is also consistent with the argument that factors other than noise are more strongly linked with school achievement.

On the other hand, there is a noise-related deficit in reading scores because of home-noise exposure. Children from noisier homes score more poorly on standard reading achievement tests. Note also that the math results, although not statistically significant, are also in the same direction. The reading results replicate earlier studies by Cohen *et al.* (1973) and Green *et al.* (1980) that showed that children residing in noisy homes had reading deficits. The data also lend validity to other studies showing that observer ratings of high home noise levels were linked with inhibited cognitive development (Wachs *et al.*, 1971).

The home and school noise findings on reading raise important questions about the importance of exposure to environmental stressors at home and at school (or on the job). The amount and type of exposure to various environmental stressors throughout the daily life of an individual may have important consequences for changing certain cognitive processes. Furthermore, as noted earlier, the effects of multiple exposures in different places may have interactive effects. Lukas and colleagues found, for example, that children from noisy homes were most at risk for noise-related deficits in reading because of noise exposure at school (Lukas *et al.*, 1981). Because home and school noise levels in our field study are highly correlated, we do not have an adequate test of the possible interactive effects of home and school noise on children's health and behavior. More discussion of the interactive effects of environmental stressors on human health and well-being are presented in Chapter 6.

Although reading achievement was unaffected by abatement, our data suggest that math achievement was higher for children in abated classrooms than in noisy classrooms. This effect is noteworthy because it occurs, as predicted, only for those children who took the achievement test at the end of the year that was spent in the abated classroom. The reader should note, as mentioned earlier, that unlike the other tasks used in our study, the achievement tests were actually taken in the classrooms. Thus, we do not know whether noise interferes directly with test performance or if the abatement-related differences reflect an aftereffect of noise.

In summary, there is little evidence in the present study that aircraft noise at school impairs reading and math skills or that these skills are related to a selective inattention strategy. Home noise levels, however, are linked with reading ability. Auditory discrimination and signal-to-noise perception are related to school achievement as well. Furthermore, deficits in signal-to-noise perception are related to chronic noise exposure. Many other field studies, but not this one, have also shown deficits in auditory discrimination from chronic noise exposure. Furthermore, noise abatement procedures may improve achievement skills in children who previously were in noisy classrooms.

Conclusion

The evidence presented in our study on the cognitive effects of noise exposure are not strongly consistent with laboratory research on environmental stress and cognition. This inconsistency is unlike our measures of health, affect, and motivation that are generally consistent with the experimental literature.

Contrary to prediction, increased years of noise exposure led to children's becoming *more* distractable by noise rather than less. Furthermore, increased noise exposure impacted limited aspects of auditory processing. Although auditory discrimination among similar sounds was not adversely affected by noise, ability to discriminate subtle changes in signal-to-noise ratio declined as a function of exposure to noise.

School achievement was generally unaffected by chronic noise exposure at school, whereas home noise did effect reading skills. Furthermore, noise abatement in classrooms showed some marginal ameliorative effects on test scores. As discussed earlier, the experimental design of our study with noise and quiet conditions covarying with different schools may have been too insensitive to capture the full effects of noise on school achievement. In addition to trying to test children from the same schools (or even better, the same classroom), future research on noise and school achievement should track children over time from the point at which they enter a noisy school. Ideally, children from both quiet and noisy homes could also be monitored in such a longitudinal design because we have shown that home noise levels can be associated with school achievement. Furthermore, as noted earlier, noise levels at home and at school may have interactive effects in human cognition.

One noteworthy aspect of the cognitive data from the Los Angeles Noise Project is the stability of several of the findings. The distraction data and, in particular, the crossover interaction was replicated in the longitudinal study. Furthermore, the auditory discrimination and school achievement results were replicated in the new third grader study. We do not know why the new third grader data did not replicate the distraction effect.

The cognitive data also strongly indicate that the cognitive effects of noise do not habituate over time. Greater length of exposure to ambient airport noise led to greater distractibility from auditory cues and to poorer performance on the signal-to-noise ratio test. An important question that our study cannot answer is, what are the long-term consequences of the cognitive effects of noise on these young children?

Experimental studies of noise and other environmental stressors indicate the stress causes deficits in specific cognitive processes, especially those related to attentional activities. Noise produces errors in serial reaction time tests where rapid responses and continuous responses are required. Responses in noise reveal momentary lapses that appear to be caused by periods of distraction due to the onset of noise bursts. Sustained vigilance tests where long periods of continuous monitoring of infrequent signals are required are also disrupted by environmental stressors. Under continuous noise, subjects make more rapid, less cautious responses. Subjects behave as if "trigger happy," producing many false alarms. Finally, when attention is needed for multiple task cues, stressors appear to inhibit the subjects' ability to accurately monitor multiple inputs. Under stress, individuals allocate greater attention to more important aspects of the task (high-priority signals) at the neglect of subsidiary cues.

Experimental research has also found some deficits in memory as a function of stress. Memory for information incidental to the primary material to be studied in a learning task is worse under noise. Environmental stressors also appear to reduce immediate memory capacity. Persons can remember items at the end of a list as well or better in noise, but they do not remember items as well that are farther back in the list. This change in immediate storage capacity may be linked to some shifts in strategy noted when individuals are told to memorize material under stress conditions. Some preliminary research suggests that, under stress, information is processed more quickly and with more attention given to sequential order than when it is studied without stress. This shift in memory strategy may also cause poorer comprehension of information such as meaning that requires more in-depth, slower processing of current stimuli and associated materials already stored in memory.

Finally, several experimental studies have shown that performance aftereffects from stressors can occur. Individuals previously exposed to a stressor, particularly if it is uncontrollable, perform more poorly on frustration-tolerance tests, may proofread more carelessly, and apparently are less cognizant of subtle social cues indicating a person's need for help (e.g., cast on the arm of the person dropping his or her books).

Field research on environmental stressors and cognition has concentrated primarily on school achievement, particularly reading, as well as on tuning-out strategies. Several studies have found reading deficits linked to ambient noise exposure at school. These effects seem strongest for elementary-school children over the fifth grade, for low achievers, and for children who also live in noisy homes. Evidence for the tuning-out hypothesis exists as well. Some children chronically exposed to high noise levels have trouble discriminating between similar sounds and may be better at gating out auditory distractors. Our study shows, however, that the enhanced ability to gate out distractors is limited to children exposed to noise for a shorter period of time. One field study has also noted a deficit in incidental memory for children from noisy home environments.

As discussed earlier, there may be some relationship between the tuning-out process that develops under chronic noise conditions and the reallocation of attention that occurs under laboratory exposures to acute noise. Tuning out may reflect more focused attention on central aspects of whatever task is at hand. If children learn to do different tasks and to develop cognitive styles under noise, then they may tune out most stimuli unrelated to the tasks they are experiencing. This process could become overgeneralized with experience. Similarly, if children continuously cope with high density by withdrawing from one another, perhaps they will adopt a general withdrawal strategy of dealing with persons regardless of the conditions of crowding.

There are several directions for future research on stress and cognition that are suggested by our study and the current literature reviewed here. First, experimental work needs to more thoroughly examine the cognitive processes involved in daily intellectual activities such as reading. Do shifts in attentional allocation and diminished information-storage capacity cause the development of poorer reading skills? What is the relationship between auditory discrimination and resistance to distraction, and how are they related to important daily cognitive activities? Research on resistance to distraction also needs to carefully explore different types of signals and background distractors. If tuning out develops with noise exposure and becomes more specific to certain noise sources, then persons who have adapted to noise should be affected differently by ambient, noise-related distractors than by other sources of distraction. More laboratory work can also integrate field research data with experimental findings by using chronic exposure to environmental stressors as a subject-blocking factor in experimental designs. For example, would children from our noise schools respond differently than the quiet-school children to Glass and Singer's aftereffect paradigm or to Hockey's dual-task situation, with noise as a stressor in the lab?

There may be some conceptual bridging possible between the learned helplessness findings discussed in Chapter 3 and some of the cognitive effects of stressors on the frustration-tolerance task. Are these effects cognitive or motivational? Learned helplessness research suggests that the two explanations are interrelated. Poor motivation to respond after continual failure to escape from a stressor is said to occur because of learning. One learns that their escape behaviors are uneffective. Behaviors become unassociated with their outcomes (Seligman, 1975). Cognitive research on stress must carefully consider the motivational aspects of the context in which tasks are performed. Furthermore, the affective meaning of materials and its relationship to the mood (physiological?) state of the individual influences information processing. Because stressors affect mood, this aspect of research on stress and cognition needs to be studied further.

This issue of motivation and cognition leads to the final area in need of further work. The whole issue of coping is just beginning to be incorporated into cognitive views of stress and performance. We know that, for at least short periods of time, many tasks can be completed, even under extremely adverse conditions. The amount and type of effort used to maintain optimum task performance under stressful conditions is interrelated with task demands, the intensity of the stressor as well as individual factors, such as experience with the stressor. Cognitive research will have to take into account the interplay among stressors, coping, and information processing.

Contextual Analyses of Environmental Stress

If you had it to do over again, how would you redesign your research? In light of the findings from your study, how have your theoretical and methodological approaches to the problem changed?

Dissertation committee
member to doctoral candidate
during final oral exam

The Los Angeles Noise project was a complex and arduous project of relatively long duration. First, there were numerous preparatory meetings among the co-principal investigators to conceptualize and design the study. Grant proposals were submitted and research funds obtained. Then, meetings were held with approximately 10 Los Angeles elementary-school principals to recruit the participation of their staff and students in our research, and a travel trailer was acquired and remodeled to serve as a mobile testing station for our research participants. During spring 1977 and spring 1978, small armies of research assistants led by at least one principal investigator per testing session made daily excursions to the participating schools where five groups, each comprised of five children, were tested in our trailer throughout the school day. Subsequently, several months were devoted to the coding and analysis of the data from the longitudinal and replication samples. Finally, research reports were prepared for presentation at conferences and for submission to various journals.

Now as we write this book, we have the advantage of hindsight. Like participants in a dissertation orals, we can step back from the specific hypotheses, methods, and findings that have been described in earlier chapters and consider some of the broader ramifications of our research. For instance, we can try to assess the changes in theoretical perspective that have occurred as we have worked together over several years on the Los Angeles Noise Project. And we can attempt to gauge the implications of our research for future investigations of environmental stress.

An important outcome of our collaborative research over the years has been the development of a *contextual* analysis of environmental stress, behavior, and health. Although earlier chapters have examined, for the most part, the direct effects of exposure to aircraft noise at school on various criteria of children's performance and health, the present chapter explores the broader ecological context in which these processes and events occurred: for instance, the social and physical conditions of the child's family and home environment and those existing within his or her classroom. Thus, in this chapter we will examine measures of school noise, performance, and health in relation to several other aspects of the child's overall life situation.

Some evidence that events within the child's school and home domains are interdependent was briefly noted in earlier chapters. Recall, for example, that the noise abatement intervention was associated with a greater reduction of blood pressure among children living within quieter homes rather than in those living within noisier areas (see Chapter 5), and that children attending noisy schools and living in the noisiest residential areas were more likely to give up on complex puzzle tasks than those from relatively quieter homes (see Chapter 4). What we have not yet done in this volume, however, is to present a systematic framework for analyzing the ecological context of environmental stressors and their impacts on behavior and health. Thus, a major purpose of this chapter is to articulate a general conceptual framework for contextual research on environmental stress. As part of that endeavor, we will present additional data from the Los Angeles Noise Project that are relevant to various assumptions of our contextual approach. Many of these data are drawn from statistical analyses that were conducted later than those described in preceding chapters. The fact that these analyses occurred later in the project reflects the gradual shift in our thinking toward a more contextual view of environmental stress over the past few years.

Our emphasis in this chapter on the ecological context of stress is part of a broader movement within the field of psychology toward the development of contextual models of human behavior and well-being (Altman & Rogoff, 1986; Lazarus & Launier, 1978; Wapner, 1986). Before outlining our framework for analyzing the contextual dimensions of environmental stress, we trace some of the historical underpinnings of the contextualist movement in psychology and specify certain definitional and procedural challenges posed by the contextual perspective.[1]

The Emergence of a Contextual Perspective in Psychology

During the 1970s and early 1980s, a growing interest in contextualism occurred within several areas of psychological research. Psychologists within every major area of the discipline noted the deficiencies of decontextualized research and called for more holistic and ecologically grounded approaches to the study of behavior. In the fields of clinical, biological, and health psychology, for example, Schwartz (1982) proposed a "biopsychosocial" view of health and illness, which

---

[1]The framework for contextual analysis, presented in the introductory portions of this chapter, is adapted from Stokols (1983, 1986).

replaces "single-cause, single-effect" models with those that address the complex interactions among physiological, psychological, and social dimensions of well-being. Similarly, the recent volumes edited by Magnusson and Allen (1983) and by Wapner and Kaplan (1983) call for holistic approaches to the study of human development and are part of an ever-widening stream of ecologically oriented research in developmental psychology (e.g., earlier discussions of this research by Bronfenbrenner, 1979; Scarr, 1979). Within the fields of cognitive, personality, and social psychology, the recent volumes by Kaplan and Kaplan (1982) and Neisser (1982) and the articles by Evans (1980), Kelley (1983), Little (1983), McGuire (1983), Smith (1983), and Veroff (1983) are indicative of an increasing trend toward contextual analyses of cognition and social behavior. And in his presidential Address to the Division of Population and Environmental Psychology, Altman (1982) contended that we are in the midst of a full-fledged scientific revolution across all areas of psychology, involving a shift from unidirectional, mechanistic analyses of environment and behavior toward transactional and contextually oriented models. Little (1986) has referred to these developments as the "contextual revolution" in psychology.

The terms *ecological* and *contextual* are certainly not new to psychologists. Explicit concern for the ecological context of behavior was evident in the writings of Koffka (1935), Lewin (1936), Tolman and Brunswik (1935), and Murray (1938) during the mid-1930s and in the subsequent work of Kohler (1947), Gibson (1960), Barker (1968), Jessor (1958), and Chein (1954). But the emergence of areas such as population, community, and environmental psychology during the 1960s and 1970s signaled a surge of interest in ecological issues that, in recent years, has begun to pervade more established areas of the field as well (Barker & Schoggen, 1973; Craik, 1973; Fawcett, 1973; Heller & Monahan, 1977; Proshansky, Ittelson, & Rivlin, 1976; Russell & Ward, 1982; Sarason, 1976; Stokols, 1982).

The current popularity of contextual approaches in psychology appears to be rooted in both societal and intellectual developments. At the community level, concerns about global population growth, resource shortages, and environmental decay have increased the salience of ecological constraints on behavior. And at a more academic level, the growing emphasis on contextual theorizing and research strategies in psychology can be viewed as part of a conceptual shift within the behavioral sciences away from exclusively intrapersonal explanations of behavior toward those that encompass not only the immediate social environment but also the broader cultural, historical, and geographic milieus of people's day-to-day activities (Cronbach, 1975; Gergen, 1982; Manicas & Secord, 1983).

Whatever its sources and the differences in terminology that surface among its proponents, the contextual perspective in psychology seems to be associated with certain widely shared core assumptions. Among these assumptions are that (a) psychological phenomena should be viewed in relation to the spatial, temporal, and sociocultural milieus in which they occur; (b) a focus on individuals' responses to discrete stimuli and events in the short-run should be supplemented by more molar and longitudinal analyses of people's everyday activities and settings; (c) the search for lawful and generalizable relationships between environment and behavior should be balanced by a sensitivity to, and analysis of, the situation specificity of psychological phenomena (Cronbach, 1975; Gergen, 1973); and (d) the criteria of ecological and external validity of research should be explicitly considered (along with the internal validity of the research) not only when designing behavioral studies but also as a basis for judging the applicability of research findings to the development of public policies and community interventions (Brinberg & McGrath, 1982; Cook & Campbell, 1979; Winkel, 1983).

Although there has been much discussion about the virtues of contextualism and some agreement about its general assumptions, considerably less progress has been made in translating these assumptions into more specific guidelines for theory development and empirical research (see Barker & Schoggen, 1973, Bronfenbrenner, 1979, and Wicker, 1986, for notable exceptions to this trend). Lest contextualism become an empty buzzword, several difficult questions must be addressed. First, what are the distinguishing features of contextual theorizing and research? What features differentiate a contextual analysis from a noncontextual one? Second, are psychological research questions differentially suited to a contextual approach? For which psychological phenomena is a contextual analysis warranted and for which is it not? And third, in those instances where a contextual perspective is adopted, what criteria determine the scope and content of the variables included in the analysis? What considerations should guide the researcher's decisions about how broadly to draw the contextual boundaries of a phenomenon, and which concepts and methods should be used in analyzing the relationships between the phenomenon at hand and the specific contexts in which it is observed?

Our goal in this chapter is not to provide clear-cut singular answers to these rather formidable questions. Rather, our intent is to offer a set of programmatic strategies for addressing these considerations in future research on environmental stress. We turn now to an analysis of the definitional and procedural questions noted before. Subsequently, we outline a set of strategies for conducting contextual research on stress that we believe is responsive to these questions.

Assumptions and Strategies of Contextual Research

Contextual and Noncontextual Analyses of Environmental Stress

A fundamental idea underlying the notion of contextual research is the concept of *embeddedness*. That is, a particular phenomenon is thought to be embedded in (and influenced by) a surrounding set of events. The first task of contextual research is to identify the central, or "target", phenomenon to be examined—for example, the impact of exposure to aircraft noise at school on children's blood pressure at school. Once the target variables have been specified, the next step is to define a set of contextual variables that are thought to exert an important influence on the form and occurrence of the target phenomena. We might suspect, for example, that the relationship between noise exposure at school and blood pressure at school is dependent on, or qualified by, other events in children's lives such as the levels of noise they experience at home or the degree of spatial density within their classrooms. These other variables are "contextual" in the sense that they surround (or are connected in time and space with) the child's experience of aircraft noise at school, and they are thought to influence the form or intensity of that experience.

If we hypothesize that the relationship between the target variables of school noise and blood pressure is independent of (or not influenced by) surrounding events, then there is no need to complicate our analysis by incorporating additional measures of home noise and classroom density levels. For instance, we might predict that high levels of noise exposure invariably raise levels of physiological arousal such as systolic and diastolic blood pressure, regardless of the context in which these variables are observed. If this hypothesis is correct, then the inclusion of contextual variables in our analysis would not add appreciably to our understanding of the target phenomenon.

If, on the other hand, we suspect that the impact of school noise levels on blood pressure *is* significantly qualified by situational factors, then it becomes important to include these factors in our analyses of the target variables. For, if we hypothesize and find that children's cardiovascular response to aircraft noise at school is moderated by the levels of noise they experience at home, then we have learned more about the relationship between noise exposure and physiology than if we had focused exclusively on the links between school noise and blood pressure, irrespective of home noise levels.

To summarize, *noncontextual research* focuses entirely on the relationships between target predictor and outcome variables (e.g., school noise and blood pressure levels). *Contextual research,* on the other hand, incorporates supplementary predictor variables drawn from the immediate situation (e.g., other conditions of the school environment such as classroom density) or more distant events (e.g., noise levels at the child's home) that presumably qualify the relationship between the target variables (see Figure 10).

With respect to criterion (outcome) measures, a contextual analysis might incorporate not only blood pressure measurements at school but also cross-setting assessments of emotional stress, coping strategies, and behavioral problems at home and in the classroom. (See, for example, our discussion in Chapter 1 regarding the generalization of coping strategies across different settings.) A contextual approach thereby widens the scope of analysis to include not only the target variables of school noise and blood pressure but also supplementary indexes of enviroment and behavior that qualify the relationship among these variables (see Figure 11).

Distinguishing Features of Contextual and Noncontextual Theories

Given a particular set of target variables, the selection of contextual variables for empirical analysis can proceed either in an exploratory and atheoretical fashion

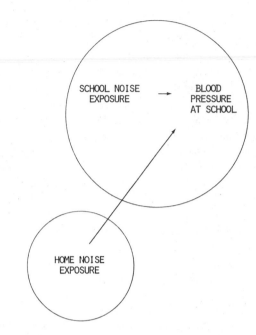

Figure 10. Hypothesized relationship between exposure to noise at school and children's blood pressure, as qualified by the degree of exposure to noise within the child's home environment.

or on the basis of theoretically derived assumptions about the target phenomenon. Lacking a well-developed theory, the researcher may begin with a tentative hunch about one or more situational moderators of the target variables. If the relevant data on these situational factors can be conveniently gathered, the researcher may pursue his or her hunch by examining the empirical relationships among the contextual and target variables. These exploratory analyses can play a useful role in the early stages of theory development by revealing situational factors that significantly influence the target variables and by excluding from further consideration those that do not.

    A more systematic and powerful form of contextual analysis occurs when the research design and the empirical assessments of situational and target variables are explicitly guided by a *contextual theory*. A distinguishing feature of contextual theories is that they specify a pattern of cross-situational variation in the target phenomenon (cf. Stokols, 1983). If, for example, the target variables are noise exposure and blood pressure, then a contextual hypothesis explicitly predicts a change in the relationship between these variables, depending on the presence or absence of certain situational factors. And a contextual theory goes on to explain *why* the hypothesized cross-situational variations in the target phenomenon occur.

    In contrast, *noncontextual theories* do not predict or explain cross-situational variation in the relationships among target variables. For instance, *environmental or situationist theories* construe behavior simply as a function of the immediate target situation (e.g., "exposure to high levels of noise at school invariably raises chil-

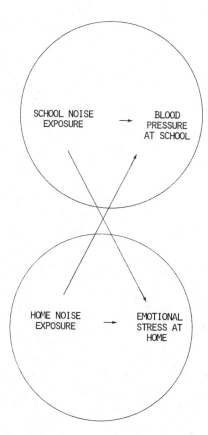

Figure 11. A contextual analysis of children's response to aircraft noise, in which the links between the target variables of school noise and blood pressure are examined in relation to supplementary indexes of noise levels and emotional stress at home.

dren's blood pressure''). *Trait theories* account for individual behavior entirely in terms of personal dispositions (e.g., ''children who are highly sensitive to noise have higher blood pressure than those who are less sensitive, regardless of the actual levels of noise to which the children are exposed''). And, *interactionist theories* account for behavior in terms of the joint influence of situational and intrapersonal factors.[2] Accordingly, Weinstein (1978) observed that dormitory residents' responses to noise were jointly influenced by their actual exposure to noise and by their personal dispositions toward noise sensitivity. In this instance, the relationship between noise exposure and behavior was moderated by an intrapersonal attribute rather than by a contextual factor. The focus of contextual theories, on the other hand, is clearly on situational rather than intrapersonal moderators of environment–

[2]See Endler and Magnusson (1976) for a more detailed discussion of situationist, trait, and interactionist theories.

behavior relationships. Thus, in contrast to environmental, trait, and interactional models, contextual theories specify a set of situational boundary conditions that qualify the relationship among targeted predictor and response variables.

The distinction between contextual and noncontextual theories is important, as it suggests a programmatic strategy for future research on environmental stress: namely the development of theories that explicitly account for the situational specificity of stress phenomena. All too often in research on environmental stress (as in other areas of behavioral study), the discovery of contextual moderators of environment–behavior relationships is treated as an "afterthought" of empirical analyses. The identification of important contextual factors tends to occur through post hoc, rather than deductive, assessments of external validity or the extent to which research findings generalize across different groups of people, settings, and times (Campbell & Stanley, 1963; Cook & Campbell, 1979; Mook, 1983; Petrinovich, 1979). Consequently, information about contextual factors is acquired in a nonprogrammatic happenstance manner as researchers gradually compare the findings from their separate and independently conducted studies. For example, early formulations of human response to crowding and noise treated these phenomena in a decontextualized manner, as if they could be understood apart from the contexts in which they occur. Accordingly, empirical studies were designed to test universalistic transsituational hypotheses about the effects of these environmental conditions on behavior and well-being. The pattern of results from these early studies, however, was far more complex than had been anticipated, revealing striking differences in people's reactions to crowding and noise depending on the situational contexts in which these events were experienced. Eventually, more contextually oriented formulations of crowding and noise were developed to account for the diverse and often-contradictory findings obtained across multiple programs of research (Cohen & Spacapan, 1984; Evans, 1982; Stokols, 1979).

Considering the progression of research on crowding and noise, it seems reasonable to suggest that future studies of environmental stress could be designed more efficiently and programmatically if contextual theories were developed prior to, or as an intended outcome of, empirical research. By making the explicit consideration of contextual factors a routine part of the research process, important aspects of the target phenomenon might be revealed that, otherwise, would have been neglected. And oversimplified assumptions about the cross-situational generality of the phenomenon might be recognized and abandoned during the early, rather than late, phases of investigation.

We therefore propose, as a general guideline for future research, that an effort be made to identify plausible contextual moderators of environmental stress in a predictive, rather than post hoc, fashion. We are not, however, suggesting that all stress phenomena will be equally amenable to contextual analysis. For instance, many studies conducted across a wide range of settings indicate that exposure to extremely high levels of noise invariably elevates cardiovascular arousal (Cohen & Weinstein, 1981). Apparently, once a certain threshold of noise intensity is exceeded, the impact of this environmental factor becomes relatively uniform across individuals and settings. Also, although the behavioral and emotional effects of certain drugs are mediated by situational factors (Schachter & Singer, 1962; Whalen & Henker, 1980), other pharmacological processes may be more exclusively dependent on intrapersonal factors and the type of drug introduced than on more remote aspects of the individual's spatial and social environments. Thus, certain stress phenomena may be relatively invariant across a wide array of situations.

The researcher's decision to adopt or not adopt a contextual view of a given problem is therefore likely to be influenced by several considerations, such as (a) existing empirical evidence for either the cross-situational variability or stability of the target phenomenon; and (b) the theoretical objectives of the research (e.g., whether the investigator is attempting to test hypotheses about intrapersonal or situational moderators of the target phenomenon). Moreover, whether or not a contextual perspective is actually translated into an operational research design may ultimately depend on more pragmatic considerations—especially the availability of sufficient research funding, personnel, and time to permit empirical study of the phenomenon across different environmental settings.

Choosing among Alternative Contextual Theories of the Same Phenomenon

We noted earlier that contextual analyses of environmental stress can have significant advantages over noncontextual approaches, particularly when there is reason to expect that the behavioral or health effects of a given stressor are mediated by situational factors. But once the researcher has opted for a contextual approach, the question then arises, Which set of contextual variables affords the greatest analytic leverage for understanding the target phenomenon? Clearly, any phenomenon can be analyzed in relation to multiple and alternative contextual factors. The key challenge in developing contextual theories is to identify, from among the myriad of potentially relevant situational factors, those that are most crucial for understanding the form and occurrence of the target phenomenon. We will refer to that subset of influential situational factors as the *effective context* of the target phenomenon (Stokols, 1983).

Assuming that all or some portion of the effective context has been identified, an important criterion for evaluating a theory is its accuracy in specifying the pattern of relationships between the target variables and one or more situational factors. This capacity to accurately account for the cross-situational variation in a phenomenon is a subcategory of predictive validity that is uniquely applicable to contextual theories.[3] Considering the earlier mentioned example of school noise levels and children's blood pressure, one theory might predict that high levels of

---

[3]Predictive validity, as it is typically defined, does not require cross-situational analysis (Carmines & Zeller, 1979). A noncontextual theory might demonstrate a high degree of predictive validity within a single setting but fail to specify important moderators of the phenomenon that would only become evident in alternative, and as yet unobserved, settings. For further discussion of the validity of contextual theories, see Stokols (1986).

noise at home intensify the cardiovascular effects of exposure to aircraft noise at school. Another theory might predict that high levels of noise at home buffer the effects of exposure to aircraft noise at school as the result of a physiological habituation process. And a third theory might suggest that home noise levels have no influence on the relationships between school noise and blood pressure but that other situational factors such as classroom density interact with school noise levels to affect children's blood pressure. Each of these theories makes a different prediction about the role of home noise levels in moderating the relationships among school noise and blood pressure. Thus, the three theories can be evaluated in terms of their relative accuracy in predicting the empirically observed pattern of covariation among the target and contextual variables.

A second criterion for gauging the adequacy of contextual theories is their relative explanatory power, or the extent to which they encompass the full range of situational factors that qualify a particular phenomenon. A contextual analysis may correctly identify some of those conditions but may exclude several others. For example, a theory may accurately account for the influence of home noise levels on the relationship between school noise exposure and blood pressure. But if other contextual variables, such as residential and classroom density, are also important in explaining the target variable relationships, then a theory that focuses only on the moderating role of home noise would be less powerful than one that explains the contextual influence of residential and classroom density as well.

Alternatively, a contextual analysis may be too inclusive, incorporating situational factors that are negligibly related to the target variables. This case suggests a third criterion for evaluating contextual theories, namely their efficiency. A contextual analysis is efficient to the extent that it includes those, and only those, situational factors that exert a significant influence on the target variables. In the preceding example, if we empirically determine that home noise, but not residential or classroom density, qualifies the relationships between school noise and blood pressure, then the theory that focuses only on the moderating role of home noise levels would be more efficient than the one that also incorporates the additional contextual factors. Both theories may be equally powerful in explaining the moderating role of home noise levels, but the former is more efficient or parsimonious than the latter because it omits the trivial variables of residential and classroom density.

Thus, it is possible to evaluate theories not only in terms of their accuracy in specifying the relationships between a particular contextual factor and the target variables but also with respect to their power and efficiency in representing the full range of situational factors that collectively exert the greatest influence on the target phenomenon. In short, the power and efficiency of a contextual theory increases to the extent that it accounts for a large, rather than small, proportion of the effective context of the target phenomenon, while excluding those situational factors that are negligibly related to the target variables.

In this discussion, we have focused on evaluative criteria that are especially useful for judging the relative adequacy of alternative contextual theories. There are, of course, several other evaluative criteria that apply equally well to assessments of noncontextual as well as contextual theories, such as the criteria of testability, construct validity, generativity, consistency with available empirical evidence, and utility or applicability to everyday problem solving. We will not discuss these more general criteria of theoretical adequacy here. For more detailed discussions of these issues, see Cook and Campbell (1979), Shaw and Costanzo (1970), Cronbach (1975), Gergen (1978), and Platt (1964).[4]

Strategies for Delimiting the Effective Context of Environmental Stress

The concept of environmental stress subsumes a wide array of physical and social stressors, on the one hand, and behavioral and physiological patterns of response, on the other. Considering the diversity of stress-related phenomena, it is highly unlikely that a singular, all-encompassing theory of environmental stress can be developed. Rather, the relationships among various categories of environmental demands (e.g., noise, spatial density, stressful life events) and response criteria (e.g., physiological arousal, emotional distress, performance decrements) are likely to be moderated by unique (although somewhat overlapping) sets of contextual factors.

Faced with a multitude of potentially relevant situational factors, identifying the effective context of various stress phenomena becomes a challenging and com-

---

[4]It is important, however, to note certain parallels between the concepts of construct validity (Cook & Campbell, 1979) and contextual validity (Stokols, 1986). *Construct validity* generally refers to the degree of match between research operations and theoretical constructs—that is, the extent to which the former adequately represent the latter. *Contextual validity* refers to the predictive accuracy of a theory in specifying the relationships among the target independent and dependent variables, as a function of one or more situational factors.

Construct validity is a broader concept that can be applied to noncontextual as well as to contextual theories. The links between construct and contextual validity become clearest when considering phenomena that are situation specific, that is, the form and occurrence of the target phenomenon vary across different contexts. From the perspective of construct validity, contextual factors that alter the relationships among target variables are viewed as situational "confoundings" of the presumed causes and effects (i.e., the target IVs and DVs). Many of the "threats to construct validity" discussed by Cook and Campbell (e.g., the interaction between testing situations and treatment, restricted sampling of treatment and response levels, and of testing intervals) relate to sources of situational confounding that can undermine the construct validity of the researcher's assessment of the target relationships. From the perspective of contextual theorizing, the identification of situational sources of variation in the target phenomenon becomes important, not so much as a basis for achieving a clean or "unconfounded" representation of the target phenomenon but rather as the basis for developing a broader theoretical understanding of the relationships between target phenomena and their situational contexts (see also Petrinovich, 1979, and Winkel, 1983, for further discussions of this perspective.)

Thus, the construct validity perspective suggests the importance of identifying sources of situational confounding and eliminating them from research assessments of the target phenomenon, whereas the latter approach attempts to identify and empirically assess sources of situational variability as the basis for developing a more integrative theory about the influence of diverse environmental contexts on the target phenomenon.

plex task. One strategy for reducing the complexity of this task is to organize the search for situational moderators of stress around certain basic dimensions of contextual analysis. In this section, we propose a general organizing framework for contextual research consisting of four basic dimensions, namely: (a) the *contextual scope* (spatial, temporal, and sociocultural) of the analysis; (b) *the objective or subjective representation* of contextual and target variables; (c) the *individual or aggregate level* at which contextual and target variable relationships are examined; and (d) the *partitive or composite representation* of contextual factors; or whether person–environment situations are represented as consisting of multiple, independently related parts, or as composites (groupings) of interdependent entities (Stokols, 1983).

Virtually any behavioral phenomenon can be analyzed in relation to these general contextual dimensions. In the ensuing discussion, we examine each of the four dimensions and the ways in which their systematic application can enhance the researcher's efforts to identify (and operationally measure) the effective context of various phenomena. We then illustrate, in a later section of the chapter, how these basic dimensions of contextual analysis have been incorporated into prior empirical work on stress, including our own study of children's response to aircraft noise.

Dimensions of Contextual Analysis

A basic starting point for contextual analyses of environmental stress is the immediate situation or setting in which the individual experiences a particular stressor. *Situations* are sequences of individual or group activities that occur at a particular time and place (Forgas, 1979; Magnusson, 1981; Pervin, 1978). *Settings* are geographic locations in which various personal or interpersonal situations recur on a regular basis (Barker, 1968; Stokols & Shumaker, 1981; Wicker, 1986). For instance, the focal situation might involve a child who is asked to complete a puzzle task that is actually insoluble. And the setting in which the situation occurs might be a testing room within an elementary school located near the Los Angeles International Airport. Although a particular situation can happen only at one place and time, multiple instances of the same type of situation can be observed within many different settings—for example, within relatively quiet versus noisy elementary schools in the Los Angeles area. (See Magnusson's, 1981, distinction between the *momentary situation* and *situation type.*) In this case, the researcher might be interested in knowing whether ambient noise levels within different school settings moderate the stressful consequences of exposure to an insoluble task.

As the preceding example suggests, the environmental contexts of stress can be described in terms of their scale or complexity. The scale of environmental units ranges from the specific stimuli and situations that occur within a given setting to the more complex *life domains* that are comprised of multiple situations and settings—the "school" domain, for example, in which students work in their classrooms, attend recess on the playground, and eat lunch in the cafeteria (see Figure 12). Life domains pertain to different spheres of an individual's life, such as family,

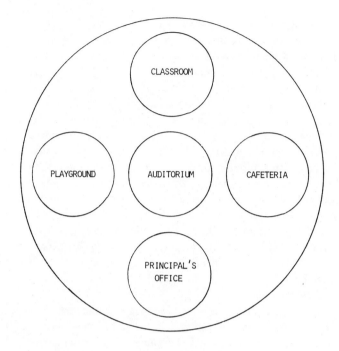

ELEMENTARY SCHOOL DOMAIN

Figure 12. Important settings within a child's elementary-school domain.

education, spiritual activities, recreation, employment, and commuting (Campbell, 1981; Stokols & Novaco, 1981). An even broader unit of contextual analysis is the individual's *overall life situation* (Magnusson, 1981), consisting of the major life domains in which a person is involved during a particular period of his or her life. The life situation of an elementary-school student, for example, might include the home, school, church, and recreational domains.

The preceding classification of environmental units suggests an important distinction between the ecological environment as it exists in reality and the environment as it is modeled in relation to a particular individual or group. For example, a comprehensive ecological description of a classroom setting would include the names and personal attributes of its members, all individual and group activities occurring during a given period, and the various furnishings and other physical features of the room itself. For purposes of contextual analysis, however, the researcher would focus only on those aspects of the setting that are presumably relevant to a particular phenomenon (e.g., the role of spatial density in moderating a child's reactions to high levels of aircraft noise). Thus, contextual variables provide highly selective representations of an individual's or group's environment. Even as the scale of the environmental units examined in an analysis increases (e.g., from a focus on situations to an analysis of life domains), the actual number of contextual variables chosen to represent the relevant environmental dimensions might remain relatively small.

Environmental Scale and Contextual Scope. Just as environmental units can be arrayed with respect to their scale or complexity, contextual analyses can be compared in terms of their relative scope. The *contextual scope* of research refers to the scale of the contextual units included in the analysis. Specifically, a set of target variables can be examined in relation to the immediate situation in which they occur or in relation to broader and more remote segments of the individual's life situation and life history. Moreover, a contextual analysis may exclude any reference to the social-structural or cultural context of the target phenomenon; or alternatively, it may examine the moderating influence of social-structural conditions within the immediate target situation as well as the influence of more distant and pervasive sociocultural conditions. Thus, theoretical and empirical analyses of stress can be compared on at least three different dimensions of contextual scope: namely spatial, temporal, and sociocultural scope.

The *spatial scope* of an analysis increases to the extent that it represents places, processes, and events occurring within a broad rather than narrow region of the individual's (or group's) geographical environment. For instance, the spatial scope of research on children's noise exposure and blood pressure at school can be broadened by considering the moderating influences of conditions within nonschool domains (e.g., noise levels at home). Here, the contextual variable of home noise levels is geographically separated from the school site at which the target variables are assessed. Also, the measurement of stress outcomes might be limited to blood pressure readings taken at school or, alternatively, could be expanded to include measures of residential experiences as well (e.g., parents' and children's reports of emotional stress at home; see Figure 11).

The *temporal scope* of an analysis increases to the extent that it represents places, processes, and events experienced by the individual (or group) within an extended rather than narrow time frame. For instance, children's reactions to noise at their present school might be analyzed solely in relation to their experiences at that school, or they could be analyzed also in relation to their experiences at previous schools (e.g., at kindergarten or elementary schools attended earlier) and with respect to their expectations about remaining at or leaving the current school in the near future (see Figure 13). In the latter case, the contextual variables refer to places, processes, and events that are temporally removed from the period during which the target variables are assessed.

Finally, the *sociocultural scope* of an analysis increases to the extent that it describes behaviorally relevant dimensions of an individual's (or group's) sociocultural environment. For instance, children's response to aircraft noise at school could be examined in relation to the normative structure (e.g., traditional vs. "open-plan" formats) or racial composition of the classroom situation. Moreover,

TEMPORAL DISTRIBUTION OF PREDICTOR VARIABLES

NOISINESS OF PREVIOUS SCHOOL ATTENDED MONTHS ENROLLED IN PREVIOUS SCHOOL	SCHOOL NOISE EXPOSURE HOME NOISE EXPOSURE	ANTICIPATED TRANSFER TO QUIETER SCHOOL ORIENTATION TOWARD FUTURE ACADEMIC ACHIEVEMENT
PREVIOUS EXPERIENCE	CURRENT SITUATION	ANTICIPATED EXPERIENCE

Figure 13. The temporal distribution of predictor variables within an analysis of children's response to noise exposure. The temporal scope of research increases to the extent that it examines places, processes, and events that are temporally remote from the period during which the target variables are assessed.

it might be hypothesized that children's response to aircraft noise at school is affected by even more pervasive, or community-wide, social conditions. For instance, a child's school might be located in a neighborhood whose residents have organized a campaign to have existing flight patterns altered by the local airport so that they are less disruptive of home and school activities. Perhaps community concern about airport noise moderates children's physiological response to school noise levels by sensitizing parents and children to the undesirable impacts of prolonged noise exposure. Here, the relationships between school noise levels and children's blood pressure are represented as being nested within, or dependent on, the social structure of the classroom and the local community as a whole.

The preceding dimensions suggest a continuum of research ranging from narrow to broad contextual scope. At the "narrow" end of the continuum are those analyses that are conceptually and methodologically reductionistic. That is, the conceptualization and measurement of the phenomena under study are limited to target events that occur within a spatially, temporally, and socioculturally restricted situation.[5] Located at the "broad" end of the continuum are analyses in which the target variables are examined in relation to conditions occurring with a wide rather than restricted region of the individual's geographical and sociocultural environment and within an extended rather than narrow interval of the individual's life experience.[6]

[5]See, for example, the critiques of conceptual and methodological reductionism in behavioral research, presented by Gergen (1973, 1982) and Schwartz (1982). See also McGuire's (1983) distinction between "convergent" and "divergent" research styles.

[6]Note that the spatial, temporal, and sociocultural dimensions of contextual scope can be considered separately for predictor and outcome variables. That is, the scope of the predictor variables might be wide, whereas that of the criterion variables is narrow, or vice versa. Alternatively, the scope of both sets of variables might be broad or narrow. To simplify the ensuing discussion, we will not examine all of these subsets of contextual analysis. Rather, we will focus on analyses of stress that vary mainly with respect to the contextual scope of their predictor variables, and whose measures of performance and stress are for the most part drawn from the immediate stressor situation.

Any attempt to discover the effective context of a phenomenon begins with some preliminary deliberation about the appropriate scope and content of the analysis. The researcher must decide how broadly to construe the relevant context of the phenomenon and which contextual factors exert a significant rather than trivial influence on the phenomenon. The broader and more complex the contextual units of analysis, the greater the potential range of factors—psychological, sociocultural, architectural, and geographic—that can affect a person's relationships with his or her surroundings. For any phenomenon, the researcher must try to determine at what point increasing or decreasing the scope of the contextual variables brings diminishing returns in terms of the explanatory power of the analysis.

When the researcher has access to extensive prior information about the situational variability of a phenomenon, decisions about the appropriate scope of the analysis become relatively straightforward. Lacking such information, however, it may be useful to adopt a broad contextual orientation during the early phases of an investigation (e.g., during the theorizing that often occurs prior to the design and implementation of the research). This approach avoids a premature narrowing of contextual scope while permitting the gradual deletion of irrelevant situational dimensions as additional insights and information about the phenomenon are acquired. Adopting a contextually narrow perspective at the outset may unduly limit the possibilities for discovering the situational moderators of the target phenomenon as the research proceeds.

The systematic assessment of spatial, temporal, and sociocultural scope does not ensure that the key contextual moderators of a phenomenon will be discovered. Nonetheless, these dimensions of contextual scope are useful in that they offer a set of analytical coordinates for mapping diverse phenomena in relation to alternate clusters of contextual variables. This exploratory mapping process often can enhance efforts to discover the effective context of a phenomenon by highlighting important geographic, temporal, and cultural aspects of the phenomenon that might otherwise be overlooked.

Having delimited the scope of analysis and selected certain contextual variables for further assessment, the researcher may then consider how best to represent those variables in operational terms. We turn now to three additional dimensions that pertain to the content and measurement of contextual variables.

Objective and Subjective Representations of Contextual and Target Variables. The environmental context of stress can be represented in objective terms, irrespective of the individual's perception and cognition (e.g., existing levels of aircraft noise in a residential area), or, alternatively, from the subjective vantage point of individual or group (e.g., residents' perceptions of the noisiness of the area). Similarly, stress reactions can be assessed objectively through physiological recordings of arousal (e.g., biochemical assays of adrenalin levels in persons exposed to noise) and records of overt behavior (e.g., observers' ratings of teacher–student interaction in noisy and quiet classrooms), or subjectively through self-report measures of stress symptoms.

For several decades, it was common practice for psychologists to construe the environment *either* in terms of its objective, material features (Brunswik, 1943; Gibson, 1960; Watson, 1913) *or* in terms of the individual's subjective impressions of those features (Freud, 1927; Lewin, 1936; Maslow, 1962; Neisser, 1967). The former approach reflects a bias toward environmental determinism or the tendency to interpret people's behavior entirely in relation to the objective properties of their physical and social environments (Franck, 1984). The latter perspective reflects a bias toward subjectivism (Sampson, 1981), whereby the direct (or nonpsychologically mediated) effects of environmental conditions on behavior are ignored. Rarely were these two orientations combined within the same analysis.

Research on human stress over the past two decades, however, indicates substantial progress toward the integration of objective and subjective representations of environment and behavior. Building upon Lazarus's (1966) psychological conceptualization of stress as distinct from purely medical or physiological models, subsequent theoretical and empirical work on crowding, noise, air pollution, commuting hassles, and a host of other stressful life events has emphasized the relationships between objective and subjective representations of the stressor and its impact on health and behavior (Chapters 1 and 5; Baum & Epstein, 1978; Cohen & Weinstein, 1981; Dohrenwend & Dohrenwend, 1974; Evans, 1982; Gore, 1981; Stokols & Novaco, 1981).

There are several theoretical and methodological reasons for combining objective and subjective perspectives in research on environmental stress. At a conceptual level, the representation of environments in both objective and subjective terms can increase researchers' sensitivity to two types of contextual effects on behavior: those that are mediated by cognitive and interpretive processes and those that are not. Although cognitive appraisal processes have been strongly emphasized in psychological studies of stress (Glass & Singer, 1972; Lazarus, 1966), certain objective features of physical environments that may be related to occupants' stress reactions have been largely ignored in that research (Evans & Cohen, 1986). For example, there is increasing evidence to suggest that certain physical features of the ambient environment, such as the presence of windows in buildings (Ulrich, 1984; Verderber, 1982) or the accessibility of trees, lakes, and other natural elements within urban areas (Kaplan, 1983; Ulrich, 1979, 1983) have important effects on behavior and well-being. By systematically considering objective features of the ambient environment as well as people's subjective interpretations of the immediate stressor, the researcher is likely to identify important aspects of the target situation that would be neglected by relying exclusively on either of these approaches.

Before psychologically salient sources of life stress can be identified for a particular individual, it is necessary to consider certain objective dimensions of the person's overall life situation. In this regard, the contextual mapping procedure mentioned earlier can be especially useful in revealing the number and spatial dispersion of the individual's important life settings (e.g., home, workplace, children's school or day-care facilities), and objectively measured levels of noise, density, poor architectural design, and other stressors within those settings. By charting the individual's typical activity patterns (or overall life situation) in objective terms, the researcher is able to define a range of settings and conditions within those settings that may be subjectively stressful for the individual (Murrell & Norris, 1983).

An objective map of a person's life situation can serve as a valuable methodological tool by highlighting important life domains and, thereby, guiding the focus and content of questionnaire or interview items designed to assess subjective stress reactions. But, there is another important methodological reason for combining objective and subjective representations of environment and behavior in research on stress. Because subjective measures of environment and behavior share common method variance (e.g., variance attributable to response sets, memory distortions, denial, and/or other psychological defense mechanisms), the degree of spurious correlation between these measures is likely to be greater than between objective and subjective (or subjective and objective) measures of environment and behavior (Derogatis, 1982; Guski & Rohrmann, 1981). By combining objective and subjective representations of both predictor and response variables, it is possible to counterbalance the respective strengths and weaknesses of these measurement approaches (Cohen, Kamarck, & Mermelstein, 1983) and to assess the degree of convergence and divergence among the various measures included in the analysis (Campbell & Fiske, 1959; Webb, Campbell, Schwartz, Sechrest, & Grove, 1981).

Just as the dimensions of spatial, temporal, and sociocultural scope provide a basis for charting the context of environmental stress, the systematic analysis of both objective and subjective features of environment and behavior can serve as an additional strategy for expanding our understanding of stress phenomena.

Individual and Aggregate Level of Analysis. The distinction between contextual and noncontextual features of the environment depends partly on whether the focus of analysis is on an individual or some aggregate of individuals. Thus, the term *contextual* does not apply to any and all attributes of real settings but, instead, only to those conditions that constitute the external environment of a particular individual or group. Conditions of crowding and noise at work, for example, are aspects of the individual's job environment that may influence his or her job satisfaction and productivity. But if the focal unit of analysis is the work organization as a whole, then the exposure of workers to crowding and noise would be viewed as an intrasystem (or nonenvironmental) factor rather than as a condition of the company's external environment. Accordingly, if we were interested in estimating the profitability and longevity of the company, it might be necessary to look beyond the physical and social conditions within the organization to more remote external events such as government monetary policies and competition from other corporations—all of which may affect the long-range prospects of the company.

Earlier, ecologically oriented conceptualizations of stress have differentiated between, and attempted to integrate, individual and aggregate levels of analysis (Dooley & Catalano, 1980; Moos, 1979; Stokols & Novaco, 1981). For instance, Stokols and Novaco described four distinct foci of research, based on a consideration of both *environmental conditions* and *stress reactions* at either *individual* or *aggregate levels:* (a) the relationships between specific environmental conditions faced by an individual and his or her experiences of, and efforts to cope with, those conditions; (b) the relationships between environmental conditions faced by an individual and manifestations of organizational or community well-being; (c) the links between community-wide environmental conditions (e.g., highly pervasive stressors such as traffic congestion, air pollution, or economic change faced by large numbers of community residents) and individual well-being; and (d) the links among community-wide environmental conditions and levels of organizational or community well-being.

The preceding research foci are somewhat overlapping and interdependent. In many instances, the same conditions that confront a particular individual are experienced by large segments of the population as well. Also, an assessment of personal stress reactions is often crucial for an understanding of community-wide health statistics. Nonetheless, because of the practical constraints of time, resources, and expertise, most theoretical and empirical analyses of stress have emphasized at most one, rather than several, of the research focuses outlined above.

Yet, the cross-level analysis of contextual and target variables from both individual and aggregate perspectives can be advantageous for several reasons. First, combining these analytical perspectives is a prerequisite for examining the interactive effects of acutely stressful events experienced by individuals (e.g., the death of a family member) and more pervasive, ambient environmental stressors (such as transportation or economic hassles) on individual behavior and well-being. The interplay between community-wide and individual-level stressors has only recently begun to receive theoretical and empirical attention (Baum, Fleming, & Davidson, 1983; Campbell, 1983; DeLongis, Coyne, Dakof, Folkman, & Lazarus, 1982; Dooley & Catalano, 1984; Evans, Jacobs, Dooley, & Catalano, 1982; Kasl & Cobb, 1982).

Moving from the community level to setting-specific analyses of stress, several recent studies have examined the interplay between aggregate indexes of organizational climate and social support and personal life events (Gore, 1981; Payne, 1980; Shumaker & Brownell, 1984). Yet, several theoretical and procedural issues remain to be addressed in future extensions of this research. Among these scarcely researched issues are the processes by which social contexts such as educational settings and work organizations promote supportive or nonsupportive interpersonal relations, and the development of reliable and valid procedures for measuring the pathogenic or health-promotive qualities of social environments (cf. Cohen & Syme, 1985).

The consideration of both individual and aggregate levels of analysis is also crucial when planning or evaluating intervention programs to reduce environmental stress. Consider, for example, the development of corporate ride-sharing programs to alleviate commuter stress and to improve organizational effectiveness (Stokols & Novaco, 1981). The proposed intervention can be analyzed in relation to alternative representations of the environment as viewed from the perspective of individuals, aggregates, or both. The cost-effectiveness of corporate van-pooling programs might be evaluated quite differently depending on whether the target phenomenon of interest was commuter stress and well-being, organizational effectiveness and profitability, or the quality of life at a community level. At the first level, an evaluation of the proposed program would involve an individually oriented analysis of the travel conditions and stress levels exhibited by participants in the van-pool program and among a comparable sample of automobile commuters. At an organizational level of analysis, the cost-effectiveness of the van-pool program might be assessed in relation to aggregate levels of employee morale, productivity, illness-related absence from work, and attrition. And at the community level, the effectiveness of the program could be measured in terms of its impact on residents' aggregate perceptions of traffic congestion and ambient noise levels in their neighborhood (e.g., Appleyard, 1981). Only by considering the proposed intervention in relation to individual as well as aggregate levels and subjective as well as objective descriptors of the environment can the appropriate criteria of cost-effectiveness be identified and understood.

Thus, for both theoretical and pragmatic reasons, the systematic analysis of stress phenomena at individual and aggregate levels can serve as another valuable strategy of contextual research.

Partitive and Composite Representations of Contextual Factors. In this section, we differentiate between two theoretical strategies for representing the contextual dimensions of stress, namely *partitive* and *composite* representations of context (Stokols, 1983). *Partitive representations* view the multiple predictor variables included in the analysis (e.g., classroom density and the number of months spent in the classroom by a particular child) as independent features of a situation that interactively affect one or more criterion measures. Moreover, the interactive relationships among these predictor variables are assumed to hold across a diversity of settings. For instance, the variables of density level and exposure time would be viewed as independent parts of the research setting (e.g., a Los Angeles elementary school), and the interactive effects of these variables on performance or health, observed in the school setting, would be viewed as representative of the relations among these variables in a host of other situations and settings (e.g., at a children's summer camp, in an overcrowded office, or at a residential facility for elderly persons).

In *composite representations* of the context of stress, predictor variables are defined so as to reflect the interdependence rather than independence among multiple components of the situation. An example of a composite term in the crowding literature is the distinction between *primary* and *secondary* environments (Stokols, 1976). This concept differentiates among high-density situations in terms of multiple attributes (time of exposure, importance of personal activities, degree of participants' anonymity). Rather than incorporating all of these dimensions as independent variables in a regression equation, a composite analysis would dichotomize situations into the two proposed types and then sample reactions to high densities within the two kinds of settings. Thus, the unit of analysis in a composite representation of context is the situation or setting (units characterized by an organizational structure that is "greater" or qualitatively different than the sum of its parts) rather than separate and independent parts of the situation. The development of concepts that treat the situation as the unit of analysis can provide theoretical criteria for comparing diverse settings in terms of their crucial differences and similarities. Such criteria are essential for implementing the "representative sampling" (of situations) advocated by Brunswik (1956) and for gauging the ecological validity of research findings and proposed environmental interventions (Petrinovich, 1979).

Note that composite concepts of context involve an intentional "chunking" of situational components into summary concepts and measures that represent the interdependence of environmental and personal components of the setting. Whereas partitive theories expand the analysis of a target phenomenon numerically (by incorporating supplementary situational predictors of criterion measures), composite concepts essentially reduce the number of independent predictor variables by substituting summary measures of the interrelations among multiple elements of the situation. Thus, composite terms often provide a more efficient representation of an environment than partitive descriptors because they consolidate interdependent environmental attributes into a single unifying construct.

Another difference between partitive and composite approaches to the study of contextual dimensions of stress is reflected in the selection and analysis of criterion variables. In the former approach, multiple criteria of performance, physiology, and affect, measured at different locations and points in time, are incorporated into the analysis on the basis of theoretical assumptions regarding a particular psychological phenomenon (e.g., the occurrence of stressful "aftereffects" following exposure to acute or chronic stressors). To the extent that the phenomenon is construed as situationally nonspecific, the adaptive costs or aftereffects of exposure to stressors should occur irrespective of situational dimensions. Although multiple measures of aftereffects might be gathered in geographically separate settings (e.g., at the source and away from the source of stress), the inclusion of these assessments would not be guided by a situation-specific view of stress.

If, on the other hand, the aftereffects phenomenon is hypothesized to depend on situational dimensions such as environmental predictability, controllability, or other structural qualities, then stress reactions would be measured across settings that are thought to differ on the relevant dimensions. Although both partitive and composite analyses of stress may incorporate criterion variables of broad contextual scope, the crucial difference between the two approaches is that in the latter, the criteria are specified on the basis of hypothesized interrelations between situation structure and stressor effects, whereas in the former, they are included to assess the target phenomenon irrespective of situation structure.

For instance, a researcher might be interested in the severity of the stressor (e.g., noise amplitude and duration) and the persistence of aftereffects. Following noise exposure in the laboratory, the researcher might assess the individual's social behavior shortly after leaving the lab and the individual's affect while at home on the evening of the experiment. Here, cross-setting measures of stress reactions are gathered to assess the persistence of aftereffects, but the measures are not in any way tied to a theory about the structure of the experimental or postlaboratory situations. Rather, they are included to gauge the interactive effects of independent variables (noise amplitude and duration) on the persistence of stress. On the other hand, a study that was designed to assess the stress-buffering effects of "restorative environments" (Kaplan, 1983) might hypothesize differential persistence of stress as a function of type of environment occupied by the individual following his or her exposure to the stressful laboratory situation. In that case, the prediction and measurement of stressor aftereffects would be explicitly derived from a theory about the restorative versus nonrestorative aspects of poststressor environments.

The utility of partitive versus composite perspectives on the context of stress depends largely on the level of interdependence that exists among situational components. Whereas "transactional" theories of situations (e.g., Altman & Rogoff, 1986; Barker, 1968; Wapner, 1986) treat interdependence as a constant or a given, the present analysis views interdependence as a variable (Weick, 1979). For instance, many person–environment encounters such as those that occur in temporary short-term situations involve minimal interdependence among individuals and the events comprising those situations (e.g., public transportation settings). In other instances, stressors are experienced within the context of highly structured settings (e.g., within home, school, or workplace). In transitory and unstructured situations, composite concepts of situational structure would be irrelevant and superfluous. In such instances, environmental and personal moderators of stressors can be viewed as relatively independent elements of the setting. In more structured settings, however, concepts that articulate stress-relevant dimensions of situational structure may be more valuable than analyses of the interactions among separate environmental and personal variables. In more complex and organized contexts, then, composite terms potentially offer a more efficient and powerful approach to the study of stress because they identify important structural features of settings that influence occupants' reactions to stressors, and they consolidate multiple environmental and personal attributes into a smaller number of summary variables reflecting the interdependence among those attributes.

In summary, the selective and appropriate use of partitive and composite terms can help avoid a haphazard atheoretical approach to the description of environments and can encourage, instead, a more systematic theoretically based assessment of the cross-situational variability of stress phenomena.

Applying Principles of Contextual Analysis to Empirical Research on Stress

In earlier sections of this chapter, we characterized a contextual theory as one that predicts and explains the cross-situational variability of a phenomenon. We then discussed criteria for judging the adequacy of contextual analyses and described a set of basic dimensions on which various behavioral phenomena can be contextually mapped. These mapping dimensions were presented as a framework for organizing contextual theorizing and research on stress.

To what extent have these assumptions and strategies of contextual research been implemented in empirical studies of environmental stress? In this section, we summarize recent empirical evidence concerning contextual dimensions of stress. This overview is by no means exhaustive, but it does suggest that empirical analyses of contextual variables within particular situations and across multiple settings and time periods are prevalent in the stress literature. Also, some research attention has been given to the relationships among objective and subjective predictors of stress reactions and the links between individual and aggregate levels of analysis. Less explicit consideration has been given to the development of composite versus partitive analyses of stressful situations. But, by and large, the recent literature on environmental stress reflects an increasing emphasis on the importance of contextual factors.

Following our review of earlier empirical research on stress, we will present more recent findings from the Los Angeles Noise Project that are relevant to contextual dimensions of stress. Some of these contextual analyses have been alluded to in preceding chapters of this volume, but most have not been reported previously. We will attempt to integrate the findings from these analyses with previous contextual research on environmental stress.

Our discussion of empirical research on stress will be organized in relation to the four dimensions of contextual analysis discussed in the preceding section of this chapter. This organizational strategy is somewhat oversimplified in the sense that the four dimensions are inherently interrelated. For example, any attempt to examine the role of multiple life domains in the etiology of stress necessarily is linked to a series of explicit or (usually) implicit decisions about whether to represent these domains in objective and/or subjective terms, at individual and/or aggregate levels, and from a partitive or composite perspective. Thus, many studies may be simultaneously relevant to multiple dimensions of contextual analysis.

On the other hand, our strategy of grouping individual studies and research programs according to their relative emphasis on each of the four contextual dimensions is heuristically useful, for it reveals certain theoretical emphases and gaps in the existing literature on environmental stress. For instance, most theories and empirical studies of stress do not specifically consider all four dimensions of contextual analysis but, instead, focus on a small subset of them (e.g., spatial and temporal scope) while treating other aspects of contextual representation (e.g., partitive vs. composite perspective) in an implicit, rather than explicit, manner. The concluding section of this chapter will address the theoretical and empirical emphases suggested by our review of the literature and highlight important directions for future contextual research on stress.

Analyses of the Spatial, Temporal, and Sociocultural Scope of Stress

A broadening of contextual perspective in stress research is especially evident in recent analyses of the intrasituational and intersetting moderators of stress experiences and outcomes. These assessments of multiple conditions within the immediate stressor situation as well as those within geographically remote life domains illustrate the increasing spatial scope of stress research.

Intrasituational Moderators of Stress. In research on density, noise, and other environmental demands, earlier work focused on the health and behavioral impact of isolated acute stressors while neglecting to consider the situational context in which these events are embedded. This overemphasis on the impact of isolated stimuli and events precluded an analysis of issues such as the ratio of uncontrollable/controllable events within a situation, the relative importance of personal and group goals with which a stressor interferes, and the extent to which the impact of a stressor is ameliorated by the availability of concurrent, compensatory rewards (Stokols, 1979; Wortman & Brehm, 1975). Subsequent studies, however, have given considerable attention to situational moderators of stress reactions.

Research on residential density, for example, has documented the specificity of crowding stress as a function of several situational factors including family structure and activity patterns (Baldassare, 1981; Gillis, 1979), group cohesiveness (Baum, Harpin, & Valins, 1975), opportunities for privacy regulation (Schmidt, Goldman, & Feimer, 1979; Verbrugge & Taylor, 1980), building height and other aspects of architectural design (Aiello & Baum, 1979; Baum & Valins, 1977; Gillis, 1974; Saegert, 1984). Also, in nonresidential settings, the stressful consequences of high density have been found to be mediated by situational factors such as the presence or absence of room decorations (Worchel & Teddlie, 1976), the degree of room partitioning (Stokols, Smith, & Prostor, 1975), the complexity of tasks being performed by occupants of the area (Evans, 1979), the degree of cooperation or competition among group members (Stokols, Rall, Pinner, & Schopler, 1973), and the gender composition of the group (cf. Epstein & Karlin, 1975; Freedman, 1975).

The stressful consequences of exposure to high-density situations has been one of the most extensively studied topics in the field of environmental stress. Thus, the previously mentioned studies of density and crowding are useful in illustrating the diversity of situational factors that affect people's reactions to environmental stressors. Although the importance of specific contextual factors varies from one type of stressor to another, people's responses to environmental demands in general are likely to be influenced by a host of situational factors. Recent reviews of people's reactions to chronic noise exposure, uncomfortable temperature, travel restraints, air pollution, and acute life events all have underlined the importance of situational variables (e.g., the kinds of activities disrupted by the stressor) as moderators of stress reactions (Bell & Greene, 1982; cf. Cohen & Weinstein, 1982; Dohrenwend & Dohrenwend, 1981; Evans & Jacobs, 1982; Stokols & Novaco, 1981).

Intersetting Moderators of Stress. The expanded geographical scope of stress research is suggested by the large number of recent studies that have explicitly assessed the role of multiple settings and life domains in the etiology of stress. For example, the impact of environmental stressors experienced in one setting on the manifestation of stress reactions in other settings has been documented in studies of crowding and noise. Baum and Valins (1977) observed that experiences of unwanted social interaction and crowding in dormitory settings were associated with more competitive tendencies and increased interpersonal distance in laboratory settings. Saegert (1984) found that elementary-school students from high-density apartments obtained lower scores on vocabulary tests and exhibited higher levels of behavioral disturbance at school than their low-density counterparts. Similarly, Cohen, Glass, and Singer (1973) found that prolonged exposure to high levels of traffic noise in a high-rise apartment building was associated with lowered reading achievement scores and impaired auditory discrimination in children's elementary-school settings. Also, Lukas, Dupree, and Swing (1981) found that children's exposure to high levels of noise at home increased their sensitivity to noise exposure at school. And, in a study of the physiological consequences of work demands, Frankenhaeuser (1981) observed that periods of increased "overtime" at work were associated with elevated urinary adrenalin levels among workers at their homes during evening hours of relaxation. These findings suggest that the behavioral and physiological aftereffects of exposure to stressors in one life domain may persist beyond the confines of that situation into other life domains (Cohen, 1980).

The relationships among multiple life domains and their joint contributions to the etiology of stress have received considerable attention in the recent literature. Many theorists have suggested that the sheer number of life domains or social identities possessed by an individual is positively associated with psychological and physical well-being (McKennell & Andrews, 1983; Murrell & Norris, 1983). In keeping with this perspective, Thoits (1983) found that the loss of social identities through various life events was associated with increased psychological distress. Verbrugge (1983) found that employed married parents displayed higher levels of mental health than individuals possessing fewer or none of these social attributes. And Gore and Mangione (1983) observed that the absence of marriage and employment domains was associated with increased depressive symptomatology among both males and females.

Other researchers have contended that the crucial determinant of vulnerability to stress is not the absolute number of life domains in which a person is involved but rather the nature of the links that exist between the different areas of one's life. A number of studies have documented the negative impacts of work stressors on spousal relationships and family functioning (Brett, 1980; Hall & Hall, 1980; Moos & Moos, 1983). One mechanism that has been proposed to explain the sometimes detrimental effects of work demands on family life is the notion of role conflict (Jacobi, 1984; Michelson, 1986a; Pearlin & Schooler, 1978). Michelson found that mothers employed full-time reported higher frequencies of conflict between their maternal and job roles than those employed part-time. The full-time employed mothers also expressed less satisfaction with the balance between work and family responsibilities than the part-time employed mothers. Michelson also found that married women with full-time jobs reported the highest levels of daily tension, as compared with nonemployed housewives (who reported the lowest tension levels) and part-time employed wives (who reported moderate tension levels). This pattern also was reflected in respondents' concerns about the quality of their marriage and parenting relationships. However, on a measure of health complaints, an opposite pattern was found, with full-time employed wives reporting the lowest levels of somatic complaints.

High levels of perceived conflict across different life domains was found by Jacobi (1984) to have both direct and interactive effects with other environmental stressors on levels of well-being among female college and graduate students. Specifically, high levels of interdomain conflict (e.g., between family, school, and recreational domains) were associated with reduced levels of self-reported energy, spirits, and health. Moreover, among students reporting a higher prevalence of environmental stressors in their home or apartment (e.g., crowding, noise), those with high levels of interdomain conflict reported more overall life stress than those with lower levels of role conflict.

Other studies also suggest that conditions within various life domains have both direct and interactive effects on physical and emotional well-being. Greenberger, Steinberg, and Vaux (1982), for example, found that low levels of congruence or satisfaction within each of three life domains (home, work, and peer relationships) were associated with health and behavioral disorders among adolescents. Also, Klassen (1981) observed that low levels of social support within the same three domains were interactively linked to increased incidence of violent behavior among adult psychiatric patients. And a study of commuting stress by Stokols and Novaco (1981) found that the effects of travel distance on drivers' blood pressure and task performance were mediated by conditions within the home and workplace (e.g., perceived choice in moving to the current residence and degree of job involvement).

On balance, the preceding studies highlight the interdependence of events and experiences that occur within the multiple domains of one's life situation. Environmental demands experienced in one setting may be exacerbated or buffered by the availability of coping resources in other domains (Bronfenbrenner, 1979). Moreover, the aftereffects of environmental stressors are not always confined to the settings in which the demands occur, and they sometimes "spill over" into seemingly unrelated life domains. Thus, the behavioral and health impacts of stressful stimuli and events can be better understood within the context of the individual's multiple life domains than in isolation, or apart from, those settings.

Temporal Aspects of Environmental Stress. Recent theoretical and empirical work on stress reflects an emphasis on temporal issues. First, the focus of stress research has shifted from short-term assessments of acute environmental demands toward longitudinal analyses of coping processes and health outcomes associated with chronic stressors (Lazarus & Launier, 1978; Pearlin, Menaghen, Lieberman, & Mullan, 1981). This broadened temporal perspective is exemplified by recent theoretical distinctions between ambient stressors, daily hassles, and acute life events (Campbell, 1983; DeLongis *et al.,* 1982), and technological versus natural disasters (Baum, *et al.,* 1983).

Second, the emphasis on chronic stressors has coincided with an increasing concern about the processes of reminiscence and anticipation, by which individuals evaluate their immediate situation in relation to previous and anticipated life experiences (Albert, 1977; Caplan, 1983; Merriam, 1980). For instance, Rowles (1981) suggested that immediate or imagined proximity to "autobiographically significant" places may enable elderly individuals to maintain a sense of self-continuity, thereby enhancing emotional stability. Others have hypothesized that favorable comparisons of the current life situation vis-à-vis earlier life stages enhances individuals' adaptation to stressful environmental transitions such as residential relocation and retirement (Gutek, Harris, Tyler, Lau, & Majchrzak, 1983; Stokols, Shumaker, & Martinez, 1983; Wapner, 1981).

Several studies suggest that optimistic attitudes about the future can reduce the stressful consequences of current stressors. Kobasa (1979, 1982) found a significant relationship between individuals' commitment to current activities and future goals and their resistance to health problems following stressful life events. Stokols *et al.* (1983) observed that among individuals expressing dissatisfaction with their current residence, health problems were more prevalent among those who perceived attractive residential options to be inaccessible rather than available to them in the future. Jacobi (1984) found that high levels of optimism about the future were directly associated with enhanced health status and also reduced the severity of stress associated with cross-domain role conflict among female university students. Moreover, Kasl and Cobb (1982) and Cohen and Hoberman (1983) reported that the perceived availability of social support buffered the detrimental impact of negative life events on physical and emotional well-being.

Research on the temporal context of environmental stress reflects objective as well as subjective conceptions of time. The previously mentioned studies of retrospective and future-oriented appraisals exemplify subjective representations of time. At an objective level, the flow of time as measured by recurring cycles of natural and human events exerts direct effects on emotional and physical well-being. For instance, objectively measured levels of the temporal and geographical constraints on people's daily routines have been found to be associated with increased irritability and social strain among family members (Cullen, 1978; Hall & Hall, 1980; Michelson, 1985).

The duration of exposure to physical stressors and negative life events is another objective index of time that is closely related to the intensity of stress reactions. Cohen, Krantz, Evans, and Stokols (1981) found that the negative effects of school-noise exposure on children's cognitive performance were more pronounced among those who had been enrolled at the school for longer periods (but this interactive pattern was not evident on measures of blood pressure; see Chapters 3, 4, & 5). Stokols *et al.* (1983) reported that length of residence was associated with increased illness symptoms among individuals who had exercised little choice in deciding where to live and were dissatisfied with their current dwelling. Other studies, however, have found that increased time of exposure to certain stressors such as urban air pollution (Evans & Jacobs, 1982), involuntary unemployment (Kasl & Cobb, 1982), and the demands of returning to college after several years away from the university (Jacobi, 1984) are associated with reduced, rather than elevated, stress levels over time. Still other studies suggest that the relationship between time of exposure to stressors such as high density or noise and the severity of stress reactions is mediated by personal dispositions such as internal/external control expectancies (cf. Baron *et al.*, 1976) and the Type A behavior pattern (cf. Krantz *et al.*, 1974).

It is clear that the relationships between the nature and intensity of the stressor, time of exposure, and situational and personal moderators of stress remain to be more fully elucidated in future experimental studies. For instance, the link between increased time of exposure to the stressor and reduced severity of stress, observed in certain studies, may be attributable to a self-selection factor. Perhaps, those who remain in the situation are those who can best cope with it, whereas those who leave are those who were most distressed and least able to cope during their exposure to the stressor. This attrition factor has not been adequately assessed in many longitudinal analyses of stress (see Chapter 2 for a more extensive discussion of this methodological issue).

The Sociocultural Context of Environmental Stress. From a contextual perspective, social structural and cultural factors are thought to play an important role in the etiology of stress. Studies of crowding, for example, indicate that group structure, composition, and cohesiveness moderate the intensity of stress reactions to high-density situations (Baldassare, 1981; Baum *et al.*, 1975; Cassel, 1974; Freedman, 1975; Epstein & Karlin, 1975; Stokols *et al.*, 1973). Also, research on territoriality and personal space suggests that cultural norms influence the nature and intensity of people's reactions to territorial infringements and interpersonal proximity with strangers (Aiello & Thompson, 1980; Altman & Chemers, 1979).

In the literature on coping and stressful life events, some theoretical attention has been given to the role of cultural and social-structural factors in the etiology of stress. Aldwin (1984) and Mechanic (1974) emphasize the role of cultural beliefs and institutions in facilitating people's efforts to cope with stressful situations. Also, Syme (1984) cites several studies indicating significant links among processes of cultural change and social mobility, stress, and disease etiology. Helson, Mitchell, and Moane (1984) suggest that normative beliefs about the appropriate timing of personal events such as marriage, college education, and employment are culturally prescribed and that deviations from these norms (e.g., return to college or employment after several years as a housewife) are associated with greater stress than if the same activities occur in a culturally prescribed sequence. Also, Pearlin and Schooler (1978) emphasize the links among role structures within family, work, and other life domains, and the ways in which these social networks influence the efficacy of coping efforts prompted by stressful life events.

Some studies suggest that environmental stress and negative life events are differentially distributed across various subgroups of the population, with low-status groups receiving a disproportionate share of these events (Pearlin & Schooler, 1978; Syme & Berkman, 1976). Evidence that social factors moderate people's exposure and response to enviromental stressors is provided by the Los Angeles Noise Project (Cohen, *et al.*, 1980; see also earlier chapters of this volume). In our study, the noise-school and quiet-school samples had nearly equal percentages of whites and did not differ significantly on levels of parents' education or family size. Additional analyses examined variations in children's blood pressure as a function of race and school-noise exposure. Blacks and Chicanos attending noisy schools had higher systolic and diastolic blood pressure than their quiet-school counterparts. For whites, there were no main effects of noise, but an interaction between noise and length of school enrollment revealed that initial elevations of blood pressure for noise-school children disappeared as length of enrollment increased.

The previously mentioned studies strongly suggest that social and cultural factors influence people's exposure to stressors, the effectiveness of coping efforts, and the severity of stress outcomes. Accordingly, future theoretical and empirical work should build upon this data base by explicitly considering the sociocultural as well as the spatial and temporal underpinnings of environmental stress.

Objective and Subjective Predictors of Stress

Affective and cognitive processes have always been accorded a central role in psychological models of stress (Lazarus, 1966; McGrath, 1970). Recent theoretical formulations portray stress as a complex transaction between individuals and their environments, whereby sequential processes of appraisal, affect, and coping play a crucial role in mediating the behavioral and physiological outcomes of environmental demands (Coyne & Holroyd, 1982; Garber & Seligman, 1980; Lazarus & Launier, 1978; Pearlin *et al.,* 1981).

Recent studies of stressful life events have identified several subjective factors that moderate the severity of stress reactions to major life changes. Both the perceived desirability and controllability of life events are associated with less severe stress outcomes following the event (Dohrenwend & Dohrenwend, 1981; Sarason, Levine, & Sarason, 1982; Vinokur & Selzer, 1975). Also, the sense of coherence (or perceived comprehensibility and meaningfulness of life events) has been identified as an important moderator of stress associated with undesirable life changes (Antonovsky, 1979). Similarly, Jacobi (1984) and Pearlin *et al.* (1981) found that feelings of self-enhancement and mastery, prompted by efforts to cope with environmental demands, were associated with enhanced health status. Jacobi also observed that feelings of self-enhancement buffered the negative health impacts associated with role conflict across multiple life domains.

Additional evidence for the health-promotive value of pleasant experiences and beliefs is reported by Cohen and Hoberman (1983) who found that both the perceived availability of social support and an increased number of positive life events reduced the impact of negative life events on physical and depressive symptomatology. Zautra and Reich (1983) present further evidence for the stress-reducing function of positive life events, whereas Dixon (1980) emphasizes the stress-buffering effects of humor.

The crucial role of environmental controllability and predictability in mediating the stressfulness of noise and high density was consistently documented in several laboratory studies (Cohen, 1980; Glass & Singer, 1972; Sherrod, 1974). Subsequent field experimental studies corroborated the laboratory findings with additional evidence that perceived control and environmental predictability buffer the negative impact of stressful life events such as residential relocation and institutionalization of the elderly (Krantz & Schulz, 1979; Langer & Rodin, 1976; Pastalan, 1980; Rowland, 1977; Schulz & Hanusa, 1978; Stokols *et al.*, 1983).

An important theoretical issue that has been examined in recent empirical studies is the relative power of objective and subjective predictors of stress. Cohen *et al.* (1983) found that a global measure of perceived life stress was a better predictor of depressive and physical symptoms than either the number or perceived impact of normatively negative life events. In a study of the determinants of perceived environmental quality, Carp (1986; Carp & Carp, 1982) found that technical (objective) environmental indexes accounted for variance in perceptions of environmental quality and the impact of environmental conditions on well-being, when closely related perceptual measures of the environment were statistically controlled. Carp also noted that the objective and subjective indexes were complementary rather than redundant, in that their relative predictive strengths varied across different dimensions and criteria of environmental quality.

In a survey study of residential density, Baldassare (1979) found that a measure of subjective crowding contributed significantly to the prediction of housing dissatisfaction, after statistically controlling for objective crowding (number of persons per room) and other socioeconomic factors. The measure of subjective crowding, however, was not significantly associated with other indexes of behavioral and social disorders. Booth (1976) also examined the relationships among survey measures of objective and subjective crowding and several dimensions of stress, poor health, and family dysfunction. Out of 33 significant relationships between the density and stress measures, 21 involved the subjective measure of crowding, whereas 12 pertained to objective density levels. The statistical tests of the subjective measures, however, did not control for objective levels of density (or vice versa), so it is impossible to assess the independent contributions of the two crowding measures to behavioral and social disorder.

The possible links between subjective reactions to traffic noise and blood pressure were examined among the residents of both noisy and quiet neighborhoods in a recent study by Neus, Ruddel, and Schulte (1983). Within the quiet area, residents who reported high levels of noise sensitivity and annoyance exhibited higher increases in blood pressure and were more likely to have sought medical treatment for hypertension than residents with lower levels of noise sensitivity. These relationships were not observed, however, among residents of the noisy area. The authors, therefore, suggested that subjective measures of noise may be more predictive of stress outcomes when objective levels of noise are low rather than high. A related set of analyses from the Los Angeles Noise Project, examining the independent contributions of subjective noise measures to stress outcomes, will be presented in a subsequent section of this chapter.

The existing literature suggests that the empirical relationships among objective and subjective measures of the environment and levels of emotional and physical stress are quite complex and vary across different categories of stressors and situations. Among the theoretical issues that remain to be examined in future studies is the manner by which nonsalient features of the physical environment (e.g., windows, natural elements, spatial arrangements, lighting) are linked to psychological and physical well-being (Ulrich, 1983, 1984; Verderber, 1982). The findings from Saegert's (1984) study of residential crowding highlights the complexity of the relationships between objective and subjective environmental indexes. Although the children in her study from high- and low-density apartments did not report different levels of perceived crowding, they did report higher levels of anger and more frequent feelings of being bothered while doing homework. And, as noted earlier, Saegert found significant differences among the high- and low-density children on measures of behavioral disorder and reading comprehension at school. Perhaps among children, the stressful consequences of residential density are less likely to be mediated by perceived crowding than among adults, or, perhaps it is more difficult to reliably measure perceived stress in children than in adults. Also, the relative primacy of affective and cognitive processes in mediating reactions to environmental stressors is an issue that has received little empirical attention (Lazarus, 1984; Zajonc, 1984). These developmental and cognitive-process issues warrant further consideration in future stress research.

Individual and Aggregate Analyses of Environmental Stress

Whereas studies of environmental stress typically emphasize either individual or aggregate levels of analysis, the recent literature reflects a trend toward cross-level theorizing and research (cf. Dooley & Catalano, 1980; Moos, 1979; Stokols & Novaco, 1981). For instance, the behavioral and physiological consequences of community-wide stressors such as air pollution (Campbell, 1983; Evans & Jacobs, 1982), economic change (Dooley & Catalano, 1984), and technological disasters (Baum *et al.*, 1983) have been assessed. Interactive effects of community-wide stressors and negative personal events also have been reported. Evans *et al.* (1982a) found that objectively measured levels of air pollution interacted with stressful life events in predicting mental health problems. Also, Kasl and Cobb (1982) observed that levels of physiological stress among unemployed men were greater following the closing of their factory for those who reported high rather than low levels of social support. Additional analyses suggested that job loss following a plant closing had a more disruptive impact on the social relationships among former employees within an urban rather than rural setting. The potential relevance of the urban-rural continuum as a moderator of economic stress also has been mentioned by Catalano, Dooley, and Jackson (1981). Catalano *et al.* suggest that levels of under- and overstaffing (cf. Barker, 1968) may vary between urban and rural settings and that these differences in the relative supply of and demand for workers within a community may moderate the stressful consequences of economic problems experienced by individuals.

Recent time-geographic studies also reflect attempts to link individual and aggregate analyses of environmental stress. Time geography involves the systematic study of peoples' daily activity patterns, both at personal and societal levels (Carlstein, Parkes, & Thrift, 1978; Michelson, 1986b). People's daily routines are examined with respect to the levels of time pressure and spatial constraints that they impose. These inventories of daily activities reveal important sources of psychological tension and physical fatigue (Cullen, 1978; Michelson, 1986a). Moreover, the individual-level data can be aggregated across various groups within the community to suggest potential environmental interventions for reducing personal stress, community health problems, and social disorder. Michelson, for example, found that the use of day-care centers was associated with lower tension scores among single and married working mothers. Accordingly, he suggests that the number and placement of child-care facilities within a community could be planned so as to reduce personal and family stress among large segments of the population. Michelson also mentions transportation planning, "flex-time" working schedules, and strategic land use planning as additional targets for stress-reducing interventions at organizational and community levels. Corporate policies and benefit structures that may reduce the family stress resulting from work demands are also discussed by Gutek, Nakamura, and Nieva (1981).

Relative to other dimensions of contextual analysis (e.g., spatial and temporal scope; objective vs. subjective representations of the environment), the linkage between individual and aggregate perspectives on stress has received only recent and limited empirical attention. Nonetheless, the studies of individuals' response to community-wide stressors mentioned before provide exemplars for future cross-level research on stress. Among the issues that warrant further consideration in future studies are the impacts of personal stress reactions on group and organizational well-being, and the development of community settings and organizations that serve to reduce the negative impact of personal and community-wide stressors.

Partitive and Composite Analyses of Environmental Stress

As noted earlier, partitive analyses represent multiple environmental conditions and the people exposed to them as functionally independent, whereas composite analyses portray environments and their occupants as interdependent components of a common system. Thus, partitive variables denote isolated and independent elements of a situation, whereas composite terms summarize various structural or organizational properties of situations and settings.

The preceding sections of this chapter mentioned several examples of partitive and composite analyses of stress; hence, the specific findings from those analyses will not be reiterated here. The use of technical environmental indexes to predict perceptions of environmental quality exemplify a partitive representation of the environment. Objective indexes of spatial density, noise, temperature, and other environmental features also reflect partitive representations of settings. On the other hand, the use of concepts such as group cohesiveness, residential choice, job involvement, social network, and role conflict reflects a composite orientation because each of these terms highlights some aspect of the interdependence among people and their environments.

As psychological models of stress have adopted a more transactional and longitudinal perspective, the theoretical and empirical attention given to composite dimensions of situations and settings has increased. Although the stress literature reveals a strong emphasis on the interdependence of experiences and events within multiple settings and life domains, it does not yet offer theoretical guidelines for deciding when a partitive or composite approach may be better suited to the analysis of particular stressors. The impacts of high-intensity stressors occurring within transient situations, for example, may be more efficiently analyzed from a partitive perspective due to the low degree of interdependence among individuals and the environment. Thus, the classification of stressor situations according to the degree of interdependence among their components may be a basic prerequisite for matching partitive or composite orientations with the complexity of the target phenomena.

To the extent that a composite analysis is undertaken, the researcher develops some basic hunches about which structural features of situations are most relevant for understanding the impacts of a particular stressor. To date, several different composite dimensions have been examined as potential moderators of the multiple outcomes resulting from diverse stressors. Little attention has been given to organizing and integrating the available information on composite dimensions that mediate stress reactions. What may prove useful as a tool for future research is the development of a taxonomy that classifies composite concepts according to the scope of the person–environmental units that they describe, their objective or subjective representation of those units, and the individual or aggregate level of analysis that they reflect. This taxonomy of contextual terms would provide a concise summary of what is currently known about the relationships between various structural features of situations, settings, and human behavior. Moreover, the taxonomy might reveal certain potentially important facets of situations and settings that have been neglected in previous research. It might also suggest ways of consolidating existing theoretical concepts into more efficient and powerful representations of person–environment transactions. Finally, the classification of composite terms in relation to their contextual scope, objective or subjective emphasis, and individual or aggregate level could assist researchers in their efforts to identify theoretically useful dimensions of situations and settings that are commensurate with the designated scope, level, and substantive focus of their investigation.

Contextual Analyses of Los Angeles Noise Project Data

As an extension of the preceding overview of empirical research, this section of the chapter presents findings from the Los Angeles Noise Project that are relevant to contextual dimensions of stress. As we mentioned earlier, our study was not designed to address the full range of contextual issues highlighted in this chapter.

The primary focus of our research was on the interrelation among aircraft noise levels at school and various criteria of performance and physiology within that environment. At the same time, however, we were interested in exploring some potential interactions between the principal target variables and certain contextual factors. Therefore, we incorporated supplementary measures of the duration of children's enrollment in their schools, levels of noise at home, children's, parents', and teachers' subjective reactions to noise, classroom and residential density, and the presence or absence of windows in the classroom.

The interactive effects of school noise levels and the duration of children's enrollment in their schools have been described earlier (Cohen *et al.*, 1980; Cohen, Krantz, Evans, & Stokols, 1981; Cohen, Evans, Krantz, Stokols, & Kelly, 1981; and the preceding chapters of this volume) and will not be reviewed here. Other findings, such as those reflecting the interactive effects of home and school noise levels, have been alluded to in previous chapters (see Chapters 3, 4, & 5) and will be summarized later, along with the data from other contextual analyses that have not been reported previously.

Before presenting the results of our contextual analyses, certain limitations of these data should be emphasized. As in many other studies of environmental stress, our assessment of the relationships between target and contextual variables was largely exploratory rather than predictive. The conceptual framework described in the preceding section of this chapter was developed only after the Los Angeles Noise Project was completed. This framework grew partly out of our efforts to integrate the data from our supplementary analyses with the findings from other studies. Because of this sequence of events, data that would have been relevant to certain dimensions of the proposed framework were not gathered during the course of our research. For instance, our analyses are most germane to the spatial dimension of contextual scope and to the relationships between objective and subjective predictors of stress reactions. We have relatively fewer findings illustrating the links between individual and aggregate levels of analysis, and none that afford a comparison of partitive and composite measures of stressful situations. All of our statistical analyses are partitive in nature, treating the specified environmental predictors of stress as independent variables. Accordingly, our discussion of the Los Angeles Noise Project data will give greater attention to the dimensions of contextual scope, and objective versus subjective predictors of stress, than to the individual—aggregate and partitive—composite dimensions.

A second limitation concerns the restricted contextual scope of the dependent variables examined in our study. Specifically, all criteria of performance and physiology used in the analyses reported next were gathered within elementary schools at a single point in time. These criteria include children's reading and math achievement scores, number of errors on the Wepman Auditory Discrimination Test, performance speed and accuracy on the crossing-out *e*'s task under conditions of auditory distraction, and systolic and diastolic blood pressure. Although children's and parents' subjective reactions to home noise levels were assessed in our study, these measures were examined as possible moderators of stress reactions at school rather than as additional criteria of stress. Thus, our supplementary analyses do not address certain contextual issues such as the persistence of stress reactions from school to residential domains or the generalizability of children's strategies for coping with prolonged noise exposure across different settings (see Chapter 1 for a theoretical analysis of these issues). We did find, however, that certain residential conditions (e.g., home noise levels, residential density) were associated with main effects on physiology and performance at school. The relationships between environmental conditions at home and behavior at school are examined in greater detail later.

Unless otherwise noted, the Los Angeles Noise Project analyses reported here employed the standard sets of covariates specified in Chapter 2 (i.e., months enrolled at school, number of children living at home, grade, and race; height and ponderal indexes were added as covariates in the analyses of blood pressure; and nondistraction performance levels were added in analyses of children's speed and accuracy on the crossing-out-*e*'s task). Also, wherever possible, analyses were replicated across two different samples: all third and fourth graders tested at Time 1, and the additional third graders tested at Time 2. All analyses are cross-sectional rather than longitudinal.

Intrasituational Moderators of Noise Stress

Our review of empirical studies earlier in this chapter suggests that the behavioral and health effects of stressors such as noise are often mediated by other environmental conditions within the same situation. Also, on the basis of previous studies linking spatial density and the presence of windows to various measures of well-being, we suspected that these features of children's classrooms might influence their reactions to aircraft noise at school. To assess the possible interactions between school noise and other classroom variables, separate 2 × 2 analyses of covariance (ANCOVAs) were performed, each incorporating school noise level (quiet vs. noisy) as one factor and either classroom density (low vs. high) or the number of classroom windows (0–4 vs. 5–9) as the other factor. Children's scores on these factors correspond to the overall noisiness of their school and the level of density and number of windows in their classroom. Density scores were computed by dividing the number of children in each class by the square footage of their classroom. Median splits on classroom density and windows were used to assign children to either the low or high level of each factor, within both the Time 1 and Time 2 samples.

The School Noise × Windows ANCOVAs revealed a significant interactive effect on the number of lines completed in the crossing-out-*e*'s task, $F(1, 117) = 9.12$, $p < .003$, *and a marginally significant interaction on the percentage of e's found in the same task*, $F(1, 117) = 3.75$, $p < .055$. These effects suggest that the number of classroom windows is not associated with performance differences among children attending quiet schools, but, among noise-school children the number of windows is inversely related to performance speed and accuracy (see

Table 21). These effects were found within the third grade sample at Time 2 but not among the third and fourth graders tested at Time 1.

To assess the degree of multicollinearity among the ANCOVA factors, chi-square and correlational analyses incorporating school noise and classroom windows were performed. School noise levels and the number of classroom windows were not significantly correlated within either the Time 1 or Time 2 sample.

Additional correlational analyses were performed on the Time 2 data to examine the possible relationships between the number of classroom windows and various behavioral, perceptual, and physiological measures. The number of classroom windows was found to be positively associated with children's perceptions of classroom noisiness, $r(161) = .16$, $p<.02$ and with teachers' estimates of the frequency at which they were interrupted, $r(130) = .58$, $p<.001$, or had to raise their voice, $r(130) = .48$, $p<.001$, due to the intrusion of exterior noise into the classroom. Also, the number of classroom windows was inversely related to the percent of e's found during the crossing-out-e's task under auditory distraction, $r(162) = -.21$, $p<.004$, and while working on the same task without distraction, $r(164) = -.14$, $p<.035$. A positive correlation between the number of classroom windows and the number of seconds taken to complete the first of two difficult but soluble puzzles was also found, $r(51) = .35$, $p<.006$. (The last analysis applies only to those children who were assigned a soluble rather than insoluble puzzle during the first portion of the puzzle task). Despite the lack of a significant correlation between *school* noise levels and classroom windows, noted earlier, the results reported here indicate a strong relationship between the number of windows, perceived noise disruptions, and performance deficits at the *classroom* level.

The preceding findings must be viewed with caution as they are based only on the sample of third graders tested at Time 2. Another reason for caution is the relatively large number of statistical tests performed and the possibility that some of

Table 21. Crossing-Out-E's Task Performance under Auditory Distraction as a Function of School Noise and Number of Classroom Windows (for Third Grade Sample at Time 2)

School noise	Number of lines completed[a]		Percentage of E's found[b]	
	M	SD	M	SD
Quiet				
0–4 Windows	60.32	13.08	86.73	9.44
5–9 Windows	60.39	12.80	86.56	8.73
Noisy				
0–4 Windows	70.32	16.01	89.37	5.38
5–9 Windows	63.41	11.87	83.84	9.80

[a]Noise × windows interaction, $p < .003$.
[b]Noise × windows interaction, $p < .055$.

the reported findings are statistically significant by chance, alone. Nonetheless, the pattern of results is provocative and suggests some interesting hypotheses for future contextual research. Previous research in educational settings found no significant relationships between the presence or absence of classroom windows and various measures of academic performance and social behavior (cf. Weinstein, 1979, for a review of this work). When compared with the findings from earlier studies, our results suggest that the links between windows and well-being may be highly dependent on contextual factors. For instance, in certain settings such as hospitals, the presence of windows may enhance well-being by providing a source of pleasant diversion from one's health problems—especially if the windows afford views of tranquil areas containing trees and other natural elements (cf. Ulrich, 1984; Verderber, 1982). In school settings, however, where a premium is placed on concentration and academic achievement, the presence of windows may interfere with teacher–student communication and children's concentration on learning tasks by increasing the potential for visual and auditory distraction. The immediate and delayed impact of these distractions on performance and subjective well-being may be accentuated in schools located within areas that are unusually noisy and distracting to begin with.

Aside from the interactive effects of school noise and classroom windows on task performance, no other significant interactions of these factors were found (i.e., on blood pressure, math, reading, and Wepman test scores). Also, the School Noise × Classroom Density ANCOVAs revealed no significant interaction effects or main effects of classroom density within either the Time 1 or Time 2 sample. However, significant Noise main effects on diastolic and systolic blood pressure were found within the Time 1 sample. These effects, indicating higher blood pressures among noise-school children at Time 1, are consistent with the regression effects of School Noise reported earlier (cf. Cohen *et al.*, 1980 and earlier chapters of this volume). Also, as in our earlier analyses, the significant noise main effects found for the Time 1 sample were not observed among the third graders tested at Time 2 (although significant School-Noise-×-Exposure-Time interactions on diastolic and systolic blood pressure were found for that sample; see Chapter 5).

Intersetting Moderators of Environmental Stress

A series of 2 × 2 ANCOVAs assessed the possible influence of residential density on children's response to aircraft noise at school. Earlier research by Baum and Valins (1977) and Saegert (1984) found that the behavioral and health effects of residential density persisted beyond the confines of a crowded apartment or dormitory setting to nonresidential situations. Saegert's findings are particularly relevant to our own study because her research specifically examined the impact of residential density on children's behavior at elementary schools. Therefore, we wanted to assess whether cross-setting effects of residential density also occurred among the elementary students tested in our study. Furthermore, we suspected that the main effects of density might be qualified by levels of children's noise exposure at school. Although the main effects of prolonged exposure to either density or noise have been documented in earlier studies, previous research has not assessed the interactive effects of these stressors within residential or academic settings.

Additional ANCOVAs examined the possible effects of home noise levels on children's behavior and physiology at school. Also, we tested for possible interactions between home noise levels and either residential density or classroom noise exposure in separate 2 × 2 ANCOVAs. Though previous research has not examined the interactive effects of density and noise in naturalistic settings, at least one earlier study assessed the influence of home noise exposure on children's reactions to noise at school (Lukas *et al.*, 1981). Lucas *et al.* found that children from noisy homes and attending noisy schools were more likely to show reading deficits than those from quieter homes. It should be noted that the study by Lucas *et al.* focused on traffic noise, which was of lower intensity than the levels of aircraft noise examined in our research.

The residential density index was computed by dividing the number of children within each household by the number of rooms in the home. Information about the number of rooms within each child's home was available only for the Time 2 sample. Therefore, the main and interactive effects of residential density were tested within that sample only. Children's assignment to either the low or high levels of residential density was based on a median split of the density scores. The median value of residential density within the Time 2 sample was .6 children per room.

Archival information about home noise levels was available for both the Time 1 and Time 2 samples but only among those children attending noisy schools. Community Noise Equivalent Level (CNEL) contours for areas adjacent to the Los Angeles International Airport and the location of children's homes within those areas determined each child's home noise score. Children's assignment to the low or high level of home noise was based on a median split of the residential CNEL scores. The median levels of home noise for the children attending noise schools were 76 CNEL for the Time 1 sample and 74 CNEL for the Time 2 sample.

Because home noise data were available only for the children attending noisy schools, the possible interactions between residential and school noise levels could not be tested directly. We did, however, assess the interactive effects of residential noise levels and the extent to which classrooms at the noisy schools had been architecturally treated for noise abatement. Children's standing on the noise-abatement factor was based on whether or not their classroom had been treated for sound abatement prior to their assignment to that room at the beginning of the school year.

Initial multicollinearity checks revealed a lack of significant association between school noise and residential density and between the home noise and classroom-abatement factors. A marginally significant correlation between home noise and residential density was found, $r(89) = .15, p < .085$.

The School Noise × Residential Density ANCOVAs indicated a lack of interactive effects on any of the dependent measures. Main effects of residential density, however, were found on the measures of reading achievement, $F(1, 52) = 4.86$, $p<.032$, math achievement, $F(1, 52) = 4.63$, $p<.036$, and Wepman test error scores, $F(1, 52) = 6.02$, $p<.018$. These effects indicated that children from the higher density homes had lower reading and math achievement scores and made a larger number of auditory discrimination errors on the Wepman test (see Table 22). The pattern of these results is consistent with Saegert's (1984) finding that high residential density was associated with lower vocabulary test scores at school. These data are also consistent with the possibility mentioned in Chapter 1 that coping strategies, such as tuning out noise at home, may generalize to other settings (e.g., school environments) and may be associated with secondary adaptive costs (e.g., reading deficits) within those settings.

The reduced degrees of freedom in the preceding analyses are due to missing data on certain variables within the Time 2 sample of third graders. In particular, reading scores were unavailable for 44% of the children, and math scores were unavailable for 12% of the children within that sample. The subjects lacking reading and math achievement scores also were omitted from the analysis of the Wepman error scores due to a listwise deletion procedure within the Statistical Package for the Social Sciences (SPSS) program. In view of the large percentage of missing data, the reported main effects of residential density are suggestive at best. Nonetheless, the pattern of results warrants consideration because of its consistency with the findings from earlier research and with information overload models of crowding and stressor aftereffects (Baum & Epstein, 1978; Cohen, 1980; Glass & Singer, 1972; Chapter 1).

The Home Noise × Residential Density ANCOVAs also indicated no significant interactions. The same pattern of significant density effects (on reading, math, and Wepman scores) reported before also was evident in the Home Noise × Residential Density ANCOVAs. Recall that in these ANCOVAs, the density effects

Table 22. Reading Achievement, Math Achievement, and Auditory Discrimination Errors as a Function of School Noise Levels and Residential Density (For Third Grade Sample at Time 2)

School noise	Reading[a]		Math[b]		Wepman test errors[c]	
	M	SD	M	SD	M	SD
Quiet						
Low density	52.00	32.39	58.10	19.59	6.10	4.15
High density	50.25	25.28	51.90	21.28	6.95	4.18
Noisy						
Low density	49.33	30.11	68.23	20.42	6.22	3.70
High density	33.07	26.78	51.86	18.79	7.79	4.05

[a]Density effect, $p < .032$.
[b]Density effect, $p < .036$.
[c]Density effect, $p < .018$.

apply only to the children attending noisy schools for whom the residential CNEL data were available, and only to those for whom reading and math scores were available.

The Home Noise × Classroom Abatement ANCOVAs on children tested at Time 1 revealed significant main effects of residential noise on systolic and diastolic blood pressure, $F(1, 104) = 7.35$, $p<.01$; $F(1, 104) = 4.48$, $p<.01$, respectively. Generally, children from the quieter homes had higher blood pressures. The residential noise effect on systolic pressure was modified by a Home-Noise-×-Classroom-Abatement interaction, $F(1, 104) = 8.44$, $p<.005$. Among the children from quieter homes, higher levels of systolic pressure were evident for those assigned to noisy classrooms than for those assigned to noise-abated rooms. This pattern was reversed among the children from noisier homes, with those assigned to noise-abated rooms exhibiting higher systolic blood pressure (see Table 23).

The Home Noise × Classroom Abatement interaction on systolic pressure suggests the possibility of residential-school noise contrast effects on children's cardiovascular arousal. Perhaps children who spend much of their time in relatively quieter residential areas develop a lower tolerance for noise at school and, therefore, exhibit higher arousal levels in nonabated classrooms. By the same reasoning, children residing in noisier areas may establish a higher adaptation level (cf. Helson, 1948) or tolerance for noise at school and, therefore, exhibit lower arousal in noisy classrooms than in abated classrooms (where the contrast between home and school noise levels is greater). This interpretation of our findings, however, is qualified by the lack of a Home Noise × Classroom Abatement interaction on diastolic pressure, and by the fact that none of the significant main and interactive effects on blood pressure within the Time 1 sample were replicated among the third graders tested at Time 2.

The Home-Noise-×-Classroom-Abatement ANCOVAs did reveal significant main and interactive effects on reading achievement and task performance within

Table 23. Systolic and Diastolic Blood Pressure (mm Hg) as a Function of Home Noise Levels and Classroom Noise Abatement (within Time 1 Sample)

Home noise	Systolic[a]		Diastolic[b]	
	M	*SD*	*M*	*SD*
Low				
Noisy classroom	92.58	14.52	49.90	10.73
Noise-abated classroom	87.59	12.56	46.80	7.52
High				
Noisy classroom	86.99	10.56	46.67	8.91
Noise-abated classroom	89.98	9.76	46.68	6.40

[a]Home Noise effect, $p < .01$; Home Noise × Abatement interaction, $p < .005$.
[b]Home Noise effect, $p < .01$.

the Time 2 sample (see Table 24). A significant effect of residential noise on reading scores was found, $F(1, 66) = 6.20$, $p<.025$, indicating that children from noisier homes had lower reading scores. The same relationship between home noise levels and reading scores also was found in earlier regression analyses of the Time 1 data, $F(1, 116) = 5.23$, $p<.02$ (see Chapter 3 for a more detailed discussion of these analyses). The main effect of residential noise on reading scores within the Time 2 sample, however, was qualified by a Home Noise × Classroom Abatement interaction, $F(1, 66) = 8.41$, $p<.005$. Among children from the quieter homes, those assigned to abated classrooms exhibited higher reading scores than those assigned to nonabated rooms. This pattern was reversed among the children from noisier homes, with those assigned to nonabated rooms having higher reading scores than those assigned to the abated classrooms. The Home Noise × Classroom Abatement interaction on reading scores, although not replicated at Time 1, does correspond to the pattern of the Time 1 blood pressure data, suggesting the possibility of residential noise–school noise contrast effects on reading achievement.

Classroom Abatement main effects were observed for the number of lines completed and the percentage of e's found during the crossing-out-e's task, $F(1, 77) = 5.23$, $p<.025$; $F(1, 78) = 5.92$, $p<.025$, respectively. Generally, children assigned to the abated classrooms exhibited greater speed and accuracy of performance on the task. The abatement effect on the number of lines completed, however, was qualified by a Home Noise × Classroom Abatement interaction, $F(1, 77) = 4.66$, $p<.05$. The pattern of means for the number of lines completed indicates that decrements in performance speed were most pronounced among the children residing in the noisier homes and assigned to the noisier (nonabated) classrooms. This finding diverges from the interaction effects reported before, whereby children from the noisier homes and assigned to the nonabated classrooms exhibited lower systolic blood pressure (within the Time 1 sample) and higher reading scores (within the

Table 24. Crossing-Out-E's Task Performance under Auditory Distraction and Reading Achievement Scores as a Function of Home Noise Levels and Classroom Noise Abatement (for Third Grade Sample at Time 2)

Home noise	Number of lines completed[a]		Percentage of E's found[b]		Reading achievement[c]	
	M	SD	M	SD	M	SD
Low						
Noisy classroom	65.22	13.83	85.25	9.52	43.56	27.29
Noise-abated classroom	68.67	14.04	88.70	6.51	44.52	33.02
High						
Noisy classroom	58.80	12.70	84.40	10.13	33.77	31.01
Noise-abated classroom	63.00	13.47	88.83	5.92	32.82	25.32

[a]Abatement effect, $p < .025$; Home Noise × Abatement interaction, $p < .05$.
[b]Abatement effect, $p < .025$.
[c]Home Noise effect, $p < .025$; Home Noise × Abatement interaction, $p < .005$.

Time 2 sample) than those assigned to the abated classrooms. We do not have an explanation for the divergent pattern of the Home Noise × Classroom Abatement interaction effects, nor can we explain why the effects are evident for certain variables (i.e., systolic pressure, reading achievement, and performance speed on the crossing-out-*e*'s task) but not others (e.g., diastolic pressure, math achievement, and performance accuracy on the *e*'s task). Moreover, it is not clear why the measures on which both main and interactive effects were found vary between the Time 1 and Time 2 samples. These mixed patterns of data add to our concern mentioned earlier that the statistical significance of some portion of the reported findings may be attributable to chance.

It is clear that our exploratory analyses of the cross-setting determinants of stress raise more questions than they answer. Our results, however, do appear to support the general finding from earlier research that contextual factors within children's residential domain can significantly affect their reactions to environmental conditions at school (cf. Lukas *et al.,* 1981; Saegert, 1984). Additional contextual research is needed to better account for the specificity of intersetting stress effects across different populations and response criteria.

Objective and Subjective Predictors of Stress

All of the ANCOVAs described in this volume have employed objective measures of environmental conditions within children's school and residential domains as the main predictor variables. These analyses do not address an important theoretical issue in the field of environmental stress: namely the relative contribution of objective and subjective assessments of stressors as predictors of performance and well-being. Although psychological theories strongly emphasize the role of subjective environmental appraisals in the etiology of stress, very few empirical studies have assessed the independent contribution of environmental perceptions to the prediction of stress while controlling for objective measures of the environment (cf. Carp & Carp, 1982; Cohen *et al.,* 1983).

To examine the relationships between objective and subjective predictors of stress, we conducted a series of hierarchical regression analyses incorporating three basic steps: (1) all standard covariates for a particular dependent variable were entered at Step 1; (2) the objective index of school noise (quiet vs. noisy) was entered at Step 2; and (3) one of four subjective measures of classroom noise was entered at Step 3. The subjective measures included children's perceptions of the noisiness of their classroom and the extent to which they were bothered by noise in their classroom in addition to teachers' perceptions of the frequency of noise-related interruptions and instances in which they had to raise their voice in the classroom due to noise from exterior sources. A second series of regression equations incorporated the same variables as the first series but did not specify the order of entry for the objective and subjective noise measures at Steps 2 and 3. These stepwise equations provided an assessment of the relative power of the subjective and objective noise measures as predictors of children's performance and physiology. The two series of regression analyses were repeated using data from both the Time 1 and Time 2 samples.

Initial multicollinearity checks revealed a significant correlation between school noise levels and children's perception of classroom noise within the Time 1 sample, $r(259) = .16, p<.006$, but not among the children tested at Time 2, as well as between school noise and children's ratings of noise annoyance within both samples, $r(259) = .17, p<.003; r(163) = .13, p<.004$, respectively. Also, school noise and teachers' perception of noise-related interruptions were inversely related within both samples, $r(102) = -.16, p<.05; r(130) = -.28, p<.001$, respectively. The reduced degrees of freedom for the latter coefficients are due to the fact that teachers' ratings of noise-related interruptions were unavailable for 61% of the children tested at Time 1 and for 21% of the sample at Time 2. It is not clear why school noise levels and teachers' perceptions of noise-related interruptions would be inversely rather than positively correlated. The variables of school noise and teachers' ratings of the frequency of having to raise their voice due to outside noise were not significantly correlated within either of the samples.

Although significant associations between the objective index of school noise and three of the subjective noise measures were found, the degree of correlation among the predictor variables was not of sufficient magnitude to preclude their simultaneous inclusion within the regression equations. By removing the variance shared by the objective and subjective noise indexes, the regression analyses provide a rather conservative test of the role of subjective environmental appraisals in the prediction of stress outcomes.

Within the Time 1 sample, the regression equations incorporating childrens' perceptions of classroom noise disclosed a significant effect on diastolic blood pressure, $F(1, 237) = 5.91, p<.025$, after controlling for the effect of school noise levels, $F(1, 237) = 6.91, p<.01$. (See Table 25). Children who perceived their classroom to be noisier had higher diastolic pressures ($M = 47.42$) than those who perceived their classrooms as less noisy ($M = 42.96$). Also, analyses incorporating teacher's perceptions of noise-related interruptions disclosed significant effects on the percentage of e's found during the crossing-out-e's task, $F(1, 91) = 12.13$, $p<.01$; the number of auditory discrimination errors during the Wepman test, $F(1, 89) = 5.37, p<.025$); and a marginally significant effect on the number of lines completed during the e's task, $F(1, 91) = 3.39, p<.07$. All of these effects indicate higher performance levels among the children whose teachers perceived a lower frequency of noise-related interruptions within their classroom (see Table 26). Also, all of these effects obtain after controlling for the objective index of school noise.

The analyses incorporating teachers' perception of having to raise their voice due to noise also disclosed significant effects on the percentage of e's found, $F(1, 91) = 19.00, p<.001$, and the number of lines completed, $F(1, 91) = 6.89, p<.01$, during the e's task, and for the number of errors on the Wepman test, $F(1, 89) =$

Table 25. Prediction of Diastolic Blood Pressure as a Function of Children's Perception of Classroom Noise, Controlling for Objective School Noise Levels and Standard Covariates (for Time 1 Sample)[a]

Predictor variable	Step	Multiple r	Cumulative r^2	Simple r	Beta	Significance level
Months enrolled	1	.013	.000	−.013	.035	n.s.
Number of children at home		.032	.001	−.030	.012	n.s.
Grade		.032	.001	−.005	−.122	n.s.
Race		.033	.001	−.008	.007	n.s.
Ponderal Index		.076	.006	.067	.072	n.s.
Height		.218	.048	.179	.220	<.001
Objective school noise level	2	.288	.083	.168	.170	<.010
Perceived noisiness of classroom	3	.324	.105	.211	.155	<.025

[a]Multiple regression, $F(8, 237) = 3.48$, $p < .01$, at Step 3.

8.25, $p<.01$, after controlling for school noise levels. The pattern of these effects corresponds to that reported for teachers' perceptions of classroom interruptions (see Table 26), indicating higher levels of performance on the crossing-out-e's and auditory discrimination tests among children whose teachers reported having to raise their voice less often due to the intrusions of outside noise.

In the second series of regression analyses, where the order of the objective and subjective noise indexes was left unspecified, the school noise measure preceded children's perception of classroom noise in predicting diastolic blood pressure, as shown in Table 25. In all of the analyses incorporating teachers' estimates of classroom noise problems, however, the perceptual measure entered the equation prior to the objective index of school noise. In none of the regression analyses incorporating teachers' ratings of noise problems was the effect of school noise

Table 26. Crossing-Out-E's Task Performance and Auditory Discrimination Errors as a Function of Teachers' Perceptions of Noise-Related Interruptions in the Classroom (for Time 1 Sample)

Perceived frequency of interruptions due to exterior noise	Number of lines completed[a]		Percentage of E's found[b]		Wepman test errors[c]	
	M	SD	M	SD	M	SD
Low	65.57	16.60	92.20	4.70	5.93	3.41
High	61.58	15.35	84.15	11.48	7.21	3.49

[a]$p < .010$, controlling for objective school noise levels.
[b]$p < .010$, controlling for objective school noise levels.
[c]$p < .025$, controlling for objective school noise levels.

significant, regardless of the order of entry for the objective and subjective noise measures at Steps 2 and 3 of the equation.

For the analyses incorporating children's ratings of noise annoyance (i.e., how bothered they were by classroom noise), a significant effect on diastolic blood pressure was found when the subjective measure was forced into the equation prior to the school noise index, $F(1, 232) = 4.21$, $p<.041$, indicating greater diastolic pressures among the children reporting higher rather than lower annoyance levels ($M = 47.51$ vs. $M = 45.69$). However, this effect becomes nonsignificant when the school noise index enters the equation prior to the noise annoyance measure: school noise $F(1, 231) = 8.11$, $p<.01$.

In the analyses of data from the third grade sample at Time 2, the two series of regression equations described previously disclosed no significant effects of either the subjective noise measures or the objective index of school noise. It is not clear why the regression effects reported for the Time 1 sample were not found within the Time 2 sample. One possibility is that the sensitivity of the regression analyses performed on the Time 2 data was reduced by the smaller size of the sample ($N = 165$ vs. 262 at Time 1). Another factor that may be related to the nonreplication of the objective and subjective noise effects within the Time 2 sample is the nature of the school noise index employed in the regression equations. As noted earlier, school noise was defined as a two-level factor (quiet school vs. noise school) rather than as a three-level factor (quiet school vs noise school, nonabated classroom vs. noise school-abated classroom) in these analyses. Also, recall that significant main effects of classroom abatement on children's task performance were found within the Time 2 sample (see Table 24 for a summary of these ANCOVA results). Considering these effects, it is plausible that the use of a two-level index of school noise (which combined children from the abated and nonabated classrooms within a single level of the noise factor) may have obscured differences in performance scores attributable to variations in classroom noise and noise-annoyance levels among the children and teachers from noisy schools.

Despite the nonreplication of the regression effects across the two samples, we believe that the significant effects of subjective noise appraisals found within the Time 1 sample are important for at least three reasons. First, our results are consistent with psychological models of stress and with the findings from earlier studies, both of which suggest that subjective environmental assessments make a significant contribution to the prediction of stress outcomes after controlling for objective measures of the stressor (Carp & Carp, 1982; Cohen *et al.*, 1983; Neus, Ruddel, & Schulte, 1983). Second, our regression analyses employed subjective environmental appraisals to predict objective indexes of performance and physiology. This strategy reduces the likelihood of statistical artifacts due to shared method variance when perceptual variables are used to predict subjective measures of well-being.

Finally, the fact that teachers' perceptions of noise-related problems predicted children's performance and auditory discrimination scores is especially interesting because the links between adults' and children's reactions to environmental stressors have not been assessed in prior research. On the one hand, teachers' assessments may simply mirror children's perceptions of classroom noise, and thus, these two categories of subjective noise ratings may be redundant. We did find, for example, that children's perception of classroom noise was significantly correlated with teachers' ratings of noise-related interruptions, $r(102) = .19, p<.026$, and the frequency of having to raise their voice, $r(102) = .17, p<.047$, within the Time 1 sample (but not within the Time 2 sample). On the other hand, these findings raise the further possibility that adults' reactions to environmental stressors directly influence children's reactions to the same conditions (cf. Bronfenbrenner, 1979). Similarly, in attempting to explain the attrition of high blood pressure subjects from noisy schools between the two phases of our study, we speculated earlier that the parents of these children may have been stressed and that their stress led, in turn, to higher stress among their children and to subsequent residential relocation (see Chapter 5, and Cohen, Krantz, Evans, & Stokols, 1981). This possibility suggests at least two useful directions for future contextual research on stress: (1) an empirical assessment of the independent and interactive contributions of children's, teachers', and parents' appraisals of classroom and residential noise to the prediction of stress outcomes, after controlling for objective noise levels; and (2) the development of composite theoretical terms that account for the ways in which the structure of interpersonal relations (e.g., among children and teachers and among children and their families) account for the behavioral and physiological effects of stressors within (and across) various life domains.

Individual and Aggregate Analyses of Stress

All of the analyses described in this chapter employed individual-level measures of children's performance and physiology. Certain earlier analyses reported by Cohen *et al.* (1980) and Cohen, Krantz, Evans, and Stokols (1981), however, employed classrooms rather than individual children as the units of analysis. In general, the effects of school noise on various clusters of dependent measures were found to be consistent between the individual and classroom levels of analysis.

Due to the small number of classrooms within the Time 1 and Time 2 samples ($N = 37$ and 27, respectively) and because of missing data on certain variables (e.g., achievement scores) for some of the classrooms in these samples, it was not possible to reliably assess the main and interactive effects of school noise and other environmental factors at the aggregate level. This limitation of our research introduces a further qualification of the significant results reported in earlier sections of this chapter because some of these effects (particularly those on reading achievement levels) may be mediated by classroom-level conditions.

Summary and Conclusions

We began this chapter with several objectives in mind: namely (a) to identify some of the distinguishing features of contextual theorizing and research; (b) to present a conceptual framework and a set of procedural strategies for organizing future contextual research on environmental stress; and (c) to review the existing empirical evidence regarding contextual dimensions of stress, including our own findings from the Los Angeles Noise Project. In this concluding section of the chapter, we will step back from these theoretical and empirical efforts and, like the participants in a dissertation orals, try to assess what has been accomplished and what remains to be done.

First, our review of the literature and the analyses from our study suggest that people's reactions to environmental stressors are moderated by diverse situational factors and that these factors can be usefully grouped according to the basic contextual dimensions that we outlined earlier. Admittedly, our overview of the literature on environmental stress was highly selective. We did not focus on the methodological details of the studies reviewed, nor did we discuss several complex methodological issues that pertain to the validity of contextual research. Some of these issues relate to the difficulties of detecting statistical interactions between environmental stressors and contextual moderators of stress (House, 1981; LoRocco, 1983; Thoits, 1982) and the relative advantages and disadvantages of using time series analyses to assess the temporal patterning of behavior and well-being (Glass, Wilson, & Gottman, 1975). Other methodological issues that are particularly relevant to contextual research are the difficulties of combining individual and aggregate data and of drawing inferences about environment–behavior relationships that span these different analytical levels (Boyd & Iverson, 1979; Firebaugh, 1978) and the relative advantages and disadvantages of representing environmental events and stressful experiences in either objective or subjective terms (Ericsson & Simon, 1980; Evans & Cohen, 1986; Marans, 1976; Weinstein, 1976).

The selectivity of our literature review was partly determined by the theoretical rather than methodological emphases of this chapter. The broad sweep of our review was intended to assess the degree to which contextual factors have been explicitly considered in previous stress research and the extent to which the findings from this research offer preliminary clues about the contextual determinants of environmental stress. Clearly, the previously mentioned statistical and methodological issues should be explicitly addressed in future studies of stress. But, for purposes of the present chapter, the main conclusions that can be drawn from our review of the literature are (a) that contextual analyses of stress have become (and in all likelihood will continue to be) quite prevalent in the field of environmental stress; and (b) that there is growing, though largely preliminary, empirical evidence that the general contextual dimensions identified in the early sections of this chapter are valuable as a basis for analyzing and understanding the cross-situational variability of stress phenomena.

Our contextual analyses of the Los Angeles Noise Project data were clearly exploratory rather than derived from an *a priori* theory about the situational determinants of environmental stress. Moreover, our findings are qualified by the nonreplication of various significant effects across the two independent samples of children tested at Time 1 and Time 2, by the constraints of missing data for certain of the variables considered in our analyses, and by the focus of our analyses on stress outcomes measured within school settings, alone, rather than across children's multiple life domains. However, the findings from our analyses of the intersetting predictors of performance and physiology at school are consistent with the data from previous studies suggesting significant links between residential conditions (i.e., home noise and density levels) and measures of children's performance at school (Lukas *et al.*, 1981; Saegert, 1984). Also, our findings offer empirical support for earlier theories that assign a central and independent role to subjective environmental appraisals in the etiology of stress (Lazarus, 1966; Lazarus & Launier, 1978; McGrath, 1970; Pearlin *et al.*, 1981). Finally, our findings suggest some important directions for subsequent empirical research, including further examination of the situational factors that qualify the relationship between proximity to windows and occupants' performance and well-being and the possible role of adults' reactions to environmental problems in mediating children's experiences of stress.

Although many empirical studies have examined situational moderators of stress, few have been guided by a systematic framework for contextual theorizing and research. For the most part, situational factors have been considered in piecemeal fashion rather than in relation to more general dimensions of contextual representation (e.g., spatial, temporal, and sociocultural scope). And, even when basic contextual dimensions have been considered, they usually have been examined singly rather than as a multidimensional matrix for charting the effective context of a phenomenon.[7] Thus, our conceptualization of stress is unique in that it offers a set of strategies for organizing contextual research and highlights the importance of developing contextual theories as a basis for predictive rather than post hoc assessments of situational moderators of stress.

Certainly, the proposed analytical strategies are ambitious in their scope and intent. Often, it is not feasible for a single study or program of research to empirically assess multiple contextual dimensions of a particular phenomenon, let alone the degree of overlap or divergence among the contextual moderators of diverse stress phenomena (e.g., circumstances affecting the intensity of people's reactions to crowding, noise, and residential relocation). These integrative research tasks are perhaps best accomplished through comparative analyses of the findings from several separate studies of stress (Glass, 1977; Hunter, Schmidt, & Jackson, 1983; Kulik, Kulik, & Cohen, 1979). Nonetheless, we suggest that the accumulation of information about contextual aspects of stress can be systematized and,

---

[7]Exceptions to this trend are recent time-geographic studies (cf. Carlstein *et al.*, 1978; Michelson, 1986a) that assess people's daily activities and experiences across spatial and temporal dimensions, at both individual and aggregate levels, and in terms of both objective and subjective representations of stressors and stress experiences. Also, Hornstein and Wapner (1986) and Jacobi (1984) developed questionnaire and interview procedures for mapping situational determinants of stress across individuals' multiple life domains and life stages. Similarly, Schwartz (1982) proposed the use of a "Patient Evaluation Grid" by physicians as a basis for organizing diagnostic information about health problems in relation to multiple contextual dimensions of the patient's overall life situation.

perhaps, hastened to the extent that researchers orient their theoretical and empirical work in relation to the basic dimensions and analytical strategies proposed in earlier sections of this chapter and in Chapter 1.

Our contextual analysis of environmental stress and our review of previous empirical studies suggest several important issues that warrant further attention in future research. First, more consideration should be given to alternative ways of conceptualizing and measuring the individual's overall life situation and the various settings and domains of activity that comprise it. For example, are individuals' life domains (e.g., home, school, and work) better construed as separable and self-contained entities or, alternatively, as overlapping and interdependent systems? Assuming that certain life domains are subjectively more important to the individual than others, how can this differential salience of domains be measured empirically as a potential mediator of emotional and physical reactions to environmental demands? Furthermore, to what extent do the mental and physiological costs of exposure to stressors in one domain generalize to other life domains, and in what ways might stressful experiences within a particular setting be ameliorated by "compensatory benefits" in other domains (Campbell, 1983; Cohen & Hoberman, 1983; Frankenhaeuser, 1981; Pearlin *et al.*, 1981; Stokols, 1982)? The explanatory power and efficiency of contextual analyses of stress will be enhanced to the extent that these questions are adequately addressed in future research.

Second, more effort should be devoted to the development of taxonomies for describing and comparing different situations and settings, and to the development of composite terms for describing their important structural dimensions (cf. Magnusson, 1981; Pervin, 1978; Stokols, 1986). Although composite terms have been incorporated into earlier stress research, little attention has been given to the integration and consolidation of these terms into more efficient and powerful representations of person–environment transactions. Also, when composite terms are used in analyses of stress, the assumption is usually made that the situational structure is relatively stable and enduring. Only minimal theoretical attention has been given to the social, cultural, and environmental antecedents of structural change within settings (Stokols & Shumaker, 1981; Wicker, 1986).

Third, the interrelations among cognitive and affective processes in mediating individuals' reactions to environmental stressors and the differential salience of physical and social stressors within different age groups remain to be addressed in subsequent research on stress.

Finally, more consideration should be given to sociocultural moderators of stress and to the measurement of stress outcomes across individual and aggregate levels. In these areas, important issues for future investigation include the possible impact of personal coping difficulties and stress reactions on group and organizational well-being and the structural properties of social systems that promote supportive, rather than dysfunctional, interpersonal relationships among their members.

Summary and Implications

7

In previous chapters, we have reviewed existing literatures and presented new evidence in regard to the relationship between environmental stress and cognitive performance, personal control, and health. We have suggested the advantages of viewing this work in the framework of "costs of coping" and "contextual" analyses of stress. Finally, we have presented data from the Los Angeles Noise Project that in many cases helps to exemplify and clarify our views and at the same time raises other questions about the process of coping with stressors in the natural environment. As noted earlier, our theoretical proposals are often not directly tested by the data we present. Instead, they reflect a growth in our conceptualization of the stress and coping process that has occured over the 8 years since we began collecting data for the project. Next we will summarize our major arguments and conclusions and discuss their implications for future research and for public policy.

Costs of Coping

In Chapter 1, we propose that detrimental effects of stress are often attributable to the costs of coping rather than to direct effects of stressors on behavior or physiology. The role of coping in behavioral deviation and pathology has been discussed elsewhere (see review in Chapter 1); however, we have gone beyond previous work in specifying conditions under which a wide range of specific costs will occur. Our proposal can be viewed as including three major categories of effects for those who actively cope with a stressor: cumulative fatigue effects, coping side effects, and overgeneralization. *Cumulative fatigue* refers to situations in which adaptive processes drain one's physiologic or cognitive (psychic) energies. Deficits in cognitive energies may result in poststress deficits in task performance, poststress insensitivity to others, and increased deficits over time on demanding tasks performed under stress. Deficits in physiologic energy may result in inability of a physiologic system to respond (or continue to respond) to a stressor, and, if the system is taxed for a prolonged period, to the total breakdown of that system's functioning. *Coping side effects* are effects that occur when coping behaviors, which were employed in an attempt to ameliorate the possible impact of the stressor, are detrimental to individuals in other ways. These include physiologic responses that are above the level that is efficient for the body's metabolic needs and behavioral adaptations that interfere with health maintenance. *Overgeneralization* can take two forms. First, a strategy that is successfully employed in coping with a stressor can persist even in situations where a person is not confronted with the offending stressor. Deleterious effects on task performance, interpersonal behavior, and health maintenance behavior occur to the extent that the coping responses are inappropriate in other situations in which they are employed. Second, in cases where a person is unsuccessful in coping with a stressor, a lack of successful coping may result in an overgeneralization of the expectation that one cannot effectively cope. This expectation may persist even in situations where coping is possible (learned helplessness) and consequently undermine the motivation to initiate further instrumental responses, interfere with learning that other outcomes are controllable, and cause depression of mood (cf. Seligman, 1975). Finally, we discuss situations in which *persons are passive* (do not engage in an active coping behavior) in the face of stressful events. We propose that when not responding is a chosen (preferred) state there are no secondary effects. When not responding is not the preferred state, effects associated with helplessness are expected. Several aspects of the noise project data provide evidence consistent with our cost-of-coping analysis. (Again, the project does not provide definitive tests, only suggestive evidence). These include evidence that when tested in a quiet setting, children living in noisy schools (a) take longer to solve and are more likely to give up on moderately difficult cognitive tasks; (b) demonstrate (at least during the first few years of exposure) an attentional focusing strategy characteristic of those exposed to a stressor; and (c) show inflated blood pressure levels.

The first two effects mentioned above are consistent with the category of coping costs that we have labeled *generalization*. First, let us consider the finding that children from noisy schools took longer to solve puzzles and were more likely to "give up" on a puzzle before their time had elapsed. As noted earlier, if the failure to effectively cope with a stressful event is attributed to a lack of skill, this expectation of ineffectiveness may generalize to other difficult coping situations where effective coping is actually possible. Second, during the first few years of exposure, children attending noise-impacted schools used a typical coping strategy, attentional focusing (or "tuning out"), outside of the stressful situation. It seems likely that this strategy was found to be effective in noise and as a result became part of the children's response repertoire used both inside and outside of stress situations. Noise project results are also consistent with the cognitive fatigue and costs-of-coping categories. For example, poorer problem-solving performance may be attributed to children's having insufficient additional psychic energy to deal with difficult cognitive tasks, and inflated blood pressure level may be attributed to a long-term cost of the sympathetic activation that occurs in response to active/effortful coping.

A Contextual Perspective

The influence of contextual factors on human behavior and health has become an increasingly popular focus of psychological research in recent years. All too often, however, the links between situational variables, behavior, and well-being are examined in a post hoc, atheoretical fashion, rather than in a more systematic manner. What has been lacking in many earlier studies is a conceptual framework for organizing theoretical and empirical research on the situational moderators of environmental stress. In Chapter 6, we offer a framework for organizing conceptual and empirical analyses of stress. This framework incorporates four basic dimensions that serve as analytical coordinates for mapping diverse behavioral phenomena in relation to alternative clusters of contextual variables: (a) the contextual scope (spatial, temporal, and sociocultural) of the analysis; (b) the use of both objective and subjective representations of contextual and target variables; (c) the analysis of stress phenomena from both individual and aggregate levels of analysis; and (d) the appropriate matching of partitive or composite forms of analysis according to the type of stress situation under investigation. We have suggested that in the initial phases of theoretical and empirical research, it is often useful to examine the target phenomenon (e.g., the relationship between school noise exposure and children's blood pressure levels) in relation to contextual frames (or groupings of situational factors) of varying contextual scope, as well as in relation to both objective and subjective, individual and aggregate, and partitive and composite representations of the contextual and target variables. The exploratory mapping of stress phenomena in relation to these four basic dimensions, particularly during the early stages of research, can reveal contextual influences on behavior and well-being that otherwise may be neglected by the researcher.

The conceptual framework, presented early in Chapter 6, provided a basis for organizing our review of previous contextual analyses of stress as well as those conducted as part of the Los Angeles Noise Project. Although our understanding of the contextual underpinnings of environmental stress is still at an early stage, our literature review and the analyses from our study suggest that people's reactions to stressors are moderated by a wide range of situational factors and that these factors can be usefully grouped according to the four basic dimensions outlined earlier. The data from the noise project, for example, are consistent with earlier studies suggesting that conditions within children's residential environment, such as home noise and density levels, exert a significant influence on children's performance and physiology at school. Also, conditions within the child's classroom, such as the number of windows and the presence or absence of treatments for sound abatement, play an important role in moderating physiological and behavioral responses to school noise. Finally, we have identified several priorities for future contextual research on environmental stress, including the development of alternative conceptualizations and measurement strategies for representing the individual's overall life situation; the development of more adequate taxonomies of situations and environmental settings and the ways in which the structure of these settings induce, buffer, or intensify stress reactions; and the conduct of further research on the interplay between cognitive, affective, and sociocultural moderators of stress reactions among individuals and groups.

Environmental Stress and Health

Chapters 1 and 4 outlined a number of different ways in which environmental stressors may influence health and well-being. We discussed several physiological mechanisms that may either be directly stimulated by stressor intensity, or may be stimulated by cognitively mediated emotional reactions to stressors. These included neural and endocrine effects attributable to activation of the sympathetic-adrenal-medullary system and the pituitary-adrenal-cortical axis, neurochemical changes in the brain relevant to mental health consequences, and suppression of immune functioning. We also discussed the possibility that stressors affect health by causing changes in behavior that may be inimical to health (e.g., coping side effects discussed previously) or by influencing affective responses linked to psychological stability.

Although existing research on environmental stress and health does not generally focus on mediating mechanisms, the literature we reviewed does provide strong evidence for the increased risk for physical illness for those exposed to intense environmental stress on a daily basis. For example, the community and industrial noise literatures consistently indicate elevations in coronary heart disease and risk for coronary heart disease among exposed groups. There is a smattering of data on various other physical health outcomes that suggests a wide range of pathogenic effects, but the sparseness and poor quality of the data leave us unconvinced at this point. There is some evidence that increased illness among exposed populations is more likely to occur in certain susceptible groups including the young, old, and institutionalized. We have noted that these groups are all characterized by a lack of control over their lives that may be exacerbated by a further lack of control over their physical environment. Data from studies on population density similarly indicate some increased risk for illness but only among susceptible groups crowded with strangers (as opposed to family). Data on stress-induced physiological effects of air pollution are insufficient at this time.

Data for the effects of environmental stressors on mental health were less compelling. Work on effects of community noise indicates the possibility that mild disorders may be aggravated by stress exposure, but that daily stressor exposure probably does not increase the incidence of (serious) clinical mental disorder. Crowding and air pollution studies generally indicate small correlations between stressor exposure and self-reported distress but do not address more serious disorder nor attempt to distinguish between new incidence of distress and aggravation of existing distress. There is, however, more than adequate evidence that daily exposure to environmental stressors can create annoyance and decrease quality of life for those who view it as unnecessary and uncontrollable. To a great extent, these outcomes appear to be based on the *meaning* of the stressor for the exposed person.

Noise Project Data on Blood Pressure

Our own data on blood pressure can be viewed as consistent with previous work. It is clear from both the initial study and replication that during the initial 2 years of exposure, children attending noise-impacted schools had higher systolic and diastolic blood pressure than those attending quiet control schools. These data could be attributed to a direct impact of the noise on cardiovascular response or to an indirect impact of the noise as mediated by an appraisal of the situation by the child. In fact, data reported in Chapter 6 indicates that children's perceptions of the noise were significant predictors of diastolic blood pressure after controlling (partialling out) for school noise levels.

There is additional evidence that suggests an even more indirect pathway by which noise may have influenced blood pressure levels. In an analysis of who moved out of the noise-impacted area over a 2-year period following our initial testing of Study I children, we found that, in noise-impacted schools, the children with high blood pressure were the most likely to move. This was not true in quiet control schools. It is important to note that blood pressure levels among noise-school children were still within the normal range, that elevated blood pressure is asymptomatic, and that parents (not children) make decisions to move. A number of alternative explanations for this result are outlined in Chapter 4. One salient alternative is that the children's elevated blood pressure was a response to their parents' noise-elicited distress rather than to their own distress. Although speculative, this interpretation clearly emphasizes the importance of the context in which persons are exposed to stressors in understanding their effects.

Additional research is clearly needed to establish the relationship between environmental stress and cardiovascular response, and between environmental stress and health in general. First, there is an almost total lack of prospective research looking at changes in health status as a function of changes in stressor exposure. Situations allowing such research are rare and hence should be capitalized on when they are found. Second, we need to design studies that take into account that people manifest stress impact in very different ways (Cassel, 1975). Hence, future work should include studies measuring a wide range of illness outcomes at the same time, and/or examine populations that are at particularly high risk for the specific disease(s) under study. Finally, future work should look closely at the process by which stress is linked to ill health. Such studies should include measures of the objective stressor, subjective response, mediating links such as health behaviors, emotional and physiologic responses, and measures of disease end points including onset, disease course, and recovery.

Environmental Stress and Control

Multiple functions of control including enhanced feelings of self-efficacy, greater instrumental coping responses, reduced uncertainty, and establishment of boundaries of aversiveness were discussed in Chapter 3. The research on control and environmental stress generally found reduced stress when behavioral control is provided. Insufficient research has been completed on other types of controls (e.g., cognitive control) to summarize their possible effects on responses to environmental stressors. Availability of behavioral control in institutional settings is an important component of healthful living conditions, particularly for persons with low levels of autonomy (e.g., the elderly in nursing homes, patients in hospitals, children in schools). Finally, research indicates that exposure to chronic, uncontrollable environmental stressors may lead to learned helplessness.

Persons who are chronically exposed to uncontrolled, adverse environmental conditions may learn that they have little or no control over their environs. When efforts to cope with chronic stressors fail, feelings of diminished competence and mastery may ensure. Children from noisy schools were significantly less likely to solve moderately difficult puzzles. Of particular interest, many of these failures were associated with giving up on the puzzle before the total time for solution (3 minutes) had elapsed. A similar pattern of results occurred from noise exposure in the home.

Children from the new third grader study were also given the opportunity to choose a game to play at the end of the testing session. Children from noisy schools were more likely to relinquish choice to an experimenter, allowing him or her to choose which game to play. Both the helplessness-on-task performance and the choice data are consistent with other recent laboratory and field studies on the impacts of uncontrollable stress on instrumental responding.

Few main effects of noise conditions were noted for children's locus of control scores. There were some weak trends suggesting a shift toward greater externality among children from noisy homes. As discussed earlier, the locus of control instrument used may have been too specific to achievement behaviors in the classroom to have registered effects from an ambient, environmental stressor.

Because our research program was designed before recent work on the reformulated attribution theory of learned helplessness, we did not include precise measures of children's attributions about the task performance situation or the importance of task success. Future research on environmental stressors and control needs to integrate the reformulated helplessness model as described in Chapter 3. This issue touches on an important developmental question as well. Children of different ages may vary in the attributions they make about uncontrollable events. Moreover, these attributions can be influenced by the type of feedback they receive in achievement situations from teachers and parents.

At this point, we can state with confidence that chronic exposure to aversive, uncontrollable environmental conditions is associated with deficits in self-initiated instrumental responses on cognitive tasks. Whether interference with early environmental mastery experiences can disrupt normal social and cognitive development is an important question still to be answered.

Environmental Stress and Cognition

The literature on stress and cognitive performance is very complex with little clear theoretical direction at this time. To be succinct, we currently have no theory that can adequately predict when stress will impair human task performance. Trends in the empirical literature suggest effects of stress on the reallocation of attention to primary task information, reduced incidental memory, and less capacity in working memory, possibly coupled with faster processing of information. There is also clear evidence of poststressor deficits on performance tasks. These postexposure deficits may be due to diminished cognitive reserves, reduced motivation to respond, or both. Two areas of theoretical research that warrant further exploration include (a) creating task typologies that are based on the number and types of cognitive strategies that can be used in performing a task; and (b) exploring the role of cognitive effort in performance on tasks with differing cognitive demands.

The cognitive data from our study, although complex, generally support the cost-of-coping approach outlined in Chapter 1. Children chronically exposed to noise reveal evidence of an overgeneralization strategy that we have called "tuning out." This is shown most clearly in children exposed to noise for less than 2 years on the distraction *e*'s task and among the new third grade sample on the signal-to-noise-ratio task. On the other hand, the Wepman auditory discrimination data were not supportive of the tuning-out hypothesis. One puzzling but stable finding was the interaction of noise and duration of exposure on the distraction *e*'s task. Although children enrolled in noisy schools for less than 2 years were less distracted by an auditory stimulus while performing a cognitive task, the opposite pattern was evident for children enrolled in noisy schools for more than 3.5 years. One tentative explanation for this reversal is that children eventually abandon the tuning-out strategy when they learn that it is ineffective.

Achievement scores (reading and math) were generally unaffected by school noise levels, whereas home noise had detrimental effects. Noise-abatement programs in the schools had marginal ameliorative impacts on achievement scores. In order to achieve more precise and statistically powerful analyses of achievement deficits from chronic environmental conditions, future research designs should monitor young children before, during, and after exposure. Cross-sectional comparisons, especially when relying on different classrooms and different schools, are very insensitive for monitoring changes in achievement scores.

More research is needed on the interrelationships among auditory stimulation levels, development of the tuning-out strategy, and cognitive performance. In addition, we need to learn more about the role of children's efforts to perform well under adverse conditions. The links between cognitive effort, adaptation, and performance also constitute a potentially important area of research. Finally, we do not know at the present time what the long-term consequences are of early chronic exposure to adverse environmental conditions on cognitive development. Now that the children from this study are in high school, do they suffer from any long-term deficits in cognitive functioning?

Influencing Public Policy

What We Would Do Differently

There are always many things to learn from conducting studies that have the scope of the Los Angeles Noise Project. Many of our conceptual and methodological insights have been discussed previously in this chapter and in the chapters of this volume. There is one issue that deserves special attention, however, because of the role it plays in creating the possibility that a study of this sort could have clearer implications for public policy. One mistake we made in our work was to not collect sound-level readings that were sufficient for establishing dose-response relationships. Although we were able to establish the effects of exposure to rather intense levels of community noise, our measurement procedures were not extensive enough or sensitive enough to establish dose-response relationships based on classroom, home, or individual exposure levels (see chapter 2 for a more detailed discussion of appropriate measures). The concept of *threshold,* which predominates in epidemiological research on environmental health effects, suggests the presence of some minimum level or range of a pollutant that is necessary for any health effects to occur. The accurate determination of a noise threshold that puts people under unacceptable risk requires a graded measurement scale. Reduction of noise levels in industry and community settings is costly, and rough guesses as to the levels that put people under risk are insuffient. Although some researchers have challenged the validity of the threshold concept in the context of environmental health (cf. Dubos, 1965; Evans & Campbell, 1983), it is important to recognize that environmental health and safety legislation and regulation currently mandate reliance on estimates of threshold effects. Specifically, an environmental impact must be of sufficient level or duration to cause debilitating physical or mental illness (Hartsough & Savitsky, 1924).

Limitations of the Bottom-Up Approach

It is also noteworthy that from a policy point of view, our research takes a bottom-up as opposed to the preferred top-down perspective (Kiesler, 1985). Investigators adopting a bottom-up view draw their hypotheses and design their studies around conceptual developments in their fields and then draw implications of their results for specific policy issues. For example, in our work, the question we ask arose from theories about the effects of noise on health and behavior and from laboratory work that itself had been theoretically based. This kind of perspective is traditional among basic scientists and provides answers to basic research questions. A top-down view asks what are the questions that we can pose that directly address the issues that are important to policymakers? For example, what specific levels of aircraft noise exposure result in increased health care costs, in increased use of community health facilities, in greater job absenteeism, or in less efficiency?

Our primary goal was to establish the ecological validity of a number of broad theoretical positions that had been tested only in laboratory settings. Establishment of a valid and broadly applicable theoretical base is essential for practical applications (e.g., decisions about noise attenuation in factories) in many situations in which studies of a specific type of noise, type of task, or situation are not available. Hence, we view a bottom-up approach as a valid one for addressing many practical as well as theoretical questions. However, looking back, it is clear that these two perspectives do not need to be pursued independently. The addition of some policy-specific measures, for example, health care utilization among students, would have added to the generality of our results in regard to basic issues while at the same time it would have increased the importance of the study for policy purposes.

The Bottom-Up Approach and Ecological Validity

Establishing that an impact of environmental stress on human health or behavior holds up in both laboratory and "real world" settings substantially increased the potential role of such results in both science and practice (Bronfenbrenner, 1974; Cohen *et al.*, 1980; DiMento, 1982; McGrath, 1982). Technical information is most influential when it emerges from a consensus among scientists (DiMento, 1982), and the replication and extension of laboratory studies in field settings heightens this concensus.

Ecological validity also relates to the issue of context. For most social scientists, the search for laws of behavior is not place specific; the goal is to achieve wide applicability in time and space. This drive toward normative rules of behavior, however, may conflict with the politician's need to know if a particular environmental condition will negatively affect his or her constituency (Altman, 1975). Thus, the contextual perspective advocated herein may in the long run prove more compatible with the political queries of place-specific impacts. Ultimately, both scientists and policymakers need to predict both *who* is most likely to be adversely effected by an environmental condition and under *what circumstances or context* that adverse effect is most likely to occur.

Some Final Comments

The Los Angeles Noise Project provided potentially important and convincing data in regard to the effects of real world environmental stressors on cardiovascular functioning, personal control, and cognitive performance. It also demonstrated the importance of adopting a broad ecological view in attempts to understand how stressors influence behavior and health. It demonstrated that both successful and unsuccessful coping can influence people's functioning outside of the stressful situation and that stressors may influence health and well-being indirectly through their effects on people's social environments rather than by directly altering behavior or bodily functioning.

We have tried to be honest in reporting our own reservations with the design and data resulting from our work. This honestly is not a reflection of any discontent on our own part. Our intent in being critical of ourselves has been to allow the shortcomings of our research to inform others interested in stress and coping in real world contexts. It is hoped that what we have reported in this volume will make others sensitive to the theoretical, methodological, and practical problems that make field studies of environmental stress such a challenge.

References

Abey-Wickrama, I., a'Brook, M. F., Gattoni, F. E. G., & Herridge, C. F. (1969). Mental-hospital admissions and aircraft noise. *Lancet, 11,* 1275–1277.

Abramson, L., Seligman, M. E. P., & Teasdale, J. (1978). Learned helplessness in humans: Critique and reformulation. *Journal of Abnormal Psychology, 87,* 49–74.

Abramson, L., Garber, J., & Seligman, M. E. P. (1980). Learned helplessness in humans: An attributional analysis. In J. Garber & M. E. P. Seligman (Eds.), *Human helplessness.* New York: Academic Press.

Ader, R. (Ed.). (1981). *Psychoneuroimmunology.* New York: Academic Press.

Ahrentzen, S. (1981). The environment and social context of distraction in the classroom. In A. Osterberg, C. Tiernan, & R. Findlay (Eds.), *Design research interactions.* Ames, IA: Environmental Design Research Association.

Ahrentzen, S., Jue, G., Skorpanich, M. A., & Evans, G. W. (1982). School environments and stress. In G. W. Evans (Ed.), *Environmental stress.* New York: Cambridge University Press.

Aiello, J. R., & Baum, A. (Eds.). (1979). *Residential crowding and design.* New York: Plenum Press.

Aiello, J. R., & Thompson, D. E. (1980). Personal space, crowding, and spatial behavior in a cultural context. In I. Altman, A. Rapoport, & J. F. Wohlwill (Eds.), *Human behavior and environment: Advances in theory and research* (Vol. 4). New York: Plenum Press.

Aiello, J. R., Epstein, Y. M., & Karlin, R. A. (1975). Effects of crowding on electrodermal activity. *Sociological Symposium, 1,* 42–53.

Albert, S. (1977). Temporal comparison theory. *Psychological Review, 84,* 485–503.

Aldwin, C. M. (1984).Culture as a determinant of coping behavior. Program in Social Ecology, Unpublished manuscript, University of California, Irvine.

Alluisi, E., & Morgan, B. (1982). Temporal factors in human performance and productivity. In E. Alluisi & E. Fleishman (Eds.), *Human performance and productivity: Stress and performance effectiveness.* Hillsdale, NJ: Erlbaum.

Altman, I. (1975). *The environment and social behavior: Privacy, territoriality, crowding and personal space.* Monterey, CA: Brooks-Cole.

Altman, I. (1982). Problems and prospects of environmental psychology. Presidential address to the Division of Population and Environmental Psychology, American Psychological Association. Annual Conference of the American Psychological Association, Washington, DC.

Altman, I., & Chemers, M. M. (1979). *Culture and environment.* Monterey, CA: Brooks-Cole.

Altman, I., & Rogoff, B. (1986). World views in psychology and environmental psychology: Trait, interactional, organismic, and transactional perspectives. In D. Stokols & I. Altman (Eds.), *Handbook of Environmental Psychology.* New York: Wiley.

American Public Health Association Committee on Hygiene of Housing. (1950). *Planning the Home for Occupancy.* Chicago: Public Administration Service.

Ando, Y., & Hattori, H. (1973). Statistical studies of the effects of intense noise during human fetal life. *Journal of Sound and Vibration, 29,* 101–110.

Ando, Y., & Hattori, H. (1977). Effects of noise on human placental lactogen (HPL) levels in maternal plasma. *British Journal of Obstetrics and Gynecology, 84,* 115–118.

Antonovsky, A. (1979). *Health, stress, and coping.* San Francisco: Jossey-Bass.

Appleyard, D. (1981). *Livable streets.* Berkeley: University of California Press.

Appley, M., & Trumbull, R. (Eds.). (1967). *Psychological stress.* New York: Appleton-Century-Crofts.

Arhlin, U., & Ohrstrom, E. (1978). Medical effects of environmental noise on humans. *Journal of Sound and Vibration, 59,* 79–87.

Averill, J. R. (1973). Personal control over aversive stimuli and its relationship to stress. *Psychological Bulletin, 80,* 286–303.

Bahrick, H., Noble, M., & Fitts, P. (1954). Extra-task performance as a measure of learning a primary task. *Journal of Experimental Psychology, 48,* 298–302.

Baldassare, M. (1979). *Residential crowding in urban America.* Berkeley: University of California Press.

Baldassare, M. (1981). The effects of household density on subgroups. *American Sociological Review, 46,* 110–118.

Barker, M. (1976). Planning for environmental indices: Observer appraisals of air quality. In K. Craik & E. Zube (Eds.), *Perceiving environmental quality.* New York: Plenum Press.

Barker, R. G. (1968). *Ecological psychology: Concepts and methods for studying the environment of human behavior.* Stanford: Stanford University Press.

Barker, R. G., Schoggen, P. (1973). *Qualities of community life.* San Francisco: Jossey-Bass.

Baron, R. (1978). Aggression and heat: The "long hot summer" revisited. In A. Baum, J. Singer, & S. Valins (Eds.), *Advances in environmental psychology* (Vol. 1). Hillsdale, NJ: Erlbaum.

Baron, R. M., & Rodin, J. (1978). Personal control as a mediator of crowding. In A. Baum, J. Singer, & S. Valins (Eds.), *Advances in environmental psychology* (Vol 1). Hillsdale, NJ: Erlbaum.

Baron, R. M., Mandel, D. R., Adams, C. A., & Griffen, L. M. (1976). Effects of social density in university residential environments. *Journal of Personality and Social Psychology, 34,* 434–446.

Basowitz, H., Persky, H., Korchin, S. J., & Grinker, R. R. (1955). *Anxiety and stress.* New York: McGraw-Hill.

Baum, A., & Davis, G. (1980). Reducing the stress of high density living: An architectural intervention. *Journal of Personality and Social Psychology, 38,* 471–481.

Baum, A., & Epstein, Y. (Eds.). (1978). *Human response to crowding.* Hillsdale, NJ: Erlbaum.

Baum, A., & Valins, S. (1977). *Architecture and social behavior: Psychological studies of social density.* Hillsdale, NJ: Erlbaum.

Baum, A., & Paulus, P. (in press). Crowding. In D. Stokols, A. Altman, & E. Willems (Eds.), *Handbook of environmental psychology.* New York: Wiley.

Baum, A., Harpin, R., & Valins, S. (1975). The role of group phenomena in the experience of crowding. *Environment and Behavior, 7,* 185–197.

Baum, A., Aiello, J., & Calesnick, L. (1978). Crowding and personal control: Social density and the development of learned helplessness. *Journal of Personality and Social Psychology, 36,* 1000–1001.

Baum, A., Fisher, J., & Solomon, S. (1981). Type of information, familiarity, and the reduction of crowding stress. *Journal of Personality and Social Psychology, 40,* 11–23.

Baum, A., Gatchel, R., Aiello, J., & Thompson, D. (1981). Cognitive mediation of environmental stress. In J. Harvey (Ed.), *Cognition, social behavior, and the environment.* Hillsdale, NJ: Erlbaum.

Baum, A., Singer, J. E., & Baum, C. S. (1981). Stress and the environment. *Journal of Social Issues, 37,* 4–35.

Baum, A., Fleming, R., Singer, J. E. (1982). Stress at Three Mile Island: Psychological impact analysis. In L. Bickman (Ed.), *Applied social psychology annual* (Vol. III). Beverly Hills, CA: Sage Publications.

Baum, A., Fleming, R., & Davidson, L. M. (1983). Natural disaster and technological catastrophe. *Environment and Behavior, 15,* 333–354.

Bechtel, R. (1977). *Enclosing behavior.* Stroudsburg, PA: Dowden, Hutchinson, & Ross.

Belgian-French Pooling Project. Assessment of Type A behavior by the Bortner scale and ischaemic heart disease. *European Heart Journal, 1984, 5b,* 440–446.

Bell, P., & Greene, T. (1982). Thermal Stress: Physiological comfort, performance, and social effects of hot and cold environments. In G. W. Evans (Ed.), *Environmental stress*. New York: Cambridge University Press.

Berenson, G. S. (1980). *Cardiovascular risk factors in children: The early natural history of atherosclerosis and hypertension*. New York: Oxford University Press.

Berglund, B., Berglund, U., & Lindval, Y. (1975). Scaling of annoyance in epidemiological studies. *Proceedings of the International Symposium on Recent Advances in the Assessment of the Health Effects of Environmental Pollution* (Vol. 1, pp. 119–137). Luxembourg: Commission of European Communities.

Berk, R. A. (1983). An introduction to sample selection bias in sociological data. *American Sociological Review, 48*, 386–98.

Berlyne, D. E. (1960). *Conflict, curiosity and arousal*. New York: McGraw-Hill.

Binik, Y. M., Theriault, G., & Shustack, B. (1977). Sudden death in the laboratory rat: Cardiac function, sensory, and experiential factors in swimming deaths. *Psychosomatic Medicine, 39*, 82–92.

Binik, Y. M., Deikel, S. M., Theriault, Shustack, B., & Balthazard, C. (1979). Sudden swimming deaths: Cardiac function, experimental anoxia, and learned helplessness. *Psychophysiology, 16*(4), 381–391.

Bleda, P., & Bleda, E. (1978). Effects of sex and smoking on reactions of spatial invasion at a shopping mall. *The Journal of Social Psychology, 104*, 311–312.

Bock, R. D. (1975). *Multivariate statistical methods in behavioral research*. New York: McGraw-Hill.

Boggs, D., & Simon, J. (1968). Differential effect of noise on tasks of varying complexity. *Journal of Applied Psychology, 52*, 148–153.

Booth, A. (1975). *Final Report: Urban Crowding Project*. Ottawa: Ministry of State for Urban Affairs. Government of Canada.

Booth, A. (1976). *Urban crowding and its consequences*. New York: Praeger.

Borsky, P. N. (1969). Effects of noise on community behavior. In W. D. Ward & J. E. Fricke (Eds.), *Noise as a public health hazard*. Washington, DC: The American Speech and Hearing Association.

Borsky, P. N. (1973). *NASA CR-221. A new field-laboratory methodology for assessing human response to noise*. Washington, DC: National Aeronautics and Space Administration.

Borsky, P. N. (1980). Research on community response to noise since 1973. In J. V. Tobias (Ed.), *The proceedings of the Third International Congress on Noise as a Public Health Problem*. Washington, DC: American Speech and Hearing Association.

Boyd, R., & Iverson, J. (1979). *Contextual analysis*. Belmont, CA: Wadsworth.

Brehm, J. (1966). *A theory of psychological reactance*. New York: Academic Press.

Brett, J. M. (1980). The effect of job transfer on employees and their families. In C. C. Cooper & R. Payne (Eds.), *Current concerns in occupational stress*. New York: Wiley.

Briere, J., Downes, A., & Spensley, J. (1983). Summer in the city: Urban weather conditions and psychiatric-emergency room visits. *Journal of Abnormal Psychology, 92*, 77–80.

Brigham, T. (1979). Some effects of choice in academic performance. In L. Perlmuter & R. Monty (Eds.), *Choice and perceived control*. Hillsdale, NJ: Erlbaum.

Brinberg, D., & McGrath, J. E. (1982). A network of validity concepts within the research process. In D. Brinberg & L. Kidder (Eds.), *New directions for methodology of social and behavioral science: Forms of validity in research*, No. 12. San Francisco: Jossey-Bass.

Broadbent, D. E. (1971). *Decision and stress*. New York: Academic Press.

Broadbent, D. E. (1978). The current state of noise research: A reply to Poulton. *Psychological Bulletin, 85*, 1052–1067.

Broadbent, D. E. (1979). Human performance and noise. In C. M. Harris (Ed.), *Handbook of noise control*. New York: McGraw-Hill.

Broadbent, D. E. (1981). The effects of moderate levels of noise on human performance. In J. Tobias & E. D. Schubert (Eds.), *Hearing: Reasearch and theory*. New York: Academic Press.

Broadbent, D. E. (1983). Recent advances in understanding performance in noise. In G. Rossi (Ed.),

Noise as a public health hazard, Proceedings of the 4th International Congress. Milano, Italy: Centro Ricerche e Studi Amplifom.

Broadbent, D. E., & Gregory, M. (1965). The effect of noise and signal rate upon vigilance analyzed by means of decision theory. *Human Factors, 7,* 155–162.

Bronfenbrenner, U. (1974). Developmental research public policy, and the ecology of childhood. *Child Development, 1,* 1–5.

Bronfenbrenner, U. (1977). Toward an experimental ecology of human development. *American Psychologist, 32,* 513–531.

Bronfenbrenner, U. (1979). *Ecology of human development.* Cambridge: Harvard University Press.

Bronzaft, A. (1981). The effect of a noise abatement program on reading ability. *Journal of Environmental Psychology, 3,* 215–222.

Bronzaft, A., & McCarthy, D. (1975). The effect of elevated train noise on reading ability. *Environment and Behavior, 7,* 517–527.

Brown, R., & Herrnstein, R. J. (1975). *Psychology.* Boston: Little, Brown.

Brunson, B., & Matthews, K. (1981). The Type A coronary prone behavior pattern and reactions to uncontrollable stress: An analysis of performance strategies, affect and attributions during failure. *Journal of Personality and Social Psychology, 40,* 906–918.

Brunswik, E. (1943). Organismic achievement and environmental probability. *Psychological Review, 50,* 255–272.

Brunswik, E. (1956). *Perception and the representative design of experiments.* Berkeley: University of California Press.

Butowsky, I., & Willows, D. (1980). Cognitive-motivational characteristics of children varying in reading ability: Evidence of learned helplessness in poor readers. *Journal of Educational Psychology, 72,* 408–422.

Calhoun, J. B. (1962). Population density and social pathology. *Scientific American, 206,* 139–148.

California Assessment Program. (1976). *Profiles of school district performance, 1975–1976: A guide to interpretation.* Sacramento: California Department of Education.

Cameron, P., Robertson, D., & Zaks, J. (1972). Sound pollution, noise pollution, and health: Community parameters. *Journal of Applied Psychology, 56,* 67–74.

Campbell, A. (1981). *The sense of well-being in America.* New York: McGraw-Hill.

Campbell, D. T., & Fiske, D. W. (1959). Convergent and discriminant validation by the multitrait-multimethod matrix. *Psychological Bulletin, 56,* 81–105.

Campbell, D. T., & Stanley, J. C. (1963). *Experimental and quasi-experimental designs for research.* Chicago: Rand McNally.

Campbell, J. (1983). Ambient stressors. *Environment and Behavior, 15,* 355-380.

Cannon, W. B. (1932). *The wisdom of the body.* New York: Norton.

Capellini, A., & Maroni, M. (1974). Clinical investigation of arterial hypertension and coronary disease and the possible relationship with the work environment in workers of the chemical industry. *Medicina del Lavoro, 65*(7-8), 297–305.

Caplan, R. D. (1983). Person–environment fit: Past, present, and future. In C. L. Cooper (Ed.), *Stress research: Issues for the eighties.* New York: Wiley.

Carlstein, T., Parkes, D., & Thrift, V. (Eds.). (1978). *Human activity and time geography.* London: Edward Arnold.

Carmines, E. G., & Zeller, R. A. (1979). *Reliability and validity assessment.* Beverly Hills: Sage Publications.

Carp, F. (1986). Environment and aging. In D. Stokols & I. Altman (Eds.), *Handbook of environmental psychology.* New York: Wiley.

Carp, F. M., & Carp, A. (1982). A role of technical environmental assessment in perceptions of environmental quality and well-being. *Journal of Environmental Psychology, 2,* 171–191.

Cassel, J. (1971). Health consequences of population density and crowding. In *Rapid population growth: Consequences and policy implications.* Baltimore, MD: Johns Hopkins University Press.

Cassel, J. (1974). Psychosocial processes and "stress": Theoretical formulation. *International Journal of Health Services, 4,* 471–482.

Cassel, J. (1975). Social science in epidemiology: Psychosocial processes and stress theoretical formulation. In E. L. Struening & M. Guttentag (Eds.), *Handbook of evaluation research* (Vol. 1). London: Sage Publications.

Catalano, R., Dooley, D., & Jackson, R. (1981). Economic predictors of admissions to mental health facilities in a nonmetropolitan community. *Journal of Health and Social Behavior, 22,* 284–297.

Chapko, M. K., & Solomon, H. (1976). Air polution and recreational behavior. *Journal of Social Psychology, 100,* 149–150.

Chein, I. (1954). The environment as a determinant of behavior. *Journal of Social Psychology, 39,* 115–127.

Chowns, R. H. (1970). Mental hospital admissions and aircraft noise. *Lancet, 1,* 467–468.

Church, R. (1964). Systematic effect of random error in the yoked control design. *Psychological Bulletin, 62,* 122–131.

Clarke, J. R. (1955). Influence of numbers on reproduction and survival in two experimental vole populations. *Proceedings of the Royal Society* (Series B), *114:*68–85.

Coates, T. J., Perry, C., Killen, F., & Slinkard, L. A. Primary prevention of cardiovascular disease in children and adolescents. In C. K. Prokop & L. A. Bradley (Eds.), *Medical psychology: Contributions to behavioral medicine.* New York: Academic Press, 1981.

Coffin, D., & Stokinger, H. (1977). Biological effects of air pollutants. In A. C. Stern (Ed.), *Air pollution.* New York: Plenum Press.

Cohen, A. (1969). Effects of noise on psychological state. In *Noise as a public health hazard.* Washington, DC: American Speech and Hearing Association.

Cohen, A. (1973). Industrial noise and medical, absence, and accident record data on exposed workers. In W. D. Ward (Ed.), *Proceedings of the International Congress on Noise as a Public Health Problem.* Washington, DC: U. S. Government Printing Office.

Cohen, A. (1976). The influence of a company hearing conservation program on extra-auditory problems in workers. *Journal of Safety Research, 8,* 146–162.

Cohen, A. (1979, September). *Remarks as discussant at symposium: Research needs with regard to nonauditory effects of noise.* Presented at the Annual Meeting of the American Psychologial Association, New York.

Cohen, J., & Cohen, P. (1975). *Applied multiple regression: Correlational analysis for the behavioral sciences.* Hillsdale, NJ: Erlbaum.

Cohen, S. (1978). Environmental load and the allocation of attention. In A. Baum, J. Singer, & S. Valins (Eds.), *Advances in environmental psychology* (Vol. 1). Hillsdale, NJ: Erlbaum.

Cohen, S. (1980). Aftereffects of stress on human performance and social behavior: A review of research and theory. *Psychological Bulletin, 88,* 82–108.

Cohen, S., & Hoberman, H. (1983). Positive events and social supports as buffers of life change stress. *Journal of Applied Social Psychology, 13,* 99–125.

Cohen, S., & Lezak, A. (1977). Noise and inattentiveness to social cues. *Environment and Behavior, 9,* 559–572.

Cohen, S., & McKay, G. (1984). Social support, stress, and the buffering hypothesis: A theoretical analysis. In A. Baum, J. E. Singer, & S. E. Taylor (Eds.), *Handbook of Psychology and Health* (Vol. 4). Hillsdale, NJ: Erlbaum.

Cohen, S., & Sherrod, D. (1978). When density matters: Environmental control as a determinant of crowding effects in laboratory and residential settings. *Journal of Population, 1,* 189–202.

Cohen, S., & Spacapan, S. (1978). The aftereffects of stress: An attentional interpretation. *Environmental Psychology and Nonverbal Behavior, 3,* 43–57.

Cohen, S., & Spacapan, S. (1984). The social psychology of noise. In D. M. Jones & A. J. Chapman (Eds.), *Noise and society.* London: Wiley.

Cohen, S., & Syme, S. L. (1985). Issues in the study and application of social support. In S. Cohen & S. L. Syme (Eds.), *Social support and health.* San Francisco: Academic Press.

Cohen, S., & Weinstein, N. (1981). Nonauditory effects of noise on behavior and health. *Journal of Social Issues, 37,* 36–70.

Cohen, S., & Weinstein, N. (1982). Nonauditory effects of noise on behavior and health. In G. W. Evans (Eds.), *Environmental stress.* New York: Cambridge University Press.

Cohen, S., Glass, D. C., & Singer, J. E. (1973). Apartment noise, auditory discrimination, and reading ability in children. *Journal of Experimental Social Psychology, 9,* 407–422.

Cohen, S., Rothbart, M., & Phillips, S. (1976). Locus of control and the generality of learned helplessness in humans. *Journal of Personality and Social Psychology, 34,* 1049–1056.

Cohen, S., Glass, D. C., & Phillips, S. (1977). Environment and health. In H. E. Freeman, S. Levine, & L. G. Reeder (Eds.), *Handbook of medical sociology.* Englewood Cliffs, NJ: Prentice-Hall.

Cohen, S., Evans, G. W., Krantz, D. S., & Stokols, D. (1980). Physiological, motivational, and cognitive effects of aircraft noise on children: Moving from the laboratory to the field. *American Psychologist, 35,* 231–243.

Cohen, S., Evans, G. W., Krantz, D. S., Stokols, D., & Kelly, S. (1981). Aircraft noise and children: Longitudinal and cross-sectional evidence on adaptation to noise and the effectiveness of noise abatement. *Journal of Personality and Social Psychology, 40,* 331–345.

Cohen, S., Krantz, D., Evans, E., & Stokols, D. (1981). Cardiovascular and behavioral effects of community noise. *American Scientist, 69,* 528–535.

Cohen, S., Kamarck, T., & Mermelstein, R. (1983). A global measure of perceived stress. *Journal of Health and Social Behavior, 24,* 385–396.

Collins, D., Baum, A., & Singer, J. (1983). Coping with chronic stress at Three Mile Island: Psychological and biochemical evidence. *Health Psychology, 2,* 149–166.

Contrada, R. J., Glass, D. C., Krakoff, L. R., Krantz, D. S., Kehoe, K., Isecke, W., Collins, C., & Elting, E. (1982). Effects of control over aversive stimulation and Type A behavior on cardiovascular and plasma catecholamine responses. *Psychophysiology, 19,* 408–419.

Contrada, R. J., Wright, R. A., & Glass, D. C. (1984). Task difficulty, Type A behavior pattern, and cardiovascular response. *Psychophysiology, 21,* 638–646.

Conway, T. L., Vickers, R. R., Jr., Ward, H. W., & Rahe, R. H. (1981). Occupational stress and variation in cigarette, coffee, and alcohol consumption. *Journal of Health and Social Behavior, 22,* 155–165.

Cook, T. D., & Campbell, D. T. (1979). *Quasi-experimentation: Design and analysis issues for field settings.* Chicago: Rand McNally.

Corah, N., & Boffa, J. (1970). Perceived control, self-observation, and response to aversive stimulation. *Journal of Personality and Social Psychology, 16,* 1–4.

Cornelius, R. R., & Averill, J. R. (1980). The influence of various types of control on psychophysiological stress reactions. *Journal of Research in Personality, 14,* 503–517.

Cox, V. C., Paulus, P. B., & McCain, G. (1984). Prison crowding research: The relevance for prison housing standards and a general approach regarding crowding phenomena. *American Psychologist, 39,* 1148–1160.

Coyne, J., & Holroyd, K. (1982). Stress, coping and illness: A transactional perspective. In T. Millon, C. Green, & R. Meagher (Eds.), *Handbook of Clinical Health Psychology.* New York: Plenum Press.

Coyne, J., Metalsky, G. I., & Lavelle, T. L. (1980). Learned helplessness as experimenter-induced failure and its alleviation with attentional redeployment. *Journal of Abnormal Psychology, 89,* 350–357.

Craik, K. H. (1973). Environmental psychology. *Annual Review of Psychology, 24,* 403–422.

Craik, F., & Lockhart, R. (1972). Levels of processing: A framework for memory research. *Journal of Verbal Learning and Verbal Behavior, 11,* 671–684.

Crandall, V. C. (1969). Sex differences in expectancy of intellectual and academic reinforcement. In C. Smith (Ed.), *Achievement-related motives in children.* New York: Russell Sage.

Crandall, V. C. (1973). *Differences in parental antecedents of internal-external control in children and in young adulthood.* Paper presented at the meeting of the American Psychological Association, Montreal.

Crandall, V. C. (1978). *New developments with the intellectual achievement responsibility scale.* Paper presented at the meeting of the American Psychological Association, Toronto.

Crandall, V. C., Katovsky, W., & Crandall, V. (1965). Children's beliefs in their own control of reinforcements in intellectual-academic achievement situations. *Child Development, 36,* 91–109.

Creer, R., Gray, R., & Treshow, M. (1970). Differential responses to air pollution as an environmental health problem. *Journal of the Air Pollution Control Association, 20,* 814–818.

Cronbach, L. J. (1975). Beyond the two disciplines of scientific psychology. *American Psychologist, 30,* 116–127.

Croog, S. H., Shapiro, D. S., & Levine, S. (1971). Denial among male heart patients. *Psychosomatic Medicine, 33,* 385–397.

Crook, M., & Langdon, F. (1974). The effects of aircraft noise in schools around London airport. *Journal of Sound and Vibration, 34,* 221–232.

Crosby, F. (1976). A model of egoistical relative deprivation. *Psychological Review, 83,* 85–113.

Cullen, I. G. (1978). The treatment of time in the explanations of spatial behavior. In T. Carlstein, D. Parkes, & V. Thrift (Eds.), *Human activity and time geography.* New York: Wiley.

Cunningham, M. (1979). Weather, mood, and helping behavior: Quasi-experiments with the sunshine samaritan. *Journal of Personality and Social Psychology, 37,* 1947–1956.

Daee, S., & Wilding, J. (1977). Effects of high intensity white noise on short-term memory for position in a list and sequence. *British Journal of Psychology, 68,* 335–349.

D'Atri, D. A. (1975). Psychophysiological responses to crowding. *Environment and Behavior, 7,* 237–250.

Dean, L. M., Pugh, U. M., & Gunderson, E. K. E. (1975). Spatial and perceptual components of crowding: Effects on health and satisfaction. *Environment and Behavior, 7,* 335–336.

de Charms, R. (1972). Personal causation training in the schools. *Journal of Applied Social Psychology, 2,* 95–113.

DeLongis, A., Coyne, J. C., Dakof, G., Folkman, S., & Lazarus, R. S. (1982). Relationship of daily hassles, uplifts, and major life events to health status. *Health Psychology, 1,* 119–136.

Dembroski, T. M., Weiss, S. M., Shields, J., Haynes, S. G., & Feinleib, M. (Eds.). (1978). *Coronary-prone behavior.* New York: Springer-Verlag.

Dembroski, T. M., MacDougall, J. M., Williams, R. B., Haney, T., & Blumenthal, J. A. (1985). Components of a Type A hostility and anger-in relationship to angiographic findings. *Psychosomatic Medicine, 47,* 219–233.

Denenberg, V. H. (1972). *The development of behavior.* Stanford, CN: Sinauer.

Derogatis, L. R. (1982). Self-report measures of stress. In L. Goldberger & S. Bresnitz (Eds.), *Handbook of stress: Theoretical and clinical aspects.* New York: Free Press.

Deutsch, C. P. (1964). Auditory discrimination and learning: Social factors. Merrill-Palmer Quarterly of Behavior and Development, 10, 277–296.

Diener, C. I., & Dweck, C. S. (1978). An analysis of learned helplessness: Continuous changes in performance, strategy, and achievement cognitions following failure. *Journal of Personality and Social Psychology, 36,* 451–462.

DiMento, J. F. (1981). Making usable information on environmental stressors: Opportunities for the research and policy communities. *Journal of Social Issues, 37,* 172–204.

DiMento, J. F. (1982). Much ado about environmental stress or research: Policy Implications. In G. W. Evans (Ed.), *Environmental stress.* New York: Cambridge University Press.

Dimond, S. (1966). Facilitation of performance through the use of the timing system. *Journal of Experimental Psychology, 71,* 181–183.

Dixon, N. F. (1980). Humor: A cognitive alternative to stress? In I. G. Sarason & C. D. Spielberger (Eds.), *Stress and anxiety* (Vol. 7). Washington, DC: Hemisphere Press.

Dohrenwend, B. S., & Dohrenwend, B. P. (Eds.). (1974). *Stressful life events: Their nature and effects.* New York: Wiley.

Dohrenwend, B. S., & Dohrenwend, B. P. (Eds.). (1981). *Stressful life events and their contexts.* New York: Prodist.

Donnerstein, D., & Wilson, D. W. (1976). Effects of noise and perceived control on ongoing and subsequent aggressive behavior. *Journal of Personality and Social Psychology, 34,* 774–781.

Dooley, D., & Catalano, R. (1980). Economic change as a cause of behavioral disorder. *Psychological Bulletin, 87,* 450–468.

Dooley, D., & Catalano, R. (1984). Why the economy predicts help-seeking: A test of competing explanations. *Journal of Health and Social Behavior, 25,* 160–176.

Dorian, B. J., Keystone, E., Garfinkel, P. E., & Brown, G. M. (1982). Aberations in lymphocyte subpopulations and functions during psychological stress. *Clinical Experimental Immunology, 50,* 132–138.

Dornic, S. (1975). Some studies on the retention of order information. In P. Rabbitt & S. Dornic (Eds.), *Attention and performance* (Vol. 5). New York: Academic Press.

Dubos, R. (1965). *Man adapting.* New Haven: Yale University Press.

Dunbar, F. (1954). *Emotions and bodily changes.* New York: Columbia University Press.

Dweck, C. S., & Elliott, E. (1983). Achievement motivation. In P. Mussen & M. Hetherington (Eds.), *Handbook of child psychology* (Vol. 4). New York: Wiley.

Dweck, C. S., & Goetz, T. (1978). Attributions and learned helplessness. In J. Harvey, W. Ickles, & R. Kidd (Eds.), *New directions in attribution research* (Vol. 2). Hillsdale, NJ: Erlbaum.

Dweck, C. S., & Licht, B. (1980). Learned helplessness and intellectual achievement. In J. Garber & M. E. P. Seligman (Eds.), *Human helplessness.* New York: Academic Press.

Easterbrook, J. A. (1959). The effect of emotion on cue utilization and the organization of behavior. *Psychological Review, 66,* 183–201.

Edmonds, L. D., Layde, P. M., & Erickson, J. D. (1979). Airport noise and tertogenesis. *Archives of Environmental Health, 34,* 243–247.

Elliot, G. R., & Eisdorfer, C. (1982). *Stress and human health: Analysis and implications of research.* New York: Springer.

Elmadjian, F. (1963, November). Excretion and metabolism of epinephrine and norepinephrine in various emotional states. *Proceedings of the 5th Pan American Congress of Endocrinology.* Lima, Peru. 341–371.

Elmadjian, F., Hope, J. M., & Larson, C. T. (1958). Excretion of epinephrine and norepinephrine under stress. *Recent Progress in Hormone Research, 14,* 513–553.

Endler, N. S., & Magnusson, D. (Eds.). (1976). *Interactional psychology and personality.* Washington, DC: Hemisphere.

Engel, G. L. (1968). A life setting conducive to illness: The giving-up-given-up complex. *Annals of Internal Medicine, 69,* 293–300.

Engel, B. T., & Bickford, A. F. (1961). Response specificity in essential hypertensives. *Archives of General Psychiatry, 5,* 478–489.

Enos, W. F., Holmes, R. H., & Beyer, J. (1953). Coronary disease among United States soldiers killed in action in Korea: Preliminary report. *Journal of the American Medical Association, 52*(1090).

Epstein, S. (1973). Expectancy and magnitude of reaction to a noxious UCS. *Psychophysiology, 10,* 100–107.

Epstein, Y., & Karlin, R. A. (1975). Effects of acute experimental crowding. *Journal of Applied Social Psychology, 5,* 34–53.

Ericsson, K. A., & Simon, H. A. (1980). Verbal reports as data. *Psychological Review, 87,* 215–251.

Evans, G. W. (1978a). Human spatial behavior: The arousal model. In A. Baum & Y. M. Epstein (Eds.), *Human response to crowding.* Hillsdale, NJ: Erlbaum.

Evans, G. W. (1978b). Crowding and the developmental process. In A. Baum & Y. M. Epstein (Eds.), *Human response to crowding.* Hillsdale, NJ: Erlbaum.

Evans, G. W. (1979). Behavioral and physiological consequences of crowding in humans. *Journal of Applied Social Psychology, 9,* 27–46.

Evans, G. W. (Ed.). (1982). *Environmental stress.* New York: Cambridge University Press.

Evans, G. W. (1980). Environmental cognition. *Psychological Bulletin, 88,* 259–287.

Evans, G. W., & Campbell, J. M. (1983). Psychological perspectives on air pollution and health. *Basic and Applied Social Psychology, 4,* 137–169.

Evans, G. W., & Cohen, S. (1986). Environmental stress. In D. Stokols & I. Altman (Eds.), *Handbook of Environmental Psychology.* New York: Wiley.

Evans, G. W., & Eichelman, W. H. (1976). Some preliminary models of conceptual linkages among proxemic variables. *Environment and Behavior, 8,* 87–116.

Evans, G. W., & Jacobs, S. V. (1982). Air pollution and human behavior. In G. W. Evans (Ed.), *Environmental stress.* New York: Cambridge University Press.

Evans, G. W., & Lovell, B. (1979). Design modification in an open-plan school. *Journal of Educational Psychology, 71,* 41–49.

Evans, G. W., Jacobs, S. V., Dooley, D, & Catalano, R. (1982). *The interaction of air pollution and stressful life events on mental health.* Paper presented at Annual Convention of the American Psychological Association, Washington, DC.

Evans, G. W., Jacobs, S. V., & Frager, N. B. (1982). Behavioral responses to air pollution. In A. Baum & J. Singer (Eds.), *Advances in environmental psychology* (Vol. 4). Hillsdale, NJ: Erlbaum. (c)

Falkner, B. (1984). Cardiovascular reactivity and psychogenic stress in juveniles. In J. M. H. Loggie, M. J. Horan, A. B. Gruskin, A. R. Hohn, J. B. Dunbar, & R. J. Havlik (Eds.), *NHLBI Workshop on Juvenile Hypertension.* Bethesda, MD: Biomedical Information Corp.

Falkner, B., Onesti, G., Angelakos, E. T., Fernandes, M., & Langman, C. (1979). Cardiovascular response to mental stress in normal adolescents with hypertensive parents: Hemodynamics and mental stress in adolescents. *Hypertension, 1,* 23–30.

Fawcett, J. T. (Ed.). (1973). *Psychological perspectives on population.* New York: Basic Books.

Finke, H. O., Guski, R., Martin, R., Rohrman, B., Schümer, R., & Schümer-Kohrs, A. (1975). Effects of aircraft noise on man. In *Proceedings of the Symposium on Noise in Transportation, Section III, Paper 1.* Southhampton: Institute of Sound and Vibration Research.

Finkelman, J., & Glass, D. (1970). Reappraisal of the relationship between noise and human performance by means of a subsidiary task measure. *Journal of Applied Psychology, 54,* 211–213.

Finkle, A. L., & Poppen, J. R. (1948). Clinical effects of noise and mechanical vibrations of a turbo-jet engine on man. *Journal of Applied Physiology, 1,* 183–204.

Firebaugh, G. (1978). A rule for inferring individual-level relationships from aggregate data. *American Sociological Review, 43,* 557–572.

Fisher, S. (1972). A distraction effect of noise bursts. *Perception, 1,* 233–236.

Fisher, C. S., Baldassare, M., & Ofshe, R. J. (1975). Crowding studies and urban life: A critical review. *American Institute of Planners Journal, 41*(6), 406–418.

Folkins, C. H. (1970). Temporal factors and the cognitive mediators of stress reaction. *Journal of Personality and Social Psychology, 14,* 173–184.

Folkman, S. (1984). Personal control and stress and coping processes: A theoretical analysis. *Journal of Personality and Social Psychology, 46,* 839–852.

Folkman, S. & Lazarus, R. S. (1980). An analysis of coping in a middle-aged community sample. *Journal of Health and Social Behavior, 21,* 219–239.

Forgas, J. P. (1979). *Social episodes: The study of interaction routines.* New York: Academic Press.

Forster, P. (1978). Attentional selectivity: A rejoinder to Hockey. *British Journal of Psychology, 69,* 505–506.

Forster, P., & Grierson, A. (1978). Noise and attentional selectivity: A reproducible phenomenon? *British Journal of Psychology, 69,* 489–498.

Franck, K. A. (1984). Exorcizing the ghost of physical determinism. *Environment and Behavior, 16,* 411–435.

Frankel, A., & Snyder, M. L. (1978). Poor performance following unsolvable problems: Learned helplessness or egotism? *Journal of Personality and Social Psychology, 36,* 1415–1423.

Frankenhaeuser, M. (1971). Behavior and circulating catecholamines. *Brain Research, 31,* 241–262.

Frankenhaeuser, M. (1980). Psychoneuroendocrine approaches to the study of stressful person–environment transactions. In H. Selye (Ed.), *Selye's guide to stress research* (Vol. 1). New York: van Nostrand.

Frankenhaeuser, M. (1981). Coping with stress at work. *International Journal of Health Services, 11,* 491–510.

Frankenhaeuser, M., & Lundberg, U. (1977). The influence of cognitive set on performance and arousal under different noise loads. *Motivation and Emotion, 1,* 139–149.

Frankenhaeuser, M., & Rissler, A. (1970). Effects of punishment on catecholamine release and efficiency of performance. *Psychopharmacologia, 17,* 378–390.

Freedman, J. L. (1973). The effects of population density on humans. In J. Fawcett (Ed.), *Psychological perspectives on population*. New York: Basic Books.

Freedman, J. L. (1975). *Crowding and behavior*. San Francisco: W. H. Freeman.

Freedman, J. L., Heshka, S., & Levy, A. (1975). Population density and pathology: Is there a relationship? *Journal of Experimental Social Psychology, 11*, 539–552.

Frerichs, R. R., Beeman, B. L., & Coulson, A. H. (1980). Los Angeles airport noise and morality: Faulty analysis and public policy. *American Journal of Public Health, 70*, 357–362.

Freud, S. (1927). *The ego and the id*. London: Hogarth Press and Institute of Psychoanalysis.

Friedman, M., & Rosenman, R. H. (1959). Association of specific overt behavior pattern with blood and cardiovascular findings: Blood cholesterol level, blood clotting time, incidence of arcus senilis and clinical coronary artery disease. *Journal of the American Medical Association, 169*, 1286–1296.

Froese, A., Hackett, T. P., & Cassem, N. H. (1974). Trajectories of anxiety and depression in denying and nondenying acute myocardial infarction patients during hospitalization. *Psychosomatic Medicine, 18*, 413–420.

Galle, O., Gove, W., & McPherson, J. (1972). Population density and pathology. *Science, 176*, 23–30.

Garber, J., & Seligman, M. E. P. (Eds.). (1980). *Human helplessness: Theory and applications*. New York: Academic Press.

Garmezy, N. (1983). Stressors of childhood. In N. Garmezy & N. Rutter (Eds.), *Stress, coping, and development in children*. New York: McGraw-Hill.

Garmezy, N., & Rutter, M. (Eds.). (1983). *Stress, coping and development in children*. New York: McGraw-Hill.

Garrity, T. F. (1975). Morbidity, mortality, and rehabilitation. In W. D. Gentry & R. B. Williams (Eds.), *Psychological aspects of myocardial infarction and coronary care*. Saint Louis: C. V. Mosby Co.

Gatchel, R. J., & Proctor, J. D. (1976). Physiological correlates of learned helplessness in man. *Journal of Abnormal Psychology, 85*, 27–34.

Gatchel, R. J., Paulus, P. B., & Maples, C. W. (1975). Learned helplessness and self-reported affect. *Journal of Abnormal Psychology, 84*, 732–734.

Gatchel, R. J., McKinney, M. E., & Koebernick, L. F. (1977). Learned helplessness, depression, and physiological responding. *Psychophysiology, 14*, 25–31.

Gattoni, F., & Tarnopolsky, A. (1973). Aircraft noise and psychiatric morbidity. *Psychological Medicine, 3*, 516–520.

Geer, J., & Maisel, E. (1972). Evaluating the effects of the prediction-control confound. *Journal of Personality and Social Psychology, 23*, 314–319.

Genest, J., Kuchel, O., Hamet, P., & Cantin, M. (Eds.). (1983). *Hypertension*. New York: McGraw-Hill.

Gentry, W. D., & Williams, R. B. (Eds.). (1979). *Psychological aspects of myocardial infarction and coronary care*. St. Louis: Mosby.

Gentry, W. D., Foster, S., & Haney, T. (1972). Denial as a determinant of anxiety and perceived health status in the coronary care unit. *Psychosomatic Medicine, 34*, 39.

Gergen, K. J. (1973). Social psychology as history. *Journal of Personality and Social Psychology, 26*, 309–320.

Gergen, K. J. (1978). Toward generative theory. *Journal of Personality and Social Psychology, 36*, 1344–1360.

Gergen, K. J. (1982). *Toward transformation in social knowledge*. New York: Springer-Verlag.

Gersten, J., Langner, T., Eisenberg, J., & Orzek, L. (1974). Child behavior and life events: Undesirable change or change *per se?* In B. S. Dohrenwend & B. P. Dohrenwend (Eds.), *Stressful life events*. New York: Wiley.

Gibson, J. J. (1960). The concept of the stimulus in psychology. *American Psychologist, 15*, 694–703.

Gillis, A. R. (1974). Population density and social pathology: The case of building type, social allowance, and juvenile delinquency. *Social Forces, 52*, 306–314.

Gillis, A. R. (1979). Coping with crowding: Television, patterns of activity, and adaptation to high density environments. *The Sociological Quarterly, 20*, 267–277.

Gillum R. F. (1979). Pathophysiology of hypertension in blacks and whites: A review of the basis of racial blood pressure differences. *Hypertension, 1,* 468–475.

Glass, D. C. (1977). Stress, behavior patterns, and coronary disease. *American Scientist, 65,* 177–187. (a)

Glass, D. C. (1977). *Behavior patterns, stress, and coronary disease.* Hillsdale, NJ: Erlbaum. (b)

Glass, D. C., & Singer, J. E. (1972). *Urban stress: Experiments on noise and social stressors.* New York: Academic Press.

Glass, D. C., Singer, J., & Friedman, L. (1969). Psychic cost of adaptation to an environmental stressor. *Journal of Personality and Social Psychology, 12,* 200–210.

Glass, D. C., Singer, J. E., Leonard, S., Krantz, D. S., Cohen, S., & Cummings, H. (1973). Perceived control of aversive stimulation and reduction of stress responses. *Journal of Personality, 41,* 577–595.

Glass, G., Wilson, V. L., & Gottman, J. M. (1975). *Design and analysis of time series experiments.* Boulder: Colorado Associated University Press.

Glass, D. C., Krakoff, L. R., Contrada, R., Hilton, W. C., Kehoe, K., Mannucci, E. G., Collins, C., Snow, B., & Elting, E. (1980). Effect of harassment and competition upon cardiovascular and plasma catecholamine responses in Type A and Type B individuals. *Psychophysiology, 17,* 453–463.

Glorig, A. (1971). Non-auditory effects of noise exposure. *Sound and Vibration, 5,* 28–29.

Gochman, I. (1979). Arousal, attribution, and environmental stress. In I. Sarason & C. Speilberger (Eds.), *Stress and anxiety* (Vol. 6). New York: Wiley.

Goldsmith, J., & Friberg, L. (1977). Effects of air pollution on human health. In A. C. Stern (Ed.), *Air pollution.* New York: Academic Press.

Gore, S. (1981). Stress-buffering functions of social support: An appraisal and clarification of research models. In B. S. Dohrenwend & B. P. Dohrenwend (Eds.), *Stressful life events and their contexts.* New York: Prodist.

Gore, S., & Mangione, T. W. (1983). Social roles, sex roles, and psychological distress: Additive and interactive models of sex differences. *Journal of Health and Social Behavior, 24,* 300–312.

Gorman, B. (1968). An observation of altered locus of control following political disappointment. *Psychological Reports, 23,* 1094.

Granati, A., Angelepi, F., & Lenzi, R. (1959). L'influenza de: rumor: sul sistema nervoso. *Folia Medica, 42,* 1313–1325.

Grandjean, E., Graf, P., Lauber, A., Meier, H. P., & Muller, H. P. (1973). A survey on aircraft noise in Switzerland. In W. D. Ward (Ed.), *Proceedings of the International Congress on Noise as a Public Health Problem.* Washington, DC: U. S. Government Printing Office.

Green, K., Pasternack, B., & Shore, R. (1980). Effect of aircraft noise on children's reading and hearing levels. *Journal of the Acoustical Society of America, 68,*(Supplement 1), 90.

Greenberger, E., Steinberg, L. D., & Vaux, A. (1982). Person–environment congruence as a predictor of adolescent health and behavioral problems. *American Journal of Community Psychology, 10,* 511–526.

Greene, T. C., & Bell, P. (1980). Additional considerations concerning the effects of warm and cool wall colours on energy conservation. *Ergonomics, 23,* 949–954.

Greene, W. A., Goldstein, S., & Moss, A. J. (1972). Psychosocial aspects of sudden death: A preliminary report. *Archives of Internal Medicine, 129,* 725–731.

Griefahn, B. (1980). Research on noise-disturbed sleep since 1973. In J. V. Tobias, G. Jansen, & W. D. Ward (Eds.), *Proceedings of the Third International Congress on Noise as a Public Health Problem.* Rockville, MD: American Speech-Language-Hearing Association.

Griffith, M. (1977). Effects of noncontingent success and failure on mood and performance. *Journal of Personality and Social Psychology, 45,* 442–457.

Griffiths, I. D., & Delauzun, F. R. (1977). Individual differences in sensitivity to traffic noise: An empirical study. *Journal of Sound and Vibration, 55,* 93–107.

Gunnar-Vongnechten, M. (1978). Changing a frightening toy into a pleasant toy by allowing the infant to control its actions. *Developmental Psychology, 14,* 157–162.

Guski, R., & Rohrmann, B. (1981). Psychological aspects of environmental noise. *Journal of Environmental Policy, 2,* 183–212.

Gutek, B. A., Nakamura, C. Y., & Nieva, V. F. (1981). The interdependence of work and family roles. *Journal of Occupational Behavior, 2,* 1–16.

Gutek, B. A., Harris, A., Tyler, T. R., Lau, R., & Majchrzak, A. (1983). The importance of internal referants as determinants of satisfaction. *Journal of Community Psychology, 11,* 111–120.

Guttman, M. C., & Benson, H. (1971). Interaction of environmental factors and systemic arterial pressure. *Medicine, 50,* 543.

Hall, D. T., & Hall, F. S. (1980). Stress and the two-career couple. In C. L. Cooper & R. Payne (Eds.), *Current concerns in occupational stress.* New York: Wiley.

Hall, E. T. (1966). *The hidden dimension.* Garden City, NY: Doubleday.

Hamilton, P., Hockey, G. R. J., Quinn, J. (1972). Information selection, arousal and memory. *British Journal of Psychology, 63,* 181–190.

Hamilton, P., Hockey, G. R. J., & Rejman, M. (1977). The place of the concept of activation in human information processing theory. In S. Dornic (Ed.), *Attention and performance* (Vol. 6). New York: Academic Press.

Hand, D. J., Tarnopolsky, A., Barker, S. M., & Jenkins, L. M. (1980). Relationships between psychiatric hospital admissions and aircraft noise: A new study. In J. V. Tobias (Ed.), *The proceedings of the Third International Congress on Noise as a Public Health Problem.* Washington, DC: U.S. Environmental Protection Agency.

Harlow, H., & Harlow, M. K. (1962). Social deprivation in monkeys. *Scientific American, 207,* 136–146.

Hartley, L. (1981a). Noise, attentional selectivity, serial reactions and the need for experimental power. *British Journal of Psychology, 72,* 101–107.

Hartley, L. (1981b). Noise and attentional selectivity: A rejoinder to Forster. *British Journal of Psychology, 72,* 123–124.

Hartsough, D., & Savitsky, J. (1984). Three-Mile Island: Psychology and environmental policy at a crossroads. *American Psychologist, 39,* 1113–1122.

Hasher, L., & Zacks, R. (1979). Automatic and effortful processes in memory. *Journal of Experimental Psychology: General, 108,* 356–388.

Haynes, S. G., Feinleib, M., & Kannel, W. B. (1980). The relationship of psychosocial factors to coronary heart disease in the Framingham study. *American Journal of Epidemiology, 3,* 37–58.

Hebb, D. O. (1972). *Textbook of psychology.* Philadelphia: W. B. Saunders.

Heckman, J. J. (1979). Sample selection bias as a specification error. *Econometrica, 47,* 153–61.

Heft, H. (1979). Background and focal environmental conditions of the home and attention in young children. *Journal of Applied Social Psychology, 9,* 47–69.

Held, R. (1965). Plasticity in sensory-motor systems. *Scientific American, 213,* 84–94.

Heller, K., & Monahan, J. (1977). *Psychology and community change.* Homewood, IL: Dorsey Press.

Helson, H. (1948). Adaptation level as a basis for a quantitative theory of frames of reference. *Psychological Review, 55,* 297–313.

Helson, R., Mitchell, V., & Moane, G. (1984). Personality and patterns of adherence and nonadherence to the social clock. *Journal of Personality and Social Psychology, 46,* 1079–1096.

Henry, J. P., & Cassel, J. C. (1969). Psychosocial factors in essential hypertension: Recent epidemiologic and animal experimental evidence. *American Journal of Epidemiology, 90,* 171.

Herd, A. J. (1978). Physiological correlates of coronary-prone behavior. In T. M. Dembroski, S. M. Weiss, J. L. Shields, S. G. Hayes, & M. Feinleib (Eds.), *Coronary-prone behavior.* New York: Spring-Verlag.

Herridge, C. F. (1974). Aircraft noise and mental health. *Journal of Psychosomatic Research, 18,* 239–243.

Herridge, C. F., & Chir, B. (1972). Aircraft noise and mental hospital admissions. *Sound, 6,* 32–36.

Hiroto, D. S. (1974). Locus of control and learned helplessness. *Journal of Experimental Psychology, 102,* 187–193.

Hiroto, D. S., & Seligman, M. E. P. (1975). Generality of learned helplessness in humans. *Journal of Personality and Social Psychology, 31,* 311–327.

Hockey, G. R. J. (1970a). Effects of loud noise on attentional selectivity. *Quarterly Journal of Experimental Psychology, 22*, 28–36.

Hockey, G. R. J. (1970b). Signal probability and spatial location as possible bases for increased selectivity in noise. *Quarterly Journal of Experimental Psychology, 22*, 37–42.

Hockey, G. R. J. (1973). Changes in information selection patterns in multi-source monitoring as a function of induced arousal shifts. *Journal of Experimental Psychology, 101*, 35–42.

Hockey, G. R. J. (1978). Attentional selectivity and the problem of replication: A reply to Forster and Grierson. *British Journal of Psychology, 69*, 499–503.

Hockey, G. R. J. (1979). Stress and the cognitive components of skilled performance. In V. Hamilton & D. Warburton (Eds.), *Human stress and cognition*. New York: Wiley.

Hockey, G. R. J., & Hamilton, P. (1970). Arousal and information selection in short-term memory. *Nature, 226*, 866–867.

Holmes, T., & Rahe, R. (1967). The social readjustment rating scale. *Journal of Psychosomatic Research, 11*, 213–218.

Hornstein, G. A., & Wapner, S. (1986). Modes of experiencing and adapting to retirement.

Horvath, F. (1959). Psychological stress: A review of definitions and experimental research. *General Systems Yearbook, 4*, 203–230.

House, J. S. (1975). Occupational stress as a precursor to coronary disease. In W. D. Gentry & R. B. Williams (Eds.), *Psychological aspects of myocardial infarction and coronary care*. St. Louis: C. V. Mosby.

House, J. S. (1981). *Work stress and social support*. Reading, MA: Addison-Wesley.

Hughes, C. W., & Lynch, J. J. (1978). A reconsideration of psychological precursors of sudden death in infrahuman animals. *American Psychologist, 33*, 419–429.

Hunter, J. E., Schmidt, F. L., & Jackson, G. B. (1983). *Meta-analysis: Cumulating research findings across studies*. Beverly Hills, CA: Sage Publications.

Hurst, J. W., Logue, R. B., Schlant, R. C., & Wenger, N. K. (Eds.). (1978). *The heart*. New York: McGraw-Hill.

Jacobi, M. (1984). *A contextual analysis of stress and health among re-entry women to college*. Doctoral dissertation, Program in Social Ecology, University of California, Irvine.

James, W. (1980). *The principles of psychology*. New York: Holt, Rinehart & Winston.

Janis, I. L. (1958). *Psychological stress: Psychoanalytic and behavioral studies of surgical patients*. New York: Wiley.

Janis, I. L. (1983). Stress inoculation in health care. In D. Meichenbaum & M. Jaremko (Eds.), *Stress reduction and prevention*. New York: Plenum Press.

Jansen, G. (1959). The occurrence of vegetative functional disturbances resulting from the influence of noise. *Gewerbepathologie und Gewerbehygiene, 17*, 238–261.

Jansen, G. (1961). Noise stress in the smelting industry. *Stahl und Eisen, 81*(4), 217–220.

Jansen, G. (1969). Effects of noise on physiologic state. In W. D. Ward & J. E. Fricke (Ed.), *Noise as a public health problem*. Washington, DC: American Speech and Hearing Association.

Jemmott, J. B., III, & Locke, S. E. (1984). Psychosocial factors, immunologic mediation, and human susceptibility to infectious diseases: How much do we know? *Psychological Bulletin, 95*, 52–77.

Jenkins, C. D. (1976). Recent evidence supporting psychologic and social risk factors for coronary disease. *New England Journal of Medicine, 294*, 1033–1038.

Jennings, J. R. (1985a). Bodily changes during attention. In M. Coles, E. Donchin, & S. Porges (Eds.), *Psychophysiology: Systems, processes, and applications*. New York: Guilford.

Jennings, J. R. (1985b). Memory, thought, and bodily response. In M. Coles, E. Donchin, & S. Porges (Eds.), Psychophysiology: Systems, processes, and applications. New York: Guilford.

Jessor, R. (1958). The problem of reductionism in psychology. *Psychological Review, 65*, 170–178.

Joffe, J., Rawson, R., & Mullick, J. (1973). Control of their environment reduces emotionality in rats. *Science, 180*, 1383–1384.

Jones, D. M. (1984). Performance effects. In D. M. Jones & A. J. Chapman (Eds.), *Noise and society*. London: Wiley.

Jones, D., Smith, A., & Broadbent, D. (1979). Effects of moderate intensity noise on the Bakan vigilance task. *Journal of Applied Psychology, 64*, 627–634.

Jones, F. N., & Tauscher, J. (1978). Residence under an airport landing pattern as a factor in teratism. *Archives of Environmental Health, 33,* 10–12.

Jones, J. W. (1978). Adverse emotional reactions of nonsmokers to secondary cigarette smoke. *Environmental Psychology and Nonverbal Behavior, 3,* 125–127.

Jones, J. W., & Bogat, G. A. (1978). Air pollution and human aggression. *Psychological Reports, 43,* 721–722.

Jonsson, E., & Sorensen, S. (1973). Adaptation to community noise—A case study. *Journal of Sound and Vibration, 26,* 571–575.

Julius, S., & Esler, M. (1975). Autonomic nervous cardiovascular regulation in borderline hypertension. *The American Journal of Cardiology, 36,* 685–689.

Julius, S., & Schork, M. A. (1977). Predictors of hypertension. *Annals of New York Academy of Sciences, 304,* 38–52.

Kahneman, D. (1973). *Attention and effort.* Englewood Cliffs, NJ: Prentice-Hall.

Kalsbeek, J. (1965). Measure objective de la surcharge mentale: Nouvelles applications de la methode des double taches. *Travail Humain, 1-2,* 122–132.

Kangelari, S. S., Abranovich-Polyakov, D. K., & Rudenkov, V. F. (1966). The effects of noise and vibration on morbidity rates. *Gigiena Truda-i Professional 'nye Zabolevaniya, 6,* 47–49.

Kannel, W. B. (1979). Cardiovascular disease: A multifactorial problem (Insights from the Framingham Study). In M. L. Pollock & D. H. Schmidt (Eds.), *Heart disease and rehabilitation.* New York: Wiley.

Kaplan, S. (1983). A model of person–environment compatibility. *Environment and Behavior, 15,* 311–332.

Kaplan, S., & Kaplan, R. (1982). *Cognition and environment.* New York: Praeger.

Karagodina, I. L., Soldatkina, S. A., Vinokur, I. L., & Klimukhin, A. A. (1969). Effect of aircraft noise on the population near airports. *Hygiene and Sanitation, 34,* 182–187.

Karasek, R. A. (1979). Job demands, job decision latitude and mental strain: Implications for job redesign. *Administrative Science Quarterly, 24,* 285–308.

Karasek, R. A., Theorell, T. G., Schwartz, J., Pieper, C., & Alfredsson, L. (1982). Job, psychological factors and coronary heart disease: Swedish prospective findings and U.S. prevalence findings using a new occupational inference method. *Advances in Cardiology, 29,* 62–67.

Karsdorf, G., & Klappach, H. (1968). The influence of traffic noise on the health and performance of secondary school students in a large city. *Zeitschrift fur die Gesamte Hygiene, 14,* 52–54.

Kasl, S. V. (1983). Pursuing the link between stressful life experiences and disease: A time for reappraisal. In C. L. Cooper (Ed.), *Stress research.* New York: Wiley.

Kasl, S. V., & Cobb, S. (1982). Variability of stress effects among men experiencing job loss. In L. Goldberger & S. Breznitz (Eds.), *Handbook of stress: Theoretical and clinical aspects.* New York: The Free Press.

Kelley, H. H. (1983). The situational origins of human tendencies: A further reason for the formal analysis of structures. *Personality and Social Psychology Bulletin, 9,* 8–30.

Kendall, P. (1983). Stressful medical procedures: Cognitive-behavioral strategies for stress management and prevention. In D. Meichenbaum & M. Jaremko (Eds.), *Stress reduction and prevention.* New York: Plenum Press.

Kerr, B. (1973). Processing demands during mental operations. *Memory and Cognition, 1,* 401–412.

Khomulo, P. S., Rodinova, L. P., & Rusinova, A. P. (1967). Changes in the lipid metabolism of man under protracted effect of industrial noise on the central nervous system. *Kardiologiia, 7,* 35–38.

Kiesler, C. A. (1985). Policy implications of research on social support and health. In S. Cohen & S. L. Syme (Eds.), *Social support and health.* New York: Academic Press.

Klassen, D. (1981). *Family, peer and work environment correlates of violent behavior.* Unpublished doctoral dissertation, Program in Social Ecology, University of California, Irvine.

Knipschild, P. G. (1976). Medische Gevolgon van Vliegtuilawaai. (Doctoral thesis with English summary, pp. 127–129). Unpublished doctoral dissertation, Amsterdam Coronel Laboratorium.

Knipschild, P. G. (1977). Medical effects of aircraft noise. *International Archives of Occupational and Environmental Health, 40,* 185–204.

Knipschild, P. G. (1980). Aircraft noise and hypertension. In J. V. Tobias, G. Jansen, W. D. Ward (Eds.), *Noise as a public health problem: Proceedings of the Third International Congress* (ASHA Report No. 10). Rockville, MD: American Speech and Hearing Association.

Knipschild, P. G., Meijer, H., & Salle, H. (n.d.). *Aircraft noise and birth weight.* the Netherlands: University of Amsterdam, Coronel Laboratory.

Knipschild, P., & Salle, H. (1979). Road traffic noise and cardiovascular disease: A population study in The Netherlands. *International Archives of Occupational and Environmental Health, 44,* 55–99.

Knipschild, P., & Oudshoorn, N. (1977). Medical effects of aircraft noise: Drug survey. *International Archives of Occupational and Environmental Health, 40,* 197–200.

Kobasa, S. C. (1979). Stressful life events, personality, and health: An inquiry into hardiness. *Journal of Personality and Social Psychology, 37,* 1–11.

Kobasa, S. C. (1982). Commitment and coping in stress resistance among lawyers. *Journal of Personality and Social Psychology, 42,* 707–717.

Koffka, J. (1935). *Principles of gestalt psychology.* New York: Harcourt, Brace and World.

Kohler, W. (1947). *Gestalt psychology.* New York: Liveright.

Kokokusha, D. (1973). *Report on investigation of living environment around Osaka International Airport.* Japan: Association for the Prevention of Aircraft Nuisance.

Krantz, D. S., & Manuck, S. B. (1984). Acute psychophysiologic reactivity and risk of cardiovascular disease: A review and methodologic critique. *Psychological Bulletin, 96,* 435–464.

Krantz, D. S., & Schulz, R. (1980). A model life crisis control, and health outcomes: Cardiac rehabilitation and relocation of the elderly. In A. Baum & J. E. Singer (Eds.), *Advances in environmental psychology* (Vol. 3): *Cardiovascular disorders and behavior.* Hillsdale, NJ: Erlbaum.

Krantz, D. S., Glass, D. C., & Snyder, M. L. (1974). Helplessness, stress level, and the coronary-prone behavior pattern. *Journal of Experimental Social Psychology, 10,* 284–300.

Krantz, D. S., Glass, D. C., Contrada, R., & Miller, N. E. (1981). *Behavior and health.* National Science Foundation's second five-year outlook on science and technology. U. S. Government Printing Office, Washington, D.C.

Krantz, D. S., Baum, A., & Singer, J. E. (Eds.). (1983). *Handbook of psychology and health* (Vol. 3): *Cardiovascular disorders and behavior.* Hillsdale, NJ: Erlbaum.

Krantz, D. S., Grunberg, N., & Baum, A. (1985). Health psychology. *Annual Review of Psychology, 36,* 349–383.

Kryter, K. D. (1970). *The effects of noise on man.* New York: Academic Press.

Kulik, J. A., Kulik, C. C., & Cohen, P. A. (1979). A meta-analysis of outcome studies of Keller's personalized system of instruction. *Americal Psychologist, 34,* 307–318.

Lacey, J. I. (1967). Somatic response patterning and stress: Some revisions of activation theory. In M. H. Appley & R. Trumbull (Eds.), *Psychological stress.* New York: Appleton-Century-Crofts.

Lane, S. R., & Meecham, W. C. (1974). Jet noise at schools near Los Angeles International Airport. *Journal of the Acoustical Society of America, 56,* 127–131.

Langer, E. J., & Rodin, J. (1976). The effects of choice and enhanced personal responsibility for the aged: A field experiment in an institutional setting. *Journal of Personality and Social Psychology, 34,* 191–198.

Langer, E. J., & Saegert, S. (1977). Crowding and cognitive control. *Journal of Personality and Social Psychology, 35,* 175–182.

Langer, E. J., Janis, R., & Wolfer, J. (1975). Reduction of psychological stress in surgical patients. *Journal of Experimental Social Psychology, 1,* 155–166.

Lauer, R. M., Connor, W. E., Leaverton, P. E., Reiter, M. A., & Clarke, W. R. (1975). Coronary heart disease risk factors in school children—The Muscatine Study. *Journal of Pediatrics, 86,* 697.

Lave, L., & Seskin, E. Air pollution and human health. *Science,* 1970, *169,* 723–733.

Lavelle, T. L., Metalsky, G. I., & Coyne, J. C. (1979). Learned helplessness, test anxiety, and acknowledgement of contingencies. *Journal of Abnormal Psychology, 88,* 381–387.

Lawrence, J. E. (1974). Science and sentiment: Overview of research on crowding and human behavior. *Psychological Bulletin, 81,* 712–720.

Lawton, M. P. (1980). *Environment and aging.* Monterey, CA: Brooks-Cole.

Lazarus, R. S. (1966). *Psychological stress and the coping process.* New York: McGraw-Hill.

Lazarus, R. S. (1977). Cognitive and coping processes in emotion. In A. Monat & R. S. Lazarus (Eds.), *Stress and coping.* New York: Columbia University Press.

Lazarus, R. S. (1980). The stress and coping paradigm. In C. Eisdorfer, D. Cohen, A. Kleinman, & P. Maxim (Eds.), *Theoretical bases for psychopathology.* New York: Spectrum.

Lazarus, R. S. (1984). On the primacy of cognition. *American Psychologist, 39,* 124–129.

Lazarus, R. S., & Cohen, J. (1977). Environmental stress. In J. Wohlwill & I. Altman (Eds.), *Human behavior and environment.* New York: Plenum Press.

Lazarus, R. S., & Folkman, S. (1984). *Stress, appraisal, and coping.* New York: Springer Publishing.

Lazarus, R. S., & Launier, R. (1978). Stress-related transactions between person and environment. In L. Pervin & M. Lewis (Eds.), *Perspectives in interactional psychology.,* New York: Plenum Press.

Lefcourt, H. (1973). The function of the illusion of control and freedom. *American Psychologist, 28,* 417–425.

Lefcourt, H. (1982). *Locus of control* (2nd ed.). Hillsdale, NJ: Erlbaum.

Lei, H., & Skinner, H. A. (1980). A psychometric study of life events and social readjustment. *Journal of Psychosomatic Research, 24,* 57–64.

Leonard, S., & Borsky, P. N. (1973). A casual model for relating noise exposure, psychological variables and aircraft noise annoyance. In W. D. Ward (Ed.), *Proceedings of the International Conference on Noise as a Public Health Problem.* Washington, DC: U.S. Government Printing Office.

Lerner, M. (1970). The desire for justice and reaction to victims. In J. Macaulay & L. Berkowitz (Eds.), *Altruism and helping.* New York: Academic Press.

Levi, L. (Ed.). (1972). *Stress and distress in response to psychosocial stimuli.* New York: Pergamon Press.

Levi, L., & Anderson, L. (1975). *Psychosocial stress: Population, environment, and quality of life.* New York: Spectrum.

Levi, L., & Herzog, A. N. (1974). Effects of population density and crowding on health and social adaptation in the Netherlands. *Journal of Health and Social Behavior, 15,* 228–240.

Lewin, K. (1936). *Principles of topological psychology.* New York: McGraw-Hill.

Lindsley, D. (1951). Emotion. In S. S. Stevens (Ed.), *Handbook of experimental psychology.* New York: Wiley.

Little, B. (1983). Personal projects: A rationale and method for investigation. *Environment and Behavior, 15,* 273–310.

Little, B. R. (1986). Personality and the environment. In D. Stokols & I. Altman (Eds.), *Handbook of environmental psychology.* New York: Wiley.

Lloyd, J. A., & Christian, J. J. (1969). Reproductive activity of individual females in three experimental freely-growing populations of house mice (Mus musculus). *Journal of Mammallogy, 50,* 49–59.

Loeb, M. (1980). Noise and performance: Do we know more? In J. Tobias, G. Jansen, & W. Ward (Eds.), *Noise as a public health problem.* Rockville, MD: The American Speech and Hearing Association.

Londe, S., Bourgoignie, J. J., Robson, A. M., & Goldring, D. (1971). Hypertension in apparently normal children. *Journal of Pediatrics, 78,* 58.

Loo, C. (1973). Important issues in researching the effects of crowding on humans. *Representative Research in Social Psychology, 4,* 219–226.

LoRocco, J. L. (1983). Theoretical distinctions between causal and interaction effects of social support. *Journal of Health and Social Behavior, 24,* 91–92.

Lukas, J. S. (1975). Noise and sleep: A literature review and a proposed criterion for assessing effect. *Journal of the Acoustical Society of America, 58b,* 1232–1242.

Lukas, J. S., DuPree, R., & Swing, J. (1981, April). *Report on a study of the effects of freeway noise on academic achievement of elementary school children, and a recommendation for a criterion level for a school noise abatement program.* Sacramento, CA: Office of Noise Control, Epidemiological Studies Section, California Department of Health Services.

Lundberg, U., & Frankenhaeuser, M. (1979). Psychophysiological reactions to noise as modified by personal control over noise intensity. *Biological Psychology, 6,* 51–60.

Mackworth, J. (1964). Performance decrement in vigilance, threshold, and high speed perceptual motor tasks. *Canadian Journal of Psychology, 18,* 209–223.

Madge, J. (1968). *International encyclopedia of the social sciences.* New York: Macmillan and Free Press.

Magnusson, D. (1981). A psychology of situations. In D. Magnusson (Ed.), *Toward a psychology of situations: An interactional perspective.* Hillsdale, NJ: Erlbaum.

Magnusson, D., & Allen, V. P., (Eds.). (1983). *Human development: An interactional perspective.* New York: Academic Press.

Maier, S., & Seligman, M. E. P. (1976). Learned helplessness: Theory and evidence. *Journal of Experimental Psychology: General, 105,* 3–46.

Malmo, R. B. (1959). Activation: A neuro-psychological dimension. *Psychological Review, 66,* 367–386.

Manicas, P. T., & Secord, P. F. (1983). Implications for psychology of the new philosophy of science. *American Psychologist, 38,* 399–413.

Manuck, S., Harvey, A., Lechleiter, S., & Neal K. (1978). Effects of coping on blood pressure responses to threat of aversive stimulation. *Psychophysiology, 15,* 544–549.

Marans, R. W. (1976). Perceived quality of residential environments: Some methodological issues. In K. H. Craik & E. H. Zube (Eds.), *Perceiving environmental quality: Research and applications.* New York: Plenum Press.

Marmot, M. G., & Syme, S. L. (1976). Acculturation and coronary heart disease in Japanese Americans. *American Journal of Epidemiology, 104,* 225–247.

Maser, A. (1978). *Effects of intrusive sound on classroom behavior: Data from a successful lawsuit.* San Francisco: Panel presentation at the Western Psychological Association Annual Meetings.

Maslow, A. (1962). *Toward a psychology of being.* New York: van Nostrand.

Mason, J. W. (1971). A re-evaluation of the concept of "non-specificity" in stress theory. *Journal of Psychiatric Research, 8,* 323–333.

Mason, J. W. (1975). A historical view of the stress field. Part 2. *Journal of Human Stress, 1,* 22–36.

Mathews, K. E., & Canon, L. (1975). Environmental noise level as a determinant of helping behavior. *Journal of Personality and Social Psychology, 32,* 571–577.

Matthews, K. A. Psychological perspectives on the Type A behavior pattern. *Psychological Bulletin,* 1982, *91,* 293–323.

Matthews, K. A., & Siegel, J. M. (1982). The Type A behavior pattern in children and adolescents: Assessment, development, and associated coronary-risk. In A. Baum & J. E. Singer (Eds.), *Handbook of psychology and health: Vol. 2: Issues in child health and adolescent health.* Hillsdale, NJ: Erlbaum.

Matthews, K. A., Manuck, S. B., Saab, P. G. (1985). *Cardiovascular responses of adolescents during a naturally occurring stressor and their behavioral and psychophysiological predictors.* Unpublished manuscript, University of Pittsburgh, Pittsburgh, PA.

McArthur, L. (1970). Luck is alive and well in New Haven. *Journal of Personality and Social Psychology, 16,* 316–318.

McCallum, R., Rusbult, C., Hong, G., Walden, T., & Schopler, J. (1979). Effects of resource availability and importance of behavior on the experience of crowding. *Journal of Personality and Social Psychology, 37,* 1304–1313.

McCroskey, R., & Devens, J. (1977). Effects of noise upon student performance in public school classrooms. *Preceedings of Technical Program.* Chicago: National Noise and Vibration Control Conference.

McCubbin, J. A., Richardson, J., Obrist, P. A., Kizer, J. S., & Langer, A. W. (1980). *Catecholaminergic and hemodynamic responses to behavioral stress in young adult males.* Paper presented at the Twentieth Annual Meeting of the Society for Psychophysiological Research.

McGrath, J. E. (Ed.). (1970). *Social and psychological factors in stress.* New York: Holt, Rinehart and Winston.

McGrath, J. E. (1982). Methodological problems in research on stress. In H. Krohne and L. Laux (Eds.), *Achievement stress and anxiety.* Washington, DC: Hemisphere Press.

McGuire, W. J. (1983). A contextualist theory of knowledge: Its implications for innovation and reform in psychological research. In L. Berkowitz (Ed.), *Advances in Experimental Social Psychology* (Vol. 16). New York: Academic Press.

McKennell, A. C. (1963). *Aircraft noise annoyance around London (Heathrow) Airport*. U.K. Government Social Survey SS 337, London: Her Majesty's Stationary Office. Central Office of Information.

McKennell, A. C., & Andrews, F. M. (1983). Components of perceived life quality. *Journal of Community Psychology, 11,* 98–110.

McKennell, A. C., & Hunt, E. A. (1966). Noise annoyance in Central London. U.K. Government Social Survey Report SS 332.

McLean, E. K., & Tarnopolsky, A. (1977). Noise discomfort and mental health. *Psychological Medicine, 1,* 19–62.

McNamara, J. J., Molot, M. A., & Stremple, J. F. (1971). Coronary artery disease in combat casualties in Vietnam. *Journal of the American Medical Association, 216,* 1185.

Mechanic, D. (1968). *Medical sociology*. New York: Free Press.

Mechanic, D. (1974). Social structure and personal adaptation: Some neglected dimensions. In G. V. Coelho, D. Hamburg, & J. Adams (Eds.), *Coping and adaptation*. New York: Basic Books.

Mechanic, D. (1977). Some modes of adaptation: Defense. In A. Monat & R. S. Lazarus (Eds.), *Stress and coping*. New York: Columbia University Press.

Meecham, W. C., & Shaw, N. (1979). Effects of jet noise on mortality rates. *British Journal of Audiology, 13,* 77–80.

Meecham, W. C., & Smith, H. C. (1977). Effects of jet aircraft noise on mental hospital admissions. *British Journal of Audiology, 11,* 81–85.

Merriam, S. (1980). The concept and function of reminiscence: A review of the research. *The Gerontologist, 20,* 604–608.

Michelson, W. (1986a). *From sun to sun: Contextual dimensions and personal implications of maternal employment*. New York: Rowman & Allanheld.

Michelson, W. (1986b). Measuring macroenvironment and behavior: The time budget and time geography. In R. Bechtel, R. Marans, & W. Michelson (Eds.), *Environmental design research methods and applications*. New York: Van Nostrand Reinhold.

Milgram, S. (1970). The experience of living in cities. *Science, 167,* 1461–1468.

Miller, J. D. (1974). Effects of noise on people. *Journal of Acoustical Society of America, 56,* 729–764.

Miller, N. E. (1980). Effects of learning on physical symptoms produced by psychological stress. In H. Selye (Ed.), *Selye's guide to stress research* (Vol. 1). New York: Van Nostrand.

Miller, S. (1979). Controllability and human stress: Method, evidence and theory. *Behavioral Research and Therapy, 17,* 287–304.

Miller, S. (1980). Why having control reduces stress: If I can stop the roller coaster, I don't want to get off. In J. Garber & M. E. P. Seligman (Eds.), *Human helplessness*. New York: Academic Press.

Miller, I., & Norman, W. (1979). Learned helplessness in humans: A review and attribution theory model. *Psychological Bulletin, 86,* 93–118.

Miller, W. R., & Seligman, M. E. P. (1975). Depression and learned helplessness in man. *Journal of Abnormal Psychology, 84,* 228–238.

Mills, J. H. (1975). Noise and children: A review of literature. *Journal of the Acoustical Society of America, 58,* 767–779.

Mills, R., & Krantz, D. S. (1979). Information, choice, and reactions to stress: A field experiment in a blood bank with laboratory analogue. *Journal of Personality and Social Psychology, 37,* 608–620.

Mitchell, R. E. (1971). Some implications of high density housing. *American Sociological Review, 36,* 18–29.

Moch-Sibony, A. (1984). Study of the effects of noise on the personality and certain psychomotor and intellectual aspects of children, after a prolonged exposure (French). *Travail Humane, 47,* 155–165.

Monat, A., Averill, J., & Lazarus, R. S. (1972). Anticipatory stress and coping reactions under various conditions of uncertainty. *Journal of Personality and Social Psychology, 24,* 237–253.

Monat, A., & Lazarus, R. S. (Eds.). (1977). *Stress and coping: An anthology.* New York: Columbia University Press.

Mook, D. G. (1983). In defense of external invalidity. *American Psychologist, 38,* 379–387.

Moos, R. H. (1976). *The human context.* New York: Wiley.

Moos, R. H. (1979). Social ecological perspectives on health. In G. C. Stone, F. Cohen, N. E. Adler, & associates (Eds.), *Health psychology: A handbook.* San Francisco: Jossey-Bass.

Moos, R. H., & Moos, B. S. (1983). Adaptation and the quality of life in work and family settings. *Journal of Community Psychology, 11,* 158–170.

Morrow, P. E. (1975). An evaluation of recent NOx toxicity data and an attempt to derive an ambient standard for NOx by established toxicological procedures. *Environmental Research, 10,* 92–112.

Moss, G. E. (1973). *Illness, immunity, and social interaction.* New York: Wiley.

Mueller, D. P., Edwards, D. W., & Yarvis, R. M. (1977). Stressful life events and psychiatric symptomatology: Change or undesirability? *Journal of Health and Social Behavior, 18,* 307–317.

Murdock, B. (1965). Effects of a subsidiary task of short-term memory. *British Journal of Psychology, 56,* 413–419.

Murray, H. A. (1938). *Explorations in personality.* New York: Oxford University Press.

Murrell, S. A., & Norris, F. H. (1983). Quality of life as the criterion for need assessment. *Journal of Community Psychology, 11,* 88–97.

Naatanen, R. (1972). The inverted-U relationships between activation and performance: A critical review. In P. Rabbitt & S. Dornic (Eds.), *Attention and performance* (Vol. 5). New York: Academic Press.

National Academy of Sciences. (1977a). *Noise abatement: Policy alternatives for transportation.* Washington, DC: National Research Council.

National Academy of Sciences. (1977b). *Medical and biological effects of environmental pollutants* Washington, DC: National Academy of Sciences.

National Heart, Lung, and Blood Institute. (NHLBI) (1978). *Cardiovascular profile of 15,000 children of school age in three communities, 1971–1975.* Washington, DC: U.S. Government Printing Office. DHEW Publication No. 78-1472.

National Air Pollution Control Administration. (1970). *Air quality criteria for carbon monoxide, AP-62.* Washington, DC: U.S. Government Printing Office.

Neisser, U. (1967). *Cognitive psychology.* New York: Appleton-Century-Crofts.

Neisser, U. (1982). *Memory observed: Remembering in natural contexts.* San Francisco: W.H. Freeman.

Neus, H., Rüddel, H., & Schulte, W. (1983). Traffic noise and hypertension: An epidemiological study of the role of subjective reactions. *International Archives of Occupational and Environmental Health, 51,* 223–229.

Neus, H., Rüddel, H., Schulte, W., & von Eiff, A. (1983). The long-term effect of noise on blood pressure. *Journal of Hypertension, 1*(Supp. 2), 251–253.

Neus, H., von Eiff, A. W., Ruddel, H., & Schulte, W. (1983). Traffic noise and hypertension. In *Proceedings of the Fourth International Congress on Noise as a Public Health Problem.* Milano, Italy: Centro Ricerche e Stud, Amplifon.

Novaco, R., Stokols, D., Campbell, J., & Stokols, J. (1979). Transportation, stress, and community psychology. *American Journal of Community Psychology, 7,* 361–380.

O'Brien, R. (1967). *Positive and negative sets in two choice discrimination learning by children.* Unpublished master's thesis, University of Illinois, Champaign-Urbana, 1967. Cited in M. E. P. Seligman, *Helplessness.* San Francisco: Freeman, 1975.

Obrist, P. A. (1981). *Cardiovascular psychophysiology.* New York: Plenum Press.

Obrist, P. A., Gaebelein, C. J., Teller, E., Langer, A., Girignolo, A., Light, K., & McCubbin, J. (1978). The relationship among heart rate, cartoid dP/dt, and blood pressure in humans as a function of the type of stress. *Psychophysiology, 15,* 102–115.

Obrist, P. A., Langer, A. W., Grignolo, A., L., Light, K. C., Hastrup, J. L., McCubbin, J. A., Koepke, J. P., & Pollack, M. H. (1983). Behavioral-cardiac interactions in hypertension. In D. S.

Krantz, A. Baum, & J. E. Singer (Eds.), *Handbook of psychology and health.* Hillsdale, N.J.: Erlbaum.

Office of Population Census and Survey. (1970). *Second survey of airport noise annoyance around London (Heathrow) Airport.* London: Her Majesty's Stationery Office.

Ostfeld, A., & Eaker, E. (Eds.). (1985). *Proceedings of the NHLBI Workshop in Measuring Psychosocial Variables in Epidemiologic Studies of Cardiovascular Disease.* Bethesda, MD: National Institutes of Health.

Parsons, J., & Ruble, D. (1977). The development of achievement-related expectancies. *Child Development, 48,* 1075–1079.

Parvizpoor, D. (1976). Noise exposure and prevalence of high blood pressure among weavers in Iran. *Journal of Occupational Medicine, 18,* 730–731.

Pasahow, R. (1980). The relation between an attributional dimension and learned helplessness. *Journal of Abnormal Psychology, 89,* 358–367.

Pasahow, R., West, S., & Boroto, D. (1982). Predicting when uncontrollability will produce performance deficits: A refinement of the reformulated learned helplessness hypothesis. *Psychological Review, 89,* 595–598.

Pastalan, L. A. (1980, September). *Relocation, mortality, and intervention.* Paper presented at the Annual Convention of the American Psychological Association, Montreal, Canada.

Patterson, M. L. (1976). An arousal model of interpersonal intimacy. *Psychological Review, 83,* 235–245.

Patterson, M. L. (1982). A sequential functional model of nonverbal exchange. *Psychological Review, 89,* 231–249.

Paulus, P. B., McCain, G., & Cox, V. C. (1978). Death rates, psychiatric commitments, blood pressure, and perceived crowding as a function of institutional crowding. *Environmental Psychology and Nonverbal Behavior, 3,* 107–116.

Payne, R. (1980). Organizational stress and social support. In C. L. Cooper & R. Payne (Eds.), *Current concerns in occupational stress.* New York: Wiley.

Pearlin, L. I., & Schooler, C. (1978). The structure of coping. *Journal of Health and Social Behavior, 19,* 2–21.

Pearlin, L. I., Menaghen, E. G., Lieberman, M. A., & Mullan, J. T. (1981). The stress process. *Journal of Health and Social Behavior, 22,* 337–356.

Pennebaker, J. W. (1982). *The psychology of physical symptoms.* New York: Springer-Verlag.

Pervin, L. A. (1978). Definitions, measurements, and classifications of stimuli, situations, and environments. *Human Ecology, 6,* 71–105.

Peterson, A. P. G., & Gross, E. E., Jr. (1972). *Handbook of noise measurement.* Concord, MA: General Radio.

Peterson, E. A., Augenstein, J. S., Tanis, D. C., & Augenstein, D. G. (1981). Noise raises blood pressure without impairing auditory sensitivity. *Science, 211,* 1450–1452.

Peterson, R. L. (1975). *Air pollution and attendance in recreation behavior settings in the Los Angeles Basin.* Chicago: American Psychological Association.

Petrinovich, L. (1979). Probabilistic functionalism: A conception of research method. *American Psychologist, 34,* 373–390.

Pittner, M. S., Houston, B. K., & Spiridigliozze, G. (1983). Control over stress, Type A behavior pattern, and response to stress. *Journal of Personality and Social Psychology, 44,* 627–637.

Platt, J. R. (1964). Strong inference. *Science, 146,* 347–353.

Pokroviskii, N. N. (1966). On the influence of industrial noise on the level of blood pressure in workers of the machine building industry. *Gigiena Truda i Professionalnye Zabolevaniia, 10,* 44–46.

Posner, M. (1969). Abstraction and the process of recognition. In G. Bower & J. Spence (Eds.), *The psychology of learning and motivation* (Vol. 3). New York: Academic Press.

Poulton, E. (1958). Measuring the order of difficulty of visual-motor tasks. *Ergonomics, 1,* 234–239.

Poulton, E. (1970). *Environment and human efficiency.* Springfield, IL: C. C. Thomas.

Poulton, E. (1977). Continuous intense noise masks auditory feedback and inner speech. *Psychological Bulletin, 84,* 977–1001.

Poulton, E. (1978). A new look at the effects of noise: A rejoinder. *Psychological Bulletin, 85,* 1068–1079.

Poulton, E. (1979). Composite model for human performance in continuous noise. *Psychological Review, 86,* 361–375.

Pribram, K. H., & McGuinness, D. (1975). Arousal, activation, and effort in the control of attention. *Psychological Review, 82,* 116–149.

Proshansky, H., Ittelson, W., & Rivlin, L. (1970). Freedom of choice and behavior in a physical setting. In H. Proshansky, W. Ittelson, & L. Rivlin (Eds.), *Environmental psychology: Man and his physical setting.* New York: Holt, Rinehart, & Winston.

Proshansky, H., Ittelson, W., & Rivlin, L. (Eds.). (1976). *Environmental psychology: People and their physical settings* (2nd ed.). New York: Holt, Rinehart, & Winston.

Raab, W., Chaplin, J. P., & Bajusz, E. (1964). Myocardial necroses produced in domesticated rats and in wild rats by sensory and emotional stress. *Proceedings of the Society of Experimental Biology and Medicine, 116,* 665–669.

Raab, W. (Ed.). (1966). *Preventive cardiology.* Springfield, IL: Charles C. Thomas.

Raab, W. (1971). Cardiotoxic biochemical effects of emotional-environmental stressors—fundamentals of psychocardiology. In L. Levi (Ed.), *Society, stress, and disease* (Vol. 1). London: Oxford University Press.

Rahe, R. H., & Arthur, R. J. (1978). Life change and illness studies: Past history and future directions. *Journal of Human Stress, 4,* 3–15.

Randolph, T. (1970). Domiciliary chemical air pollution in the etiology of ecological mental illness. *International Journal of Social Psychiatry, 16,* 243–265.

Rankin, R. E. (1969). Air pollution control and public apathy. *Journal of the Air Pollution Control Association, 19,* 565–569.

Raytheon Service Company. (1972). *The effects of a company hearing conservation program on extra-auditory disturbances in workers.* Washington, DC: U.S. Department of Health, Education, and Welfare. Contract No. CDC-99-74-28.

Reim, B., Glass, D., & Singer, J. (1971). Behavioral consequences of exposure to uncontrollable and unpredictable noise. *Journal of Applied Social Psychology, 1,* 44–56.

Reimanis, G. (1971). *Effects of experimental internal/external modification techniques and home environment variables on internality/externality.* Paper presented at the meeting of the American Psychological Association, Washington, DC.

Review Panel. (1981). Coronary-prone behavior and coronary heart disease: A critical review. *Circulation, 63,* 1199–1215.

Rholes, W., Blackwell, J., Jordan, C., & Walters, C. (1980). A developmental study of learned helplessness. *Developmental Psychology, 16,* 616–624.

Richter, C. P. (1957). On the phenomenon of sudden death in animals and man. *Psychosomatic Medicine, 19,* 191–198.

Riley, M. W., & Foner, A. (1968). *Aging and society.* New York: Russell Sage Foundation.

Rivlin, L. (1974). Personal communication. In R. Baron, D. Byrne, & W. Griffitt (Eds.), *Social psychology.* Boston: Allyn & Bacon.

Robinson, J. P. (1983). Environmental differences in how Americans use time: The case for subjective and objective indicators. *Journal of Community Psychology, 11,* 171–181.

Rodin, J. (1976). Density, perceived choice and response to controllable and uncontrollable outcomes. *Journal of Experimental Social Psychology, 12,* 564–578.

Rodin, J. & Baum, A. (1978). Crowding and helplessness: Potential consequences of density and locus of control. In A. Baum & Y. Epstein (Eds.), *Human response to crowding.* Hillsdale, NJ: Erlbaum.

Rodin, J., & Langer, E. J. (1977). Long term effects of a control-relevant intervention with the institutionalized aged. *Journal of Personality and Social Psychology, 35,* 897–902.

Rodin, J., Solomon, S., & Metcalf, J. (1978). Role of control in mediating perceptions of density. *Journal of Personality and Social Psychology, 36,* 989–999.

Rodin, J., Rennert, K., & Solomon, S. (1980). Intrinsic motivation for control: Fact or fiction. In A. Baum & J. Singer (Eds.), *Advances in environmental psychology* (Vol. 2). Hillsdale, NJ: Erlbaum.

Rodin, J., Bohm, L., & Wack, J. (1983). Control, coping, and aging: Models for research and intervention. In L. Bickman (Ed.), *Applied social psychology annual* (Vol. 3). Beverly Hills, CA: Sage.

Rosenman, R. H., Brand, R. J., Jenkins, C. D., Friedman, M., Straus, R., & Wurm, M. (1975). Coronary heart disease in the Western Collaborative Group Study: Final follow-up experience of 8½ years. *Journal of the American Medical Association, 233,* 872–877.

Roth, S. (1980). A revised model of learned helplessness in humans. *Journal of Personality, 48,* 103–133.

Roth, S., & Kubal, L. (1975). The effects of noncontingent reinforcement on tasks of differing importance: Facilitation and learned helplessness. *Journal of Personality and Social Psychology, 32,* 680–691.

Rothbaum, F., Weisz, J., & Snyder, S. (1982). Changing the world and changing the self: A two process model of perceived control. *Journal of Personality and Social Psychology, 42,* 5–37.

Rotter, J. B. (1966). Generalized expectancies for internal versus external control of reinforcement. *Psychological Monographs,* Vol. 80 *[Whole No. 609].*

Rotter, J. B. (1975). Some problems and misconceptions related to the construct of internal versus external control of reinforcement. *Journal of Consulting and Clinical Psychology, 43,* 56–67.

Rotton, J. (1983). Affective and cognitive consequences of malodorous pollution. *Basic and Applied Social Psychology, 4,* 171–191.

Rotton, J., & Frey, J. (1982). *Atmospheric conditions, seasonal trends, and psychiatric emergencies: Replications and extensions.* Washington, DC: American Psychological Association.

Rotton, J., Barry, T., Frey, J., & Soler, E. (1978). Air pollution and interpersonal attraction. *Journal of Applied Social Psychology, 8,* 57–71.

Rotton, J., Olszewski, D., Charleston, M., & Soler, E. (1978). Loud speech, conglomerate noise and behavioral aftereffects. *Journal of Applied Psychology, 63,* 360–365.

Rotton, J., Frey, J., Barry, T., Milligan, M., & Fitzpatrick, M. (1979). The air pollution experience and interpersonal aggression. *Journal of Applied Social Psychology, 9,* 397–412.

Rowkaw, S., Detels, R., Coulson, A., Sayre, J., Tashkin, D., Allwright, S., & Massey, F. (1980). The UCLA population studies of chronic obstructive respiratory disease. *Chest, 78,* 252–262.

Rowland, K. F. (1977). Environmental events predicting death for the elderly. *Psychological Bulletin, 84,* 349–372.

Rowles, G. D. (1981). Geographical perspectives on human development. *Human Development, 24,* 651–688.

Russell, J. A., & Ward, L. M. (1982). Environmental psychology. *Annual Review of Psychology, 33,* 651–688.

Saegert, S. (1984). Environment and children's mental health: Residential density and low income children. In A. Baum & J. Singer (Eds.), *Handbook of psychology and health* (Vol. 2). Hillsdale, NJ: Erlbaum.

Saegert, S., MacKintosh, I., & West, S. (1975). Two studies of crowding in urban public spaces. *Environment and Behavior, 7,* 159–184.

Sampson, E. E. (1981). Cognitive psychology as ideology. *American Psychologist, 36,* 730–743.

Sandler, I. N., & Barrera, M., Jr. (1980, September). *Social support as a stress-buffer: A multi-method investigation.* Paper presented at the meeting of the American Psychological Association, Montreal.

Sarason, S. B. (1976). *The psychological sense of community: Prospects for a community psychology.* San Francisco: Jossey-Bass.

Sarason, I. G., Levine, H. M., & Sarason, B. R. (1982). Assessing the impact of life changes. In T. Millon, C. Green, & R. Meagher (Eds.), *Handbook of clinical health psychology.* New York: Plenum Press.

Scarr, S. (1979). Psychology and children: Current research and practice. *American Psychologist, 34,* 809–811.

Schachter, S., & Singer, J. E. (1962). Cognitive, social and physiological determinants of emotional state. *Psychological Review, 69,* 379–399.

Schachter, S., Kozlowski, L. T., & Silverman, B. (1977). Effect of urinary pH on cigarette smoking. *Journal of Experimental Psychology: General, 106,* 13–19.

Schmidt, D. E., & Keating, J. P. (1979). Human crowding and personal control: An integration of the research. *Psychological Bulletin, 86,* 680–700.

Schmidt, D. E., Goldman, R. D., & Feimer, N. R. (1979). Perceptions of crowding: Predicting at the residence, neighborhood, and city levels. *Environment and Behavior, 11,* 105–130.

Schmitt, R. C. (1966). Density, health, and social disorganization. *Journal of the American Institute of Planners, 32,* 38–40.

Schulz, R. (1976). The effects of control and predictability on the physical and psychological well-being of the institutionalized aged. *Journal of Personality and Social Psychology, 33,* 563–573.

Schulz, R., & Hanusa, B. (1978). Long-term effects of control and predictability-enhancing interventions: Findings and ethical issues. *Journal of Personality and Social Psychology, 36,* 1194–1201.

Schulz, R., & Hanusa, B. (1979). Environmental influences on the effectiveness of control-and-competence-enhancing interventions. In L. Perlmuter & R. Monty (Eds.), *Choice and perceived control.* Hillsdale, NJ: Erlbaum.

Schwartz, G. E. (1982). Testing the biopsychosocial model: The ultimate challenge facing behavioral medicine? *Journal of Consulting and Clinical Psychology, 50,* 1040–1053.

Schwartz, S. (1975). *The effects of arousal on recall, recognition, and organization of memory.* Unpublished manuscript.

Seligman, M. E. P. (1975). *Helplessness: On depression, development, and death.* San Francisco: Freeman.

Seligman, M. E. P., Abramson, L., Semmel, A., & von Baeyer, C. (1979). Depressive attributional style. *Journal of Abnormal Psychology, 88,* 242–247.

Selye, H. (1956). *The stress of life.* New York: McGraw-Hill.

Shapiro, J., & Shapiro, D. (1979). The psychology of responsibility. *New England Journal of Medicine, 301,* 211–212.

Shaw, M. E., & Costanzo, P. R. (1970). *Theories of social psychology.* New York: McGraw-Hill.

Sherrod, D. R. (1974). Crowding, perceived control, and behavioral aftereffects. *Journal of Applied Social Psychology, 4,* 171–186.

Sherrod, D. R., & Cohen, S. (1978). Density, personal control and design. In S. Kaplan & R. Kaplan (Eds.), *Humanscape.* North Scituate, MA: Duxbury.

Sherrod, D. R., & Downs, R. (1974). Environmental determinants of altruism: The effects of stimulus overload and perceived control on helping. *Journal of Experimental Social Psychology, 10,* 468–479.

Sherrod, D. R., Hage, J. N., Halpern, P. L., & Moore, B. S. (1977). Effects of personal causation and perceived control on responses to an aversive environment: The more control, the better. *Journal of Experimental Social Psychology, 13,* 14–27.

Shumaker, S. A., & Brownell, A. (1984). Toward a theory of social support: Closing conceptual gaps. *Journal of Social Issues, 40,* 11–36.

Shumaker, S. A., & Reizenstein, J. (1982). Environmental factors affecting inpatient stress in acute care hospitals. In G. W. Evans (Ed.), *Environmental stress.* New York: Cambridge University Press.

Singer, J. E. (1980). Traditions of stress research: Integrated comments. In E. Sarason & C. D. Speilberger (Eds.), *Stress and anxiety* (Vol. 7). Washington: Hemisphere Press.

Singer, J. E., Lundberg, U., & Frankenhauser, M. (1978). Stress on the train: A study of urban commuting. In A. Baum, J. Singer, & S. Valins (Eds.), *Advances in environmental psychology* (Vol. 1). Hillsdale, NJ: Erlbaum.

Sklar, L. S., & Anisman, H. (1979). Stress and coping factors influence tumor growth. *Science, 205,* 513–515.

Sklar, L. S., & Anisman, H. (1981). Stress and cancer. *Psychological Bulletin, 89,* 369–406.

Smith, A. (1980). Low levels of noise and performance. In J. Tobias, G. Jansen, & D. Ward (Eds.), *Noise as a public health problem.* Rockville, MD: The American Speech and Hearing Association.

Smith, A. & Broadbent, D. (1981). Noise and levels of processing. *Acta Psychologica, 47,* 129–142.

Smith, M. B. (1983). The shaping of American social psychology: A personal perspective from the periphery. *Personality and Social Psychology Bulletin, 9,* 165–180.

Snyder, M. L., Smoller, B., Strenta, A., & Frankel, A. (1981). A comparison of egotism, negativity, and learned helplessness as explanations for poor performance after unsolvable problems. *Journal of Personality and Social Psychology, 40,* 24–30.

Sokolov, E. (1963). *Perception and conditioned reflex.* Oxford, England: Pergamon.

Solomon, S., Holmes, D. S., & McCaul, K. D. (1980). Behavioral control over aversive events: Does control that requires effort reduce anxiety and physiological arousal? *Journal of Personality and Social Psychology, 39,* 729–736.

South Coast Air Quality Management District. (1977). *Summary of air quality in South Coast Air Basin 1977.* Sacremento, CA: Author. ENP 78-1.

Southwick, C. (1955). The population dynamics of confined houdr mivr dupplirf eiyh unlimited food. *Ecology, 36,* 212–225.

Spacapan, S., & Cohen, S. (1983). Effect and aftereffects of stressor expectations. *Journal of Personality and Social Psychology, 45,* 1243–1254.

Spitz, R. (1946). Anaclitic depression. *The Psychoanalytic Study of the Child, 2,* 313–342.

State of California. (1976). *VIII.* Vol. 2: *California Air Resources Board: Technical Services Division, V.* Sacramento, CA: Author.

Steptoe, A. (1981). *Psychological factors in cardiovascular disorders.* New York: Academic Press.

Stipek, D., & Weisz, J. (1981). Perceived control and academic achievement. *Review of Educational Research, 51,* 101–137.

Stokols, D. (1972). A social-psychological model of human crowding phenomena. *The Journal of the American Institute of Planners, 6,* 72–83. (a)

Stokols, D. (1972b). On the distinction between density and crowding: Some implications for future research. *Psychological Review, 79,* 275–277.

Stokols, D. (1976). The experience of crowding in primary and secondary environments. *Environment and Behavior, 8,* 49–86.

Stokols, D. (1978). A typology of crowding experiences. In A. Baum & Y. Epstein (Eds.), *Human response to crowding.* Hillsdale, NJ: Erlbaum.

Stokols, D. (1979). A congruence analysis of human stress. In I. G. Sarason & C. D. Spielberger (Eds.), *Stress and anxiety* (Vol. 6). New York: Wiley.

Stokols, D. (1983, August). *Scientific and policy challenges of a contextually oriented psychology.* Presidential address to the Division of Population and Environmental Psychology of the American Psychological Association, Annual Conference of the American Psychological Association, Anaheim, California.

Stokols, D. (1986). Environmental psychology: A coming of age. In A. Krant (Ed.), *G. Stanley Hall Lecture Series* (Vol. 2). Washington, DC: American Psychological Association. (a)

Stokols, D. (1986). Conceptual strategies of environmental psychology. In D. Stokols & I. Altman (Eds.), *Handbook of environmental psychology.* New York: Wiley. (b)

Stokols, D. & Altman, I. (Eds.). (1986). *Handbook of environmental psychology.* New York: Wiley.

Stokols, D., & Novaco, R. W. (1981). Transportation and well-being: An ecological perspective. In J. Wohlwill, P. Everett, & I. Altman (Eds.), *Human behavior and environment—Advances in theory and research* (Vol. 5). *Transportation environments.* New York: Plenum Press.

Stokols, D., & Shumaker, S. A. (1981). People in places: A transactional view of settings. In J. Harvey (Ed.), *Cognition, social behavior and the environment.* Hillsdale, NJ: Erlbaum.

Stokols, D., Rall, M., Pinner, B., & Schopler, J. (1973). Physical, social and personal determinants of the perception of crowding. *Environment and Behavior, 4,* 87–115.

Stokols, D., Smith, T. E., & Prostor, J. J. (1975). Partitioning and perceived crowding in a public place. *American Behavioral Scientist, 18,* 792–814.

Stokols, D., Novaco, R. W., Campbell, J., & Stokols, J. (1978). Traffic congestion, Type-A behavior, and stress. *Journal of Applied Psychology, 63,* 467–480.

Stokols, D., Ohlig, W., & Resnik, S. (1978). Perception of residential crowding, classroom experiences, and student health. In A. Esser & B. Greenbie (Eds.), *Design for communality and privacy.* New York: Plenum Press.

Stokols, D., Shumaker, S. A., & Martinez, J. (1983). Residential mobility and personal well-being. *Journal of Environmental Psychology, 3,* 5–19.

Strahilevitz, M., Strahilevitz, A., & Miller, J. (1979). Air pollutants and the admission rate of psychiatric patients. *American Journal of Psychiatry, 136,* 205–207.

Sundstrom, E. (1978). Crowding as a sequential process: Review of research on the effects of population density on humans. In A. Baum & Y. Epstein (Eds.), *Human response to crowding.* Hillsdale, NJ: Erlbaum.

Syme, L. S. (1984). Sociocultural factors and disease etiology. In W. D. Gentry (Ed.), *Handbook of behavioral medicine.* New York: Guilford Press.

Syme, L. S., & Berkman, L. F. (1976). Social class, susceptibility, and sickness. *American Journal of Epidemiology, 104,* 1–8.

Syme, L. S., & Torfs, M. S. (1978). Epidemiological research in hypertension: A critical appraisal. *Journal of Human Stress, 4,* 43–48.

Tarnopolsky, A., Watkins, G., & Hand, D. J. (1980). Aircraft noises and mental health: I: Prevalence of individual symptoms. *Psychological Medicine, 10,* 683–698.

Taylor, S. (1979). Hospital patient behavior: Reactance, helplessness, or control? *Journal of Social Issues, 35,* 156–184.

Taylor, S. (1983). Adjustment to threatening events. *American Psychologist, 38,* 1161–1173.

Thoits, P. A. (1982). Conceptual, methodological, and theoretical problems in studying social support as a buffer against life stress. *Journal of Health and Social Behavior, 23,* 145–159.

Thoits, P. A. (1983). Multiple identities and psychological well-being: A reformulation and test of the social isolation hypothesis. *American Sociological Review, 48,* 174–187.

Thompson, S. C. (1981). A complex answer to a simple question: Will it hurt if I can control it? *Psychological Bulletin, 90,* 89–101.

Tolman, E. C., & Brunswik, E. (1935). The organism and the causal texture of the environment. *Psychological Review, 42,* 43–77.

Tracor, Inc. (1970). *Community reactions to airport noise.* Washington, DC: National Aeronautics and Space Administration. NASA CR-1761.

Ulrich, R. S. (1979). Visual landscapes and psychological well-being. *Landscape Research, 4,* 17–23.

Ulrich, R. S. (1983). Aesthetic and affective response to natural environment. In I. Altman & J. F. Wohlwill (Eds.), *Human behavior and the natural environment.* New York: Plenum Press.

Ulrich, R. S. (1984). View through a window may influence recovery from surgery. *Science, 224,* 420–421.

U.S. Department of Health, Education, & Welfare (USDHEW). (1979). *Healthy people: A report of the Surgeon General on health promotion and disease prevention.* Washington, DC: U.S. Government Printing Office. U. S. Public Health Services Publication No. 79-55071.

U.S. Environmental Protection Agency (USEPA). (1974). *EPA Publication No. 550/0-74004. Information on levels of environmental noise requisite to protect public health and welfare with an adequate margin for safety.* Washington, DC: U.S. Government Printing Office.

U.S. Environmental Protection Agency (USEPA). (1980). *EPA Report No. 550/9-80-101. Noise, general stress responses, and cardiovascular disease processes: Review and reassessment of hypothesized relationships.* Washington, DC: Office of Noise Abatement and Control.

U.S. Environmental Protection Agency (USEPA). (1981). *EPA Report No. 550/9-81-103(A-C). Epidemiology feasibility study: Effects of noise on the cardiovascular system.* Washington, DC: Office of Noise Abatement and Control.

Valins, S., & Baum, A. (1973). Residential group size, social interaction, and crowding. *Environment and Behavior, 5,* 421–439.

Verbrugge, L. M. (1983). Multiple roles and physical health of women and men. *Journal of Health and Social Behavior, 24,* 16–30.

Verbrugge, L. M., & Taylor, R. B. (1980). Consequences of population density and size. *Urban Affairs Quarterly, 16,* 135–160.

Verderber, S. F. (1982). Designing for the therapeutic functions of windows in hospital rehabilitation environment. In P. Bart & G. Francescato (Eds.), *Knowledge for design: Proceedings of the 1982 Conference of the Environmental Design Research Association.* College Park: University of Maryland Press.

Veroff, J. (1983). Contextual determinants of personality. *Personality and Social Psychology Bulletin, 32*, 331–343.

Verrier, R. L., DeSilva, R. A., & Lown, B. (1983). Psychological factors in cardiac arrhythmias and sudden death. In D. S. Krantz, A. Baum, & J. E. Singer (Eds.), *Handbook of psychology and health* (Vol. 3). Hillsdale, NJ: Erlbaum.

Vinokur, A., & Selzer, M. L. (1975). Desirable versus undesirable life events: Their relationship to stress and mental distress. *Journal of Personality and Social Psychology, 32*, 329–337.

Visintainer, M. A., Volpicelli, J. R., & Seligman, M. E. P. (1982). Tumor rejection in rats after inescapable shock. *Science, 216*, 437–439.

Visintainer, M. A., Seligman, M. E. P., & Volpicelli, J. R. (1983). Helplessness, chronic stress, and tumor development. *Psychosomatic Medicine, 301*, 1249–1254.

Von Eiff, A. W., Friedrich, G., & Neus, H. (1982). Traffic noise, a factor in the pathogenesis of essential hypertension. In G. M. Berlyne, S. Giovanetti, S. Thomas (Eds.), *Contributions to Nephrology, 30*, 82–86. Basel: S. Karger.

Voors, A. W., Foster, T. A., Frerichs, R. R., Weber, L. S., & Berenson, G. S. (1976). Studies of blood pressure in children, ages 5–14 years, in a total biracial community. *Circulation, 54*, 319–327.

Wachs, T. D. (1978). The relationship of infants' physical environment to their Binet performance at 2½ years. *International Journal of Behavioral Development, 1*, 51–65.

Wachs, T. D. (1979). Proximal experience and early cognitive development: The physical environment. *Merrill-Palmer Quarterly, 25*, 3–41.

Wachs, T. D., Uzgiris, I., & Hunt, J. (1971). Cognitive development in infants of different age levels and from different environmental backgrounds: An exploratory investigation. *Merrill-Palmer Quarterly of Behavior and Development, 17*, 283–317.

Wachtel, P. L. (1968). Anxiety, attention, and coping with threat. *Journal of Abnormal Psychology, 73*, 137–143.

Walster, E. (1966). Assignment of responsibility for an accident. *Journal of Personality and Social Psychology, 3*, 73–79.

Wapner, S. (1981). Transactions of persons-in-environments: Some critical transitions. *Journal of Environmental Psychology, 1*, 223–239.

Wapner, S. (1986). A holistic, developmental, systems-oriented environmental psychology: Some beginnings. In D. Stokols & I. Altman (Eds.), *Handbook of environmental psychology*. New York: Wiley.

Wapner, S., & Kaplan, B. (Eds.). (1983). *Toward a holistic developmental psychology*. Hillsdale, NJ: Erlbaum.

Ward, S. K. (1975). Overcrowding and social pathology: A re-examination of the implications for the human population. *Human Ecology, 3*, 275–286.

Ward, L., & Suedfeld, P. (1973). Human response to highway noise. *Environmental Research, 6*, 306–326.

Watkins, G., Tarnopolsky, A., & Jenkins, L. M. (1981). Use of medicines and health care services. *Psychological Medicine, 11*, 155–168.

Watson, J. B. (1913). Psychology as the behaviorist views it. *Psychological Review, 20*, 159–177.

Watson, J., & Ramey, C. (1972). Reactions to response-contingent stimulation in early infancy. *Merrill-Palmer Quarterly, 18*, 219–228.

Webb, J. W., Campbell, D. T., Schwartz, R. D., & Sechrest, L. (1966). *Unobtrusive Measures: Nonreactive research in the social sciences*. Chicago: Rand McNally.

Webb, E. J., Campbell, D. T., Schwartz, R. D., Sechrest, L., & Grove, J. B. (1981). *Nonreactive measures in the social sciences*. Boston: Houghton Mifflin.

Weick, K. E. (1979). *The social psychology of organizing*. (2nd ed.). Reading, MA: Addison-Wesley.

Weiner, B., (Ed.). (1974). *Achievement motivation and attribution theory*. Morristown, NJ: General Learning Press.

Weiner, H. (1977). *Psychobiology and human disease*. New York: Elsevier.

Weiner, B., Frieze, I., Kukla, A., Reed, L., Rest, S., & Rosenbaum, R., (1971). *Perceiving the causes of success and failure*. Morristown, NJ: General Learning Press.

Weinstein, C. (1979). The physical environment of the school: A review of the research. *Review of Educational Research, 49,* 577–610.

Weinstein, N. D. (1974). Effects of noise on intellectual performance. *Journal of Applied Psychology, 59,* 548–554.

Weinstein, N. D. (1976). Human evaluations of environmental noise. In K. H. Craik & E. H. Zube (Eds.), *Perceiving environmental quality: Research and applications.* New York: Plenum Press.

Weinstein, N. D. (1977). Noise and intellectual performance: A confirmation and extension. *Journal of Applied Psychology, 62,* 104–107.

Weinstein, N. D. (1978). Individual differences in reactions to noise. *Journal of Applied Psychology, 63,* 458–466.

Weiss, B. (1983). Behavioral toxicology and environmental health science. *American Psychologist, 38,* 1174–1187.

Weiss, J. M. (1971). Effects of coping behaviors in different warning signal conditions on stress pathology in rats. *Journal of Comparative and Physiological Psychology, 77,* 1–13.

Weiss, J. M. (1972). Psychological factors in stress and disease. *Scientific American, 226,* 104–113.

Weiss, J. M. (1977). Psychological and behavioral influences on gastrointestinal lesions in animal models. In J. D. Maser & M. E. P. Seligman (Eds.), *Psychopathology: Experimental models.* San Francisco: W. H. Freeman.

Weiss, J. M. (1985). In E. Katkin & S. B. Manuck (Eds.), *Advances in behavioral medicine.* New York: JAI.

Weiss, J. M., Stone, E. A., & Harrell, N. (1970). Coping behavior and brain norephinephrine in level in rats. *Journal of Comparative and Physiological Psychology, 72,* 153–160.

Weiss, J. M., Glazer, H. I., & Pohorecky, L. A. (1977). Coping behavior and neurochemical changes: An alternative explanation for the original ''learned helplessness'' experiments. In G. Serban (Ed.), *Psychopathology of human adaptation.* New York: Plenum Press.

Weisz, J. (1983). Can I control it? The pursuit of veridical answers across the life span. In P. Baltes (Ed.), *Life-span development and behavior* (Vol. 5). New York: Academic Press.

Welch, B. L. (1979, June). *Extra-auditory health effects of industrial noise: Survey of foreign literature.* Report of Aerospace Medical Division, Air Force Systems Command, Wright-Patterson AFB.

Wells, J. A. (1985). Chronic life situations and life events. In A. Ostfeld & E. Eaker (Eds.), *Measuring psychosocial variables in epidemiological studies of cardiovascular disease.* Washington, DC: National Institutes of Health.

Wener, R., & Kaminoff, R. (1983). Improving environment information: Effects of signs on perceived crowding and behavior. *Environment and Behavior, 15,* 3–20.

Wepmen, J. (1958). *Manual of directions: Auditory discrimination test.* Chicago: Author.

Whalen, C. K., & Henker, B. (1980). *Hyperactive children: The social ecology of identification and treatment.* New York: Academic Press.

White, R. (1959). Motivation reconsidered: The concept of competence. *Psychological Review, 66,* 297–333.

Wicker, A. W. (1973). Undermanning theory and research: Implications for the study of psychological and behavioral effects of excess populations. *Representative Research in Social Psychology, 4,* 185–206.

Wicker, A. W. (1986). Behavior settings reconsidered: Temporal stages, resources, internal dynamics, context. In D. Stokols & I. Altman (Eds.), *Handbook of environmental psychology.* New York: Wiley.

Wilding, J., & Mohindra, N. (1983). Noise slows phonological coding and maintenance rehearsal: An explanation for some effects of noise on memory. In G. Rossi (Ed.), *Noise as a public health hazard, Proceedings of the 4th International Congress.* Milano, Italy: Centro Ricerche e Studi Amplifom.

Wilkinson, R. (1969). Some factors influencing the effects of environmental stressors upon performance. *Psychological Bulletin, 72,* 260–272.

Wineman, J. D. (1982). The office environment as a source of stress. In G. W. Evans (Ed.), *Environmental stress.* New York: Cambridge University Press.

Winkel, G. (1983). Ecological validity issues in field research settings. In A. Baum & J. E. Singer (Eds.), *Advances in environmental psychology, Volume 5: Methods of environmental investigations*. Hillsdale, NJ: Erlbaum.

Winsborough, H. H. (1965). The social consequences of high population density. *Law and Contemporary Problems, 30,* 120–126.

Wohlwill, J. F. (1974). Human response to levels of environmental stimulation. *Human Ecology, 2,* 127–147.

Wohlwill, J. F., Nasar, J. L., DeJoy, D. M., & Foruzani, H. H. (1976). Behavioral effects of a noisy environment: Task involvement versus passive exposure. *Journal of Applied Psychology, 1,* 67–74.

Wolf, S., & Goodell, H. (Eds.). (1968). *Harold G. Wolff's "stress and disease."* Springfield, IL: Charles C.Thomas.

Woodhead, M. (1964). Searching a visual display in intermittent noise. *Journal of Sound and Vibration, 1,* 157–161.

Worchel, S. (1978). Reducing crowding without increasing space: Some applications of an attributional theory of crowding. *Journal of Population, 1,* 216–230.

Worchel, S., & Teddlie, C. (1976). Factors affecting the experience of crowding: A two-factor theory. *Journal of Personality and Social Psychology, 34,* 30–40.

Wortman, C. B., & Brehm, J. W. (1975). Responses to uncontrollable outcomes: An integration of reactance theory and the learned helplessness model. In L. Berkowitz (Ed.), *Advances in experimental social psychology* (Vol. 8). New York: Academic Press.

Zajonc, R. B. (1984). On the primacy of affect. *American Psychologist, 39,* 117–123.

Zautra, A. J., & Reich, J. W. (1983). Life events and perceptions of life quality: Developments in the two-factor approach. *Journal of Community Psychology, 11,* 121–132.

Zimring, C., Weitzer, W., & Knight, R. C. (1982). Opportunity for control and the designed environment: The case of an institution for the developmentally disabled. In A. Baum & J. Singer (Eds.), *Advances in environmental psychology* (Vol. 4). Hillsdale, NJ: Erlbaum.

Zlutnick, S., & Altman, I. (1972). Crowding and human behavior. In J. F. Wohlwill & D. H. Carson (Eds.), *Environment and the social sciences: Perspectives and applications*. Washington, DC: American Psychological Association.

Author Index

Subject Index